Variational and Hemivariational Inequalities. Theory, Methods and Applications

Volume I

Nonconvex Optimization and Its Applications

Volume 69

VARIATIONAL AND HEMIVARIATIONAL INEQUALITIES
Theory, Methods and Applications

Volume I: UNILATERAL ANALYSIS AND UNILATERAL MECHANICS

D. GOELEVEN
IREMIA, University of La Reunion,
FRANCE

D. MOTREANU
University of Perpignan,
FRANCE

Y. DUMONT
IREMIA, University of La Reunion,
FRANCE

M. ROCHDI
IREMIA, University of La Reunion,
FRANCE

Springer Science+Business Media, LLC

Library of Congress Cataloging-in-Publication

CIP info or:

Title: Variational and Hemivariational Inequalities: Theory, Methods and Applications
     Volume I: Unilateral Analysis and Unilateral Mechanics
Author: Goeleven, Motreanu, Dumont, Rochdi
ISBN 978-1-4613-4646-3   ISBN 978-1-4419-8610-8 (eBook)
DOI 10.1007/978-1-4419-8610-8

This book is dedicated
to the memory of Prof.
P.D. Panagiotopoulos.

Contents

List of Figures

Acknowledgments

The authors acknowledge the great assistance they have received from Prof. P.D. Panagiotopoulos in preparing Chapter 2 of the book.

Many thanks are also due to A. Goeleven who prepared the final text with the LATEX program.

We also wish to thank our editors in Kluwer for their cooperation during the project.

Chapter 1

UNILATERAL ANALYSIS

1.1   BASIC MATHEMATICAL TOOLS

The purpose of this chapter is to provide some notions and fundamental results of convex analysis which will be used throughout this book. Starting with the notion of convexity, some basic results on convex and lower semi-continuous functionals are given. Particular attention is paid to the separation theorems of convex sets. There follow some results on lower estimate of lower semi-continuous convex functions. In particular we deal with a significant result of Szulkin [405]. The famous Ekeland's variational principle is also presented. Projection operators on closed convex sets are discussed. The chapter ends with four mathematical principles of particular interest in the study of inequality problems: The KKM principle, Minty's principle, the complementarity principle and the variational principle. In preparing this Section a number of well-known works have been followed, in particular those of Aubin [24], [25], Baiocchi and Capelo [31], Bourbaki [59], Brézis [67], Ekeland [125], Ekeland and Temam [124], Granas [181], Köthe [229], Pascali and Sburlan [358] and Rockafellar [374].

1.1.1   CONVEX SETS AND FUNCTIONALS

Let X be a real Banach space and K a subset of X. We recall that K is said to be convex if

$$\lambda x_1 + (1 - \lambda)x_2 \in K, \tag{1.1.1}$$

whenever $x_1 \in K$, $x_2 \in K$ and $0 \leq \lambda \leq 1$ (Fig. 1.1.1). All linear subspaces of X (including X) are convex. By convention, the empty

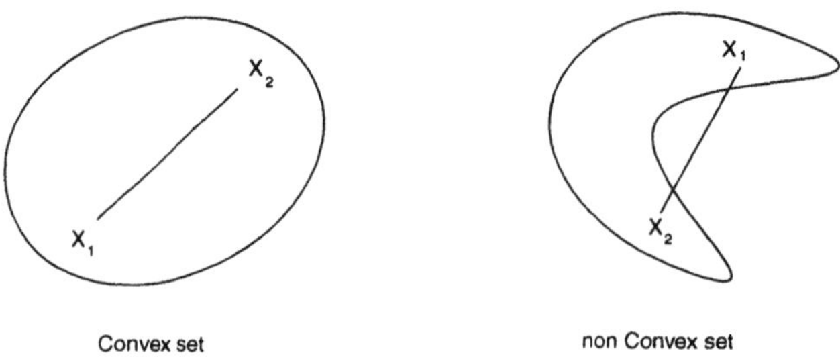

Figure 1.1.1.   Convex set - non Convex set.

set $\emptyset$ is convex. The general characteristic of convex sets is that they contain, along with any two distinct points x_1 and x_2, the line segment between x_1 and x_2 (Fig. 1.1.1). Any intersection of convex sets is convex; the same does not hold for an arbitrary union of convex sets.

Given a set $A \subset X$, then the necessarily convex set of all finite linear combinations $\sum_{i=1}^{n} \lambda_i x_i, x_i \in A, i = 1, \cdots, n$, with $\sum_{i=1}^{n} \lambda_i = 1$, is a set called the affine hull of A. If additionally, $\lambda_i \geq 0, i = 1, \cdots, n$ it is called the convex hull of A and is denoted by $conv\{A\}$ which is also convex. So

$$conv\{A\} = \left\{ \sum_{i=1}^{n} \lambda_i x_i; \lambda_i \geq 0, x_i \in A \ (i = 1, \cdots, n), \sum_{i=1}^{n} \lambda_i = 1 \right\}.$$

A real-valued functional $f : K \to \mathbb{R}$ is convex (resp. strictly convex) on K if for each $x_1 \in K$, $x_2 \in K$ and $0 < \lambda < 1$

$$f(\lambda x_1 + (1 - \lambda)x_2) \leq (\text{resp. } <) \ \lambda f(x_1) + (1 - \lambda)f(x_2). \qquad (1.1.2)$$

A geometrical interpretation of this definition is given in Fig 1.1.2. A functional F is said to be concave (resp. strictly concave) if and only if $-F$ is convex (resp. strictly convex). A linear functional is at the same time convex and concave, but not strictly.

Let us here also recall the following important result for twice differentiable functionals.

Theorem 1.1.1 Let D be a nonempty, open and convex subset of X. Suppose that f is twice differentiable in D. Then f is convex if and only if

$$f''(x)(h, h) \geq 0, \ \forall \, x \in D, \ \forall \, h \in X.$$

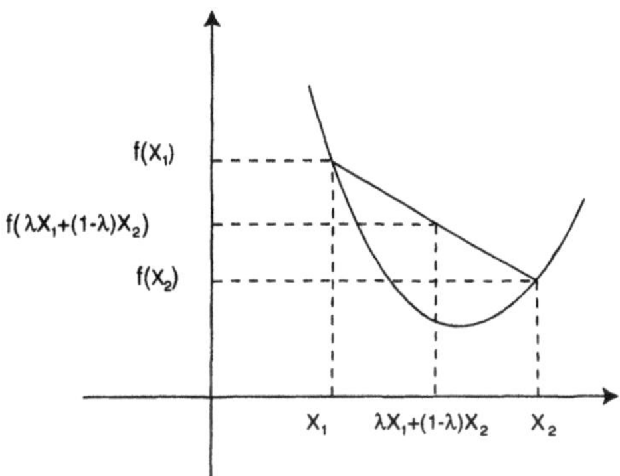

Figure 1.1.2.   Geometrical interpretation of the definition of convexity.

The convexity definition given in (1.1.2) can be extended to include the case of convex functionals with possibly infinite values. A functional $f : K \to \overline{\mathbb{R}}$ ($\equiv \mathbb{R} \cup \{\pm\infty\}$) is defined to be convex on K if for every $x_1 \in K$ and $x_2 \in K$ (1.1.2) holds whenever the right-hand side makes sense. It is obvious that the right-hand side cannot be defined if $f(x_1) = -f(x_2) = \pm\infty$. Because of the fact that a convex functional may have infinite values, we can limit our attention to functionals defined on all of X. Indeed, if $f : K \to \mathbb{R}$ is convex on K, then we can define an extension $\bar{f}$ of f to all of X by setting $\bar{f}(x) = f(x)$ for $x \in K$ and $\bar{f}(x) = +\infty$ for $x \notin K$. To every convex set K we can associate a convex functional $\Psi_K : X \to \mathbb{R} \cup \{+\infty\}$, defined by

$$\Psi_K(x) = \left\{ \begin{array}{ll} 0 & \text{for } x \in K \\ +\infty & \text{for } x \notin K \end{array} \right. \tag{1.1.3}$$

and called the indicator of K. With respect to the convex functional $f : X \to \overline{\mathbb{R}}$, the set

$$epi \ f = \{(x, \lambda) \in X \times \mathbb{R} : \ f(x) \leq \lambda\} \tag{1.1.4}$$

is defined and called the epigraph of f. An equivalent general definition of convexity results, as may be easily established, if we define $f : X \to \overline{\mathbb{R}}$

as being convex, whenever *epi* f is a convex subset of $X \times \mathbb{R}$. The effective domain $D(f)$ of a convex functional f on X is defined by

$$D(f) = \{x \in X : \mid f(x) \mid < \infty\}. \tag{1.1.5}$$

A functional f is said to be proper if $f : X \to (-\infty, +\infty]$ and $f \not\equiv +\infty$. Note here that throughout the rest of this book only proper functionals will be considered. If f is convex, λf ($\lambda \geq 0$) is convex. For f_1 and f_2 convex, $f_1 + f_2$ is convex as well (defining $(f_1 + f_2)(x) = +\infty$ for $f_1(x) = -f_2(x) = \pm\infty$).

A function $f : X \to \mathbb{R} \cup \{+\infty\}$ is said to be (sequentially) lower semi-continuous (weakly lower semi-continuous) on X if

$$x_n \to x \quad (\text{resp. } x_n \rightharpoonup x) \Rightarrow f(x) \leq \liminf f(x_n). \tag{1.1.6}$$

Similarly, f is (sequentially) upper semi-continuous (weakly upper semi-continuous) on X if

$$x_n \to x \quad (\text{resp. } x_n \rightharpoonup x) \Rightarrow f(x) \geq \limsup f(x_n). \tag{1.1.7}$$

Recall that one defines

$$\liminf_{n \to \infty} f(x_n) = \sup_{n \in \mathbb{N}} (\inf_{p \geq 0} f(x_{n+p})) = \lim_{n \to \infty} (\inf_{p \geq 0} f(x_{n+p})) \tag{1.1.8}$$

and

$$\limsup_{n \to \infty} f(x_n) = \inf_{n \in \mathbb{N}} (\sup_{p \geq 0} f(x_{n+p})) = \lim_{n \to \infty} (\sup_{p \geq 0} f(x_{n+p})). \tag{1.1.9}$$

The property of lower semicontinuity is closely related to the closeness of *epi* f. The following proposition holds (see e.g. [358], [74]).

Proposition 1.1.2   A functional $f : X \to \mathbb{R} \cup \{+\infty\}$ is lower semi-continuous (resp. weakly lower semi-continuous) if and only if *epi* f is a closed (weakly closed) subset of $X \times \mathbb{R}$.

Note also that *epi* f is a closed (weakly closed) subset of $X \times \mathbb{R}$ if and only if the level set

$$\{x \in X : f(x) \leq c\}$$

is closed (weakly closed) for any $c \in \mathbb{R} \cup \{+\infty\}$. (Note that the case $c = +\infty$ is immediate).

Important properties concerning lower semi-continuous (l.s.c.) (resp. weakly lower semi-continuous (w.l.s.c.)) functionals are listed in the following proposition.

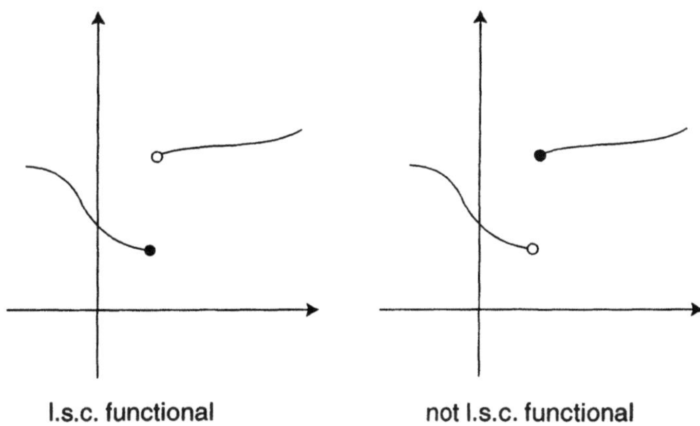

Figure 1.1.3.   l.s.c. functional - not l.s.c. functional.

Proposition 1.1.3   Let $f, g : X \to \mathbb{R} \cup \{+\infty\}$ be l.s.c. (resp. w.l.s.c) functionals. Then $f + g$, $\sup\{f, g\}$, $\inf\{f, g\}$ are l.s.c. (resp. w.l.s.c) functionals.

If in addition $f, g \geq 0$ then the functional $f.g$ (that one supposes well-defined) is l.s.c. (resp. w.l.s.c.).

Examples of l.s.c. and not l.s.c. real-valued functionals are given in Fig. 1.1.3.

It is clear that each weakly lower semi-continuous function is lower semi-continuous. If $f : X \to (-\infty, +\infty]$ is convex then the converse holds true (see e.g. [Pas78]). If $f : X \to (-\infty, +\infty]$ is a weakly lower semi-continuous function on a reflexive Banach space then the existence of a bounded minimizing sequence, i.e. a bounded sequence $\{x_n\} \subset X$ such that $f(x_n) \leq \inf_X f + \varepsilon_n$ for some sequence $\{\varepsilon_n\} \subset \mathbb{R}_+$ such that $\varepsilon_n \to 0^+$, guarantees the existence of a minimum for f (see e.g. [124]).

Proposition 1.1.4   If $f : X \to (-\infty, +\infty]$ is weakly lower semi-continuous on a reflexive Banach space X and has a bounded minimizing sequence, then f has a minimum on X.

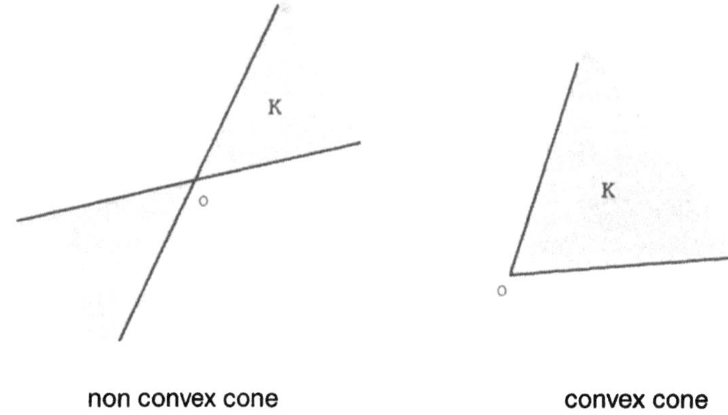

Figure 1.1.4.   non convex cone - convex cone.

The existence of a bounded minimizing sequence will be in particular insured when f is weakly coercive, i.e. such that

$$f(u) \to +\infty \ \ \text{if} \ \ \|u\| \to +\infty.$$

If in addition f is strictly convex then the minimum is unique.

Recall also that the sum of finitely many lower semi-continuous is lower semi-continuous whenever it is defined.

1.1.2   CONE, POLAR CONE, DUAL CONE, TANGENT CONE, NORMAL CONE

A set $K \subset X$ is said to be a cone if

$$0 \in K \tag{1.1.10}$$

and

$$\lambda K \subset K, \ \forall \, \lambda > 0 \tag{1.1.11}$$

It is easy to see that a cone K is convex if and only if (see Fig. 1.1.4)

$$K + K \subset K. \tag{1.1.12}$$

To every set $A \subset X$, we may associate a convex cone called the "cone generated by A", denoted by $\text{CONE}(A)$ and defined through the relation

$$\text{CONE}(A) = \bigcup_{\lambda \geq 0} \lambda \, conv\{A\}. \tag{1.1.13}$$

Let A be a subset of X. The polar cone A^+ of A is defined by

$$A^+ = \left\{ x^* \in X^* : \langle x^*, y \rangle \leq 0, \ \forall\, y \in A \right\}. \tag{1.1.14}$$

An important result in convex analysis asserts that

$$A^{++} = \overline{\mathrm{CONE}(A)}. \tag{1.1.15}$$

The dual cone of A is then defined by the relation

$$A^* = -A^+. \tag{1.1.16}$$

Let A be a nonempty subset of X. To every point $x \in X$, we can associate the set

$$T_A(x) \ = \ \left\{ y \in X : \exists \{x_k\} \subset A, \{\lambda_k\} \subset \mathbb{R}_+ \setminus \{0\}, \ x_k \to x, \atop \lambda_k(x_k - x) \to y \right\} \tag{1.1.17}$$

called the tangent cone of A at x. The normal cone of A at x is then defined as the polar cone of $T_A(x)$, i.e.

$$N_A(x) = T_A(x)^+. \tag{1.1.18}$$

Note that for $x \in X \setminus \overline{A}$, $T_A(x) = \emptyset$ while for $x \in int\{A\}$, $T_A(x) = X$ and $N_A(x) = \{0\}$. It is also easy to check that for any $x \in \overline{A}$, the set $T_A(x)$ is a closed cone.

Precising the structure of the set A, we may specify the sets $T_A(x)$ and $N_A(x)$.

Proposition 1.1.5 Let $a_i \in X^*(i = 1, \cdots, N)$ and $b_i \in \mathbb{R}(i = 1, \cdots, N)$ be given. We consider the set

$$A = \left\{ x \in X : \langle a_i, x \rangle \leq b_i, \ i \in \{1, \cdots, N\} \right\}. \tag{1.1.19}$$

Let $\overline{x} \in A$ and set

$$E(\overline{x}) = \left\{ \alpha \in \{1, \cdots, N\} : \langle a_\alpha, \overline{x} \rangle = b_\alpha \right\}.$$

Then

$$T_A(\overline{x}) = \left\{ y \in X : \langle a_\alpha, y \rangle \leq 0, \ \forall\, \alpha \in E(\overline{x}) \right\} \tag{1.1.20}$$

and

$$N_A(\overline{x}) = \left\{ \sum_{\alpha \in E(\overline{x})} \lambda_\alpha a_\alpha ; \lambda_\alpha \geq 0 \ (\alpha \in E(\overline{x})) \right\}. \qquad (1.1.21)$$

Proof. Let $y \in T_A(\overline{x})$ be given. Then there exists $\{x_k\} \subset A$, $\{\lambda_k\} \subset \mathbb{R}_+ \setminus \{0\}$ such that $x_k \to \overline{x}$ and $\lambda_k(x_k - \overline{x}) \to y$. Thus, for $\alpha \in E(\overline{x})$, we have

$$\begin{aligned}
\langle a_\alpha, y \rangle &= \langle a_\alpha, \lim_{k \to \infty} \lambda_k(x_k - \overline{x}) \rangle \\
&= \lim_{k \to \infty} \lambda_k \langle a_\alpha, x_k - \overline{x} \rangle \\
&= \lim_{k \to \infty} \lambda_k \left(\langle a_\alpha, x_k \rangle - b_\alpha \right).
\end{aligned}$$

We know that $x_k \in A$ and thus $\langle a_\alpha, x_k \rangle \leq b_\alpha$. It results that

$$\langle a_\alpha, y \rangle \leq 0$$

and

$$T_A(\overline{x}) \subset \left\{ y \in X : \langle a_\alpha, y \rangle \leq 0, \ \forall \, \alpha \in E(\overline{x}) \right\}.$$

Suppose now that $y \in X$ with $\langle a_\alpha, y \rangle \leq 0$, $\forall \, \alpha \in E(\overline{x})$. We set $\lambda_k = k$ and $x_k = \overline{x} + \frac{y}{k}$ $(k \in \mathbb{N} \setminus \{0\})$. We have $x_k = \overline{x} + \frac{1}{k}y \to \overline{x}$ and $\lambda_k(x_k - \overline{x}) = y \to y$. In addition

$$\langle a_i, x_k \rangle = \langle a_i, \overline{x} \rangle + \frac{1}{k}\langle a_i, y \rangle \ (i \in \{1, \cdots, N\}).$$

If $i \in E(\overline{x})$ then $\langle a_i, \overline{x} \rangle = b_i$ and $\langle a_i, y \rangle \leq 0$, so that

$$\langle a_i, x_k \rangle \leq b_i.$$

If $i \in \{1, \cdots, N\} \setminus E(\overline{x})$ then $\langle a_i, \overline{x} \rangle < b_i$ and for k great enough

$$\langle a_i, x_k \rangle < b_i.$$

That means that for k great enough

$$x_k \in A.$$

Thus

$$\left\{ y \in X : \langle a_\alpha, y \rangle \leq 0, \ \forall \, \alpha \in E(\overline{x}) \right\} \subset T_A(\overline{x}).$$

We have proved that

$$\begin{aligned}
T_A(\overline{x}) &= \left\{ y \in X : \langle a_\alpha, y \rangle \leq 0, \ \forall \, \alpha \in E(\overline{x}) \right\} \\
&= \left\{ a_\alpha ; \alpha \in E(\overline{x}) \right\}^+.
\end{aligned}$$

Thus using (1.1.18)

$$\begin{aligned}
N_A(\overline{x}) &= \{a_\alpha;\ \alpha \in E(\overline{x})\}^{++} \\[2mm]
&= \overline{\mathrm{CONE}\{a_\alpha; \alpha \in E(\overline{x})\}} \qquad\qquad (1.1.22) \\[2mm]
&= \{\textstyle\sum_{\alpha \in E(\overline{x})} \lambda_\alpha a_\alpha; \lambda_\alpha \geq 0\ (\alpha \in E(\overline{x})\}.
\end{aligned}$$

$\blacksquare$

A more general structure for A is now investigated.

Proposition 1.1.6   Let $\Phi_i : X \to \mathbb{R}$ $(i \in \{1, \cdots, n\})$ and $\Psi_i : X \to \mathbb{R}$ $(i \in \{1, \cdots, n\})$ be Gâteaux-differentiable functionals. We consider the set

$$\begin{aligned}
A = \Big\{ &x \in X : \Phi_i(x) \leq 0\ (i \in \{1, \cdots, N\}), \\
&\Psi_j(x) = 0\ (j \in \{1, \cdots, M\}) \Big\}.
\end{aligned} \qquad (1.1.23)$$

Let $\overline{x} \in A$ and set

$$E(\overline{x}) = \Big\{ \alpha \in \{1, \cdots, N\} : \Phi_i(\overline{x}) = 0 \Big\}.$$

We suppose that the derivatives $\Psi'_j(\overline{x})$, $1 \leq j \leq N$, are linearly independent. Moreover, we assume the existence of $y_0 \in X$ such that

$$\langle \Phi'_\alpha(\overline{x}), y_0 \rangle < 0,\ \forall\, \alpha \in E(\overline{x}). \qquad (1.1.24)$$

Then

$$\begin{aligned}
T_A(\overline{x}) = \Big\{ &y \in X : \langle \Phi'_\alpha(\overline{x}), y \rangle \leq 0\ (\alpha \in E(\overline{x})), \\
&\langle \Psi'_j(\overline{x}), y \rangle = 0\ (j \in \{1, \cdots, M\}) \Big\}
\end{aligned} \qquad (1.1.25)$$

and

$$\begin{aligned}
N_A(\overline{x}) = \Big\{ &\sum_{\alpha \in E(\overline{x})} \lambda_\alpha \Phi'_\alpha(\overline{x}) + \sum_{j=1}^{M} \lambda_j \Psi'_j(\overline{x}); \\
&\lambda_\alpha \geq 0(\alpha \in E(\overline{x})),\ \lambda_j \in \mathbb{R}\ (j \in \{1, \cdots, M\}) \Big\}.
\end{aligned} \qquad (1.1.26)$$

Note that (1.1.25) can be written as

$$T_A(\overline{x}) = \Big\{ \Phi'_\alpha(\overline{x}),\ -\Psi'_j(\overline{x}),\ +\Psi'_j(\overline{x}); \alpha \in E(\overline{x}),\ j \in \{1, \cdots, M\} \Big\}^{+}$$

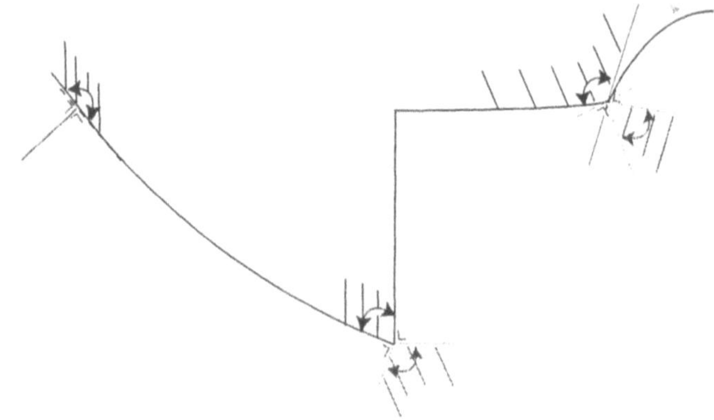

Tangent and normal cones to a set

Figure 1.1.5.   Tangent and normal cones to a set.

so that

$$N_A(\overline{x}) = \overline{\mathrm{CONE}\Big\{ \Phi'_\alpha(\overline{x}),\ -\Psi'_j(\overline{x}),\ +\Psi'_j(\overline{x}); \alpha \in E(\overline{x}),\ j \in \{1,\cdots,M\}\Big\}}$$

which gives formula (1.1.26).

Though the set defined in (1.1.23) encompasses the structure given in (1.1.19), Proposition 1.1.5 is more appropriate for this last case since formula (1.1.26) holds without the condition (1.1.24). Tangent and normal cones are depicted in Fig. 1.1.5.

Let us now suppose that A is a nonempty convex set. Then the following results holds.

Proposition 1.1.7   Let $A \subset X$ be a nonempty convex set. If $x \in A$ then

$$T_A(x) = \mathrm{CONE}(A - \{x\}) \tag{1.1.27}$$

and

$$N_A(x) = \Big\{x^* \in X^* : \langle x^*, y - x \rangle \leq 0,\ \forall\, y \in A\Big\}. \tag{1.1.28}$$

Proof.   Let us first remark that if $y \in A$ then $y - x \in T_A(x)$. Indeed, it suffices to set $x_k = \frac{1}{k}y + (1 - \frac{1}{k})x$ (which belongs to A for $k \geq 1$) and $\lambda_k = k$. Then $x_k \to x$ and $\lambda_k(x_k - x) = y - x$. Thus

$$A - \{x\} \subset T_A(x).$$

It results that

$$\mathrm{CONE}(A - \{x\}) \subset T_A(x)$$

since $T_A(x)$ is a cone. Then

$$\overline{\mathrm{CONE}(A - \{x\})} \; \subset \; \overline{T_A(x)} \; = \; T_A(x). \qquad (1.1.29)$$

Let $y \in T_A(x)$ be given. There exist $\{x_k\} \subset A, \{\lambda_k\} \subset \mathbb{R}_+ \setminus \{0\}$ with $x_k \to x$ and $\lambda_k(x_k - x) \to y$. We have

$$\lambda_k(x_k - x) \in \mathrm{CONE}(A - \{x\}), \; \forall\, k \; \in \; \mathbb{N}$$

and thus

$$y \in \overline{\mathrm{CONE}(A - \{x\})}. \qquad (1.1.30)$$

From this last result and (1.1.29), we deduce that

$$T_A(x) = \overline{\mathrm{CONE}(A - \{x\})}.$$

Let $x^* \in X^*$ with $\langle x^*, y - x \rangle \leq 0, \; \forall\, y \; \in \; A$. Then

$$\langle x^*, z \rangle \leq 0, \; \forall\, z \; \in \; A - \{x\}$$

so that

$$\langle x^*, z \rangle \leq 0, \; \forall\, z \; \in \; \overline{\mathrm{CONE}(A - \{x\})},$$

that is also

$$\langle x^*, z \rangle \leq 0, \; \forall\, z \; \in \; T_A(x).$$

That means that

$$x^* \in T_A(x)^+ = N_A(x).$$

Let now $x^* \in N_A(x)$ be given. We have

$$\langle x^*, z \rangle \leq 0, \; \forall\, z \; \in \; T_A(x).$$

If $y \in A$ then $y - x \in T_A(x)$ and thus

$$\langle x^*, y - x \rangle \leq 0, \; \forall\, y \; \in \; A.$$

It results that (1.1.28) holds.   ∎

The results of Proposition 1.1.7 are illustrated in Fig. 1.1.6.

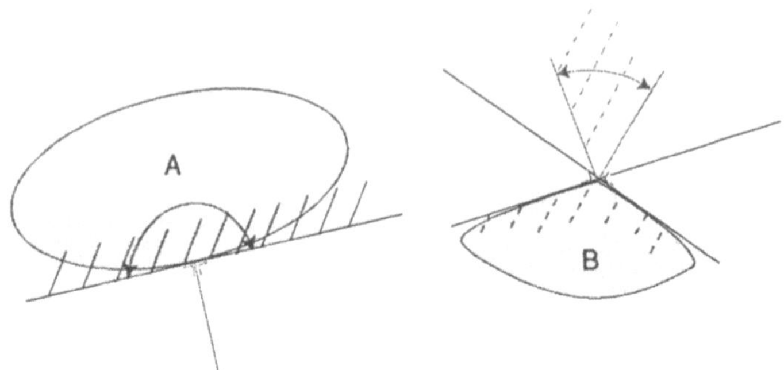

Figure 1.1.6.   Tangent and normal cones to a convex set.

1.1.3   SEPARATION OF CONVEX SETS

Let X be a vector space. We start with the analytic version of the Hahn-Banach theorem.

Theorem 1.1.8   Let $p : X \to \mathbb{R}$ be a function satisfying
(1) $p(\lambda x) = \lambda p(x)$, $\forall\, x \in X$, $\forall\, \lambda \geq 0$,
(2) $p(x + y) \leq p(x) + p(y)$, $\forall\, x,\, y \in X$.
Let $Y \subset X$ be a vector subspace of X and let $f : Y \to \mathbb{R}$ be a linear function satisfying
(3) $f(x) \leq p(x)$, $\forall\, x \in Y$.
Then there exists a linear function $\tilde{f}$ extending f to X, i.e.

$$\tilde{f}(x) = f(x), \; \forall\, x \in Y$$

and such that

$$\tilde{f}(x) \leq p(x), \; \forall\, x \in X.$$

A proof of this theorem may be found in [67]. Now let X be a topological vector space and let X^* be its dual space, i.e. the space of all linear continuous real-valued functions on X. For any $0 \neq f \in X^*$ and any $c \in \mathbb{R}$, the set

$$\mathcal{H}(c, f) = \{x \in X : \langle f, x \rangle = c\} \tag{1.1.31}$$

is called a closed hyperplane. The sets

$$f_c = \{x \in X : \langle f, x \rangle \le c\} \tag{1.1.32}$$

and

$$f^c = \{x \in X : \langle f, x \rangle \ge c\} \tag{1.1.33}$$

are called the closed half-spaces determined by $\mathcal{H}(c, f)$. Let A, B be two nonempty subsets of X. We say that A, B are separated by the hyperplane $\mathcal{H}(c, f)$ if either $A \subset f_c$ and $B \subset f^c$ or $A \subset f^c$ and $B \subset f_c$.

Let us now state the first geometric form of the Hahn-Banach theorem.

Theorem 1.1.9   Let A and B be two nonempty convex subsets of a normed vector space X, such that $A \cap B = \emptyset$. Suppose that A is open. Then there exists a closed hyperplane which separates A from B.

Proof. We set
$$C = A - B.$$

The set C is convex as the difference of two convex sets. It is open since $C = \cup_{y \in B}(A - y)$ and $0 \notin C$ since $A \cap B = \emptyset$. We claim that there exists $w \in X^*$ such that

$$\langle w, x \rangle < 0, \ \forall \, x \, \in \, C.$$

Indeed, let $x_0 \in -C$ and set

$$D = x_0 + C.$$

It is clear that $x_0 \notin D$ and $0 \in D$. We set

$$p(x) = \inf\{\alpha > 0 : \ \alpha^{-1}x \in D\},$$

that is p is the Minkowski functional of D. It is known that p satisfies (1) and (2) in Theorem 1.1.8 (see e.g. [67], [229]). Moreover, there exists $M > 0$ such that (see Lemma I.2 in [67])

$$0 \le p(x) \le M\|x\|, \ \forall \, x \, \in \, X \tag{1.1.34}$$

and

$$D = \{x \in X : \ p(x) < 1\}. \tag{1.1.35}$$

We consider the subspace $Y = \mathbb{R}\,x_0$ and the linear function $f : Y \to \mathbb{R}$ defined by

$$f(tx_0) = t, \ \forall \, t \, \in \, \mathbb{R}.$$

It is easy to see that

$$f(x) \le p(x), \ \forall \, x \, \in \, Y.$$

Using Theorem 1.1.8, we obtain a linear function $\tilde{f} : X \to \mathbb{R}$ such that

$$\tilde{f}(x) = f(x), \ \forall \, x \, \in \, Y$$

and

$$\tilde{f}(x) \leq p(x), \ \forall \, x \, \in \, X.$$

The right-hand inequality in (1.1.34) entails the continuity of $\tilde{f}$. Moreover

$$\tilde{f}(x) \leq p(x) < 1 = \tilde{f}(x_0) \, , \ \forall \, x \, \in \, D.$$

Thus there exists $w \in X^*$ such that

$$\langle w, x \rangle < \langle w, x_0 \rangle, \ \forall \, x \, \in \, x_0 \, + C.$$

That is also

$$\langle w, z \rangle < 0, \ \forall \, z \, \in \, C.$$

Thus

$$\langle w, x \rangle < \langle w, y \rangle, \ \forall \, x \, \in \, A, \, \forall \, y \, \in \, B.$$

We now choose $c \in \mathbb{R}$ such that

$$\sup_{x \in A} \langle w, x \rangle \leq c \leq \inf_{y \in B} \langle w, y \rangle \, .$$

This means that the hyperplane $\mathcal{H}(c, w)$ separates A and B.   ∎

Let X be a normed vector space, the first separation theorem asserts that, every nonempty open convex set A and every nonempty convex set B, not intersecting A, can be separated by a closed hyperplane, that is, we can find $c \in \mathbb{R}$ and $w \in X^*$ such that

$$\langle w, x \rangle \leq c \leq \langle w, y \rangle, \ \forall \, x \, \in \, A, \, \forall \, y \, \in \, B. \tag{1.1.36}$$

Now if $x \in \bar{A}$ then there exists a sequence $x_n \in A$ such that $x_n \to x$. By (1.1.36) we have

$$\langle w, x_n \rangle \leq c \leq \langle w, y \rangle, \ \forall \, y \, \in \, B$$

and taking the limit as $n \to \infty$, we obtain

$$\langle w, x \rangle \leq c \leq \langle w, y \rangle, \ \forall \, y \, \in \, B.$$

Thus (1.1.36) entails also

$$\langle w, x \rangle \leq c \leq \langle w, y \rangle, \ \forall \, x \, \in \, \bar{A}, \, \forall \, y \, \in \, B. \tag{1.1.37}$$

We say that A and B are strictly separated by $\mathcal{H}(c, f)$ if either $A \subset int\{f_c\}$ and $B \subset int\{f^c\}$ or $A \subset int\{f^c\}$ and $B \subset int\{f_c\}$.

Let us now give the second geometric form of the Hahn-Banach theorem.

Theorem 1.1.10 Let A and B be two nonempty, disjoint convex subsets of a normed vector space, such that A is closed and B is compact. Then there exists a closed hyperplane strictly separating A from B.

Proof. For $\varepsilon > 0$ we set $A_\varepsilon = A + B(0, \varepsilon)$ and $B_\varepsilon = B + B(0, \varepsilon)$ with $B(0, \varepsilon) = \{x \in X : \|x\| < \varepsilon\}$. The sets A_ε and B_ε are open, convex and, for ε small enough, disjoint. Using Theorem 1.1.9 we get the existence of $w \in X^* \backslash \{0\}$ and $c \in \mathbb{R}$ such that

$$\langle w, x + \varepsilon z \rangle \leq c \leq \langle w, y - \varepsilon z \rangle, \ \forall \, x \in A, \ \forall \, y \in B, \ \forall \, z \in B(0, 1).$$

From
$$\langle w, x \rangle + \varepsilon \langle w, z \rangle \leq c, \forall \, z \in B(0, 1)$$
one deduces that
$$\langle w, x \rangle + \varepsilon \|w\|_* \leq c,$$
where $\| \cdot \|_*$ denote the norm in the dual space X^*. On the other hand, from
$$c + \varepsilon \langle w, z \rangle \leq \langle w, y \rangle, \forall \, z \in B(0, 1)$$
one obtains that
$$c + \varepsilon \|w\|_* \leq \langle w, y \rangle.$$
Thus

$$\langle w, x \rangle + \varepsilon \|w\|_* \leq c \leq \langle w, y \rangle - \varepsilon \|w\|_*, \ \forall \, x \in A, \ \forall \, y \in B.$$

From the previous inequalities we conclude that $\mathcal{H}(w, c)$ separates strictly A from B. ∎

A fundamental consequence of this separation theorem is the following result.

Theorem 1.1.11 Let X be a real Banach space. Any convex and lower semi-continuous function $f : X \to (-\infty, +\infty]$ is bounded from below by an affine continuous function.

Proof. If f is not proper then the statement of Theorem 1.1.11 is obvious. Let us now assume that f is proper. For any fixed $y \in X$

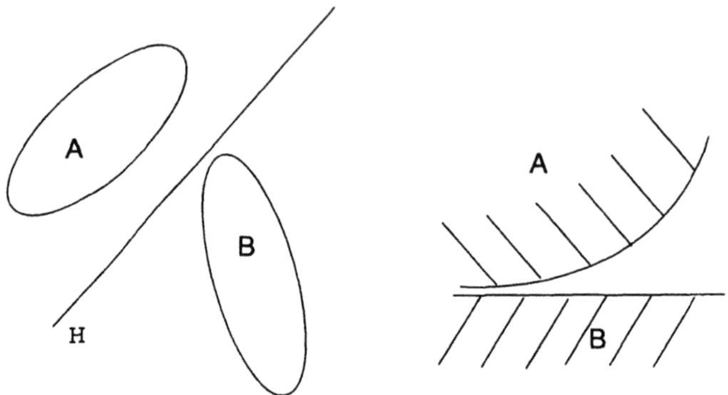

A and B are strictly separated by H

A and B cannot be strictly separated by a closed hyperplane

Figure 1.1.7.   Separation of convex sets.

choose $t \in \mathbb{R}$ such that $t < f(y)$. Since $(y,t) \notin epi\ f$ and $epi\ f$ is a closed convex set, we may apply Theorem 1.1.10 to get $w \in X^* \backslash \{0\}$, α, $\beta \in \mathbb{R}$ such that the closed hyperplane

$$\{(x,\tau) \in X \times \mathbb{R} : \langle w, x \rangle + \alpha\tau = \beta\}$$

separates (y,t) from $epi\ f$. Hence

$$\langle w, y \rangle + \alpha t < \beta < \langle w, x \rangle + \alpha\tau, \ \forall\ (x,\tau) \in epi\ f.$$

If $f(y) < +\infty$, taking $(y, f(y)) \in epi\ f$, we obtain $\alpha(f(y) - t) > 0$, i.e., $\alpha > 0$. Thus

$$\frac{\beta}{\alpha} - \frac{1}{\alpha}\langle w, y \rangle < f(y).$$

This inequality remains obviously true if $f(y) = +\infty$. Hence $l(x) = \frac{1}{\alpha}(\beta - \langle w, x \rangle)$ is an affine continuous function which is a minorant of the function f.   ∎

It results from Theorem 1.1.11 that if $f : X \to (-\infty, +\infty]$ is a convex and weakly lower semi-continuous function then there exists $\alpha \geq 0$ and $\beta \in \mathbb{R}$ such that

$$f(x) \geq -\alpha\|x\| + \beta, \ \forall\ x \in X.$$

1.1.4   THE SZULKIN LEMMA

Let X be a real Banach space and $\mathcal{X} : X \to (-\infty, +\infty]$ a lower semi-continuous convex function with $\mathcal{X}(0) = 0$. We know (Theorem 1.1.11) that $\mathcal{X}$ is bounded from below by an affine function, i.e. $\mathcal{X}(x) \geq \langle z, x \rangle - \beta$ on X, for some $z \in X^*$ and $\beta \in \mathbb{R}$. The following result of Szulkin [405] asserts that in some cases we can choose $z \neq 0$ with norm ≤ 1 and $\beta = 0$.

Lemma 1.1.12 Let X be a real Banach space and $\mathcal{X} : X \to (-\infty, +\infty]$ a convex and lower semi-continuous function with $\mathcal{X}(0) = 0$. If

$$\mathcal{X}(x) \geq -\|x\|, \ \forall \, x \in X,$$

then there exists a $z \in X^* \backslash \{0\}$ such that $\|z\|_* \leq 1$ and

$$\mathcal{X}(x) \geq \langle z, x \rangle, \ \forall \, x \in X.$$

Proof. In the space $X \times \mathbb{R}$ define

$$A = \{(x,t) : \|x\| < -t\} \quad \text{and} \quad B = \{(x,t) : \mathcal{X}(x) \leq t\}.$$

It is easy to verify that A and B are convex (in fact, B is the epigraph of $\mathcal{X}$) and A is open. Moreover, since $\mathcal{X}(x) \geq -\|x\|$ we get $A \cap B = \emptyset$. Consequently, there exists an hyperplane separating A and B, i.e., we can find $\alpha, \beta \in \mathbb{R}$ and $w \in X^*$ such that (see Theorem 1.1.9 and (1.1.36))

$$\langle w, x \rangle - \alpha t - \beta \geq 0, \ \forall \, (x,t) \in \bar{A},$$
$$\langle w, x \rangle - \alpha t - \beta \leq 0, \ \forall \, (x,t) \in B.$$

Since $(0,0) \in \bar{A} \cap B$, $\beta = 0$. Taking $t = -\|x\|$ in the first of these inequalities gives

$$\langle w, x \rangle \geq -\alpha \|x\|, \ \forall \, x \in X.$$

It follows that $\alpha \geq 0$ and $\|w\|_* \leq \alpha$. If $\alpha = 0$, then $w = 0$ and there is no hyperplane. So $\alpha > 0$. Set $z = w/\alpha$ and $t = \mathcal{X}(x)$ in the second of the above inequalities. Then $\|z\|_* \leq 1$ and $\langle z, x \rangle \leq t = \mathcal{X}(x)$. ∎

1.1.5   EKELAND'S VARIATIONAL PRINCIPLE

The aim of this Section is to present the famous Ekeland's variational principle [125] which will be used later in Chapter 4.

Theorem 1.1.13 Let (X, d) be a complete metric space, and $f : X \to (-\infty, +\infty]$ be a proper lower-semi-continuous function which is bounded from below. If $\delta > 0$, $\lambda > 0$ and $x \in X$ satisfy

$$f(x) \leq \inf_{v \in X} f(v) + \delta,$$

then there exists $y \in X$ such that
(1) $f(y) \leq f(x)$,

(2) $d(x, y) \leq \frac{1}{\lambda}$,
(3) $f(w) > f(y) - \delta\lambda d(y, w)$, $\forall\, w \neq y$.

Proof.   Let us define inductively a sequence $\{x_n\}_{n \in \mathbb{N}}$ starting with $x_0 = x$. Suppose that $x_n \in X$ is known. If

$$f(w) > f(x_n) - \delta\lambda d(x_n, w), \ \forall\, w \neq x_n, \tag{1.1.38}$$

we set $x_{n+1} = x_n$. On the contrary if there exists $w \neq x_n$ such that

$$f(w) \leq f(x_n) - \delta\lambda d(x_n, w) \tag{1.1.39}$$

then we choose $x_{n+1} \in S_n$, where S_n is the set of all w satisfying (1.1.39), such that

$$f(x_{n+1}) - \inf_{S_n} f \leq \frac{1}{2}\Big(f(x_n) - \inf_{S_n} f\Big). \tag{1.1.40}$$

The sequence $\{x_n\}$ is a Cauchy sequence. Indeed, if (1.1.38) ever occurs, it is stationary, and if not, (1.1.39) entails

$$\delta\lambda d(x_n, x_{n+1}) \leq f(x_n) - f(x_{n+1}), \ \forall\, n \in \mathbb{N}.$$

Adding the inequalities up, we get

$$\delta\lambda d(x_n, x_p) \leq f(x_n) - f(x_p), \ \forall\, n \leq p. \tag{1.1.41}$$

The sequence $\{f(x_n)\}$ is decreasing and bounded from below by $\inf f$ and thus convergent. Therefore (1.1.41) entails that $\{x_n\}$ is a Cauchy sequence. Since the space X is complete, x_n converges to some $y \in X$. We have

$$f(x) \geq f(x_1) \geq \cdots \geq f(x_n) \geq \cdots$$

and, using the fact that f is lower semi-continuous, we obtain (1) by taking the lower limit as $n \to \infty$. Setting $n = 0$ in (1.1.41) we get

$$\delta\lambda d(x, x_p) \leq f(x) - f(x_p), \ \forall\, p \geq 0. \tag{1.1.42}$$

Thus

$$\delta\lambda d(x, x_p) \leq f(x) - \inf_X f \leq \delta,$$

so that

$$d(x, x_p) \leq \frac{1}{\lambda}, \ \forall \, p \geq 0.$$

Taking the limit as $p \to +\infty$, we obtain (2). If the inequality (3) is supposed not to hold, there would be some $w \neq y$ such that

$$f(w) \leq f(y) - \delta\lambda d(y, w).$$

Letting $p \to \infty$ in (1.1.41) we obtain

$$f(y) \leq f(x_n) - \delta\lambda d(x_n, y).$$

Thus

$$f(w) \leq f(x_n) - \delta\lambda(d(y, w) + d(x_n, y)) \leq f(x_n) - \delta\lambda d(x_n, w).$$

If $w \neq x_n$, using (1.1.40), we obtain

$$2f(x_{n+1}) - f(x_n) \leq \inf_{S_n} f,$$

and thus

$$2f(x_{n+1}) - f(x_n) \leq f(w).$$

This relation remains true if $w = x_n$. When $n \to \infty$, $f(x_n) \to l$, and the previous relation yields $l \leq f(w)$. Since f is lower semi-continuous, we also have $f(y) \leq l$. Finally, we get the inequality $f(y) \leq f(w)$, contradicting the definition of w. ∎

The variational principle of Ekeland means that in a neighborhood of x we can find a point y which minimizes exactly the perturbed function

$$w \longmapsto f(w) + \frac{\delta}{\lambda} d(y, w).$$

1.1.6   PROJECTION OPERATOR ON A CLOSED CONVEX SET

Let $\{X, (.,.)\}$ be a real Hilbert space and C a nonempty closed convex subset of X. To each $x \in X$, we can associate a unique element $P_C x \in C$ satisfying (see e.g. [25])

$$\|x - P_C x\| = \min_{z \in C} \|x - z\|. \tag{1.1.43}$$

The operator $P_C : X \to C$ which is so defined is called the projection operator onto C (or best approximation operator of C) and satisfies the properties listed in the following Proposition.

Proposition 1.1.14 The operator $P_C : X \to C$ is idempotent, nonexpansive and monotone, i.e.

$$P_C^2 = P_C, \tag{1.1.44}$$

$$\|P_C x - P_C y\| \leq \|x - y\|, \ \forall \, x, \, y \, \in \, X, \tag{1.1.45}$$

$$(P_C x - P_C y, x - y) \geq 0, \ \forall \, x, \, y \, \in \, X. \tag{1.1.46}$$

Let us recall that here the duality mapping $J : X \to X^*$ from the Hilbert space X onto its dual can be defined by the formula

$$\langle Ju, v \rangle = (u, v), \ \forall \, u, \, v \, \in \, X.$$

The following examples hold.

Example 1.1.15 Let X be a real Hilbert space and let $C \subset X$ be defined by

$$C = \{y \in X : Ay = b\},$$

where $A \in \mathcal{L}(X, X^*)$ is surjective and $b \in X^*$. Then

$$P_C(x) = x - J^{-1}A^*(AJ^{-1}A^*)^{-1}(Ax - b).$$

The operator

$$A^+ = J^{-1}A^*(AJ^{-1}A^*)^{-1}$$

is the orthogonal right inverse of A.

Example 1.1.16 Let X be a real Hilbert space and let $C \subset X$ be defined by

$$C = \{y \in X : \langle p, y \rangle = 0\},$$

where $p \in X^* \backslash \{0\}$. Then

$$P_C(x) = x - \frac{\langle p, x \rangle}{\|p\|_*^2} J^{-1}p.$$

Example 1.1.17 Let $C = \mathbb{R}_+^n$ $(n \in \mathbb{N}, n \geq 1)$. Then

$$P_C(x) = x^+,$$

where x^+ denotes the vector $x^+ := (\max\{0, x_i\})_{i=1,\dots,n}$.

The relationship between the projector P_C and the variational inequality problem is now discussed.

Proposition 1.1.18 Relation (1.1.43) is satisfied if and only if

$$(P_C x - x, v - P_C x) \geq 0, \ \forall \, v \, \in \, C. \tag{1.1.47}$$

Proof. Suppose that (1.1.43) is satisfied. Then

$$\|x - P_C x\|^2 \leq \|x - z\|^2, \ \forall \, z \, \in \, C.$$

Setting $z = (1 - t)P_C x + tv$ for a given $t \in (0, 1]$ and $v \in C$, we obtain

$$\|x - P_C x\|^2 \leq \|P_C x - x\|^2 + t^2\|v - P_C x\|^2 - 2t(P_C x - x, P_C x - v).$$

Thus

$$(P_C x - x, P_C x - v) \leq \frac{t}{2}\|P_C x - v\|^2.$$

Taking the limit as $t \to 0^+$ and since v is arbitrary, we obtain (1.1.47). Suppose now that (1.1.47) is satisfied. We have

$$\begin{aligned}
\|x - P_C x\|^2 &= (x - P_C x, x - P_C x) \\
&= (x - P_C x, x - z) + (x - P_C x, z - P_C x)
\end{aligned}$$

and thus for $z \in C$, we obtain thanks to (1.1.47) the inequality

$$\|x - P_C x\|^2 \leq (x - P_C x, x - z) \leq \|x - P_C x\| \, \|x - z\|$$

This means that

$$\|x - P_C x\| \leq \|x - z\|, \ \forall \, z \, \in \, C.$$

$\blacksquare$

The previous result means that for each $f \in X$, the element $P_C f \in C$ is the unique solution of the variational inequality: Find $u \in C$ such that

$$(u - f, v - u) \geq 0, \ \forall \, v \, \in \, C.$$

Moreover, (1.1.45) implies that the mapping $f \to P_C(f)$ is Lipschitz continuous.

Let us consider the more general variational inequality:

Find $u \in C$ such that

$$\langle Au - f, v - u \rangle \geq 0, \ \forall \, v \, \in \, C, \tag{1.1.48}$$

where $A : X \to X^*$ is a possibly nonlinear operator, $f \in X^*$, X a real Hilbert space and C a nonempty closed convex subset of X. Then we see easily that the inequality (1.1.48) is equivalent to the following one

$$(u - (u - \sigma J^{-1} Au - \sigma J^{-1} f), v - u) \geq 0, \ \forall \, v \, \in \, C,$$

where $\sigma > 0$ is any positive constant. This means that the variational inequality problem is also equivalent to the fixed point problem:

Find $u \in X$ such that

$$u = P_C(u - \sigma J^{-1} Au - \sigma J^{-1} f), \tag{1.1.49}$$

the positive real number σ being fixed.

1.1.7   THE KKM PRINCIPLE

In this Section we state a fundamental result of Knaster-Kuratowski-Mazurkiewicz [227] expressing through the statement that a given intersection is nonempty.

Let X be a real vector space and $A \subset X$ a subset. A set-valued mapping $G : A \to 2^X$ is called a KKM mapping if

$$conv\{x_1, \cdots, x_n\} \subset \bigcup_{i=1}^{n} G(x_i) \tag{1.1.50}$$

for each finite subset $\{x_1, \cdots, x_n\}$ of A.

Example 1.1.19   Let X be a real Banach space, $C \subset X$ a nonempty convex subset and $F : C \to X$ a mapping. Let us define $G : C \to 2^X$ by

$$G(x) = \{v \in C : \langle F(v), v - x \rangle \leq 0\}, \ x \in C.$$

We claim that G is a KKM mapping. Indeed, suppose on to the contrary the existence of $y_0 \in conv\{x_1, \cdots, x_n\}$ such that $y_0 \notin \bigcup_{i=1}^{n} G(x_i)$. Then

$$\langle F(y_0), y_0 - x_i \rangle > 0, \ \forall \, i = 1, \ldots, n.$$

Therefore

$$x_i \in \Lambda := \{x \in C : \langle F(y_0), y_0 - x \rangle > 0\}$$

for each $i \in \{1, \cdots, n\}$. The set Λ is convex and thus contains $y_0 \in conv\{x_1, \cdots, x_n\}$. We get the contradiction $\langle F(y_0), y_0 \rangle > \langle F(y_0), y_0 \rangle$.

Let us now state the KKM principle (see e.g. [181]).

Theorem 1.1.20   Let X be a vector space and $A \subset X$ a nonempty subset. Let $G : A \to 2^X$ be a KKM mapping such that for each $x \in A$, the set $G(x) \cap L$ is closed for any finite-dimensional subspace L of X. Then the family of sets $\{G(x) : x \in A\}$ has the finite intersection property, that is every nonempty finite subfamily of $\{G(x) : x \in A\}$ has a nonempty intersection.

Proof.   Suppose by contradiction that

$$\bigcap_{i=1}^{n} G(x_i) = \emptyset,$$

for some $x_1, \cdots, x_n \in A$.

Let $L = span\{x_1, \cdots, x_n\}, C = conv\{x_1, \cdots, x_n\}$ and d the metric distance on L. Let $h : C \to \mathbb{R}$ be defined by

$$h(x) = \sum_{i=1}^{n} d(x, L \cap G(x_i)).$$

The set $\bigcap_{i=1}^{n} L \cap G(x_i)$ is empty and therefore $h(x) \neq 0$ for all $x \in C$. Let now $f : C \to C$ be defined by

$$f(x) = \frac{1}{h(x)} \sum_{i=1}^{n} d(x, L \cap G(x_i))x_i, \ x \in C.$$

The map f is continuous and using Brouwer's fixed point Theorem for simplexes (Bohl's Theorem [53]), we get $x_0 \in C$ such that $x_0 = f(x_0)$. If $d(x_0, L \cap G(x_i)) \neq 0$ then $x_0 \notin G(x_i)$. Thus if we set $I = \{i \in \{1, \ldots, n\} : d(x_0, L \cap G(x_i)) \neq 0\}$ then

$$x_0 \notin \bigcup_{i \in I} G(x_i).$$

However, by construction,

$$f(x_0) \in conv\{x_i : i \in I\}$$

and since f is a KKM mapping, we obtain

$$f(x_0) \in \bigcup_{i \in I} G(x_i),$$

which is a contradiction since $x_0 = f(x_0)$.   ∎

Note that if E is a compact topological space and $\mathcal{F}$ a family of closed subsets of E then $\bigcap_{F \in \mathcal{F}} F \neq \emptyset$ provided that $\mathcal{F}$ has the finite intersection property [219]. Using this last property we obtain the following analytic version of the KKM principle.

Theorem 1.1.21 Let X be a real reflexive Banach space and C a non-empty closed convex and bounded subset of X. Let $\Phi : X \to \mathbb{R} \cup \{+\infty\}$ be a convex and l.s.c. function such that $C \cap D(\Phi) \neq \emptyset$. Let $f, g : C \times C \to \mathbb{R}$ be two functions satisfying

(1) $g(x, y) \leq f(x, y)$, $\forall\, x, y \in C$;

(2) $x \mapsto f(x, y)$ is concave on C, for each fixed $y \in C$;

(3) $y \mapsto g(x, y)$ is w.l.s.c. on C, for each fixed $x \in C$.

Then for each $\lambda \in \mathbb{R}$, the following alternative holds: either

(a) there exists $y_0 \in C \cap D(\Phi)$ such that

$$g(x, y_0) + \Phi(y_0) - \Phi(x) \leq \lambda, \ \forall\, x \in C \cap D(\Phi),$$

or

(b) there exists $x_0 \in C$ such that

$$f(x_0, x_0) > \lambda.$$

Proof. Let $\lambda \in \mathbb{R}$ be given and let $G : C \cap D(\Phi) \to 2^X$ be defined by

$$G(x) = \{v \in C \cap D(\Phi) : \Phi(v) - \Phi(x) + g(x, v) \leq \lambda\}.$$

Assumption (3) together with our assumptions on C imply that each set $G(x)$ is weakly compact. It entails that $G(x) \cap L$ is closed for any finite dimensional subspace of X. Moreover we know that the set $\bigcap_{x \in C \cap D(\Phi)} G(x)$ is nonempty provided that the family $\{G(x) : x \in C \cap D(\Phi)\}$ has the finite intersection property. Thus if G is a KKM mapping we obtain the existence of $y_0 \in C$ such that $y_0 \in \bigcap_{x \in C \cap D(\Phi)} G(x)$

and thus satisfying (a). If G is not a KKM mapping then there exists $\{x_1, \cdots, x_n\} \subset C \cap D(\Phi)$ and $x_0 \in conv\{C \cap D(\Phi)\} = C \cap D(\Phi)$ such that $x_0 \notin \bigcup_{i=1}^{n} G(x_i)$, that is

$$\lambda < \Phi(x_0) - \Phi(x_i) + g(x_i, x_0), \ \forall \ i = 1, \cdots, n.$$

Using assumption (1), we obtain

$$x_i \in A := \{x : f(x, x_0) + \Phi(x_0) - \Phi(x) > \lambda\}$$

Assumption (2) entails that A is convex and therefore $x_0 \in A$, i.e. (b) is satisfied. ∎

A direct consequence of Theorem 1.1.21 is the analytic form of the famous Ky-Fan principle [129].

Corollary 1.1.22 Let X be a reflexive Banach space and C a nonempty closed convex and bounded subset of X. Let $f : C \times C \to \mathbb{R}$ be a function satisfying

(1) $f(x, x) \le 0, \ \forall \ x \in C$;

(2) $x \mapsto f(x, y)$ concave on C, for each fixed $y \in C$;

(3) $y \mapsto f(x, y)$ w.l.s.c. on C, for each fixed $x \in C$.

Then there exists $y_0 \in C$ such that

$$f(x, y_0) \le 0, \ \forall \ x \in C.$$

Proof. Apply Theorem 1.1.21 with $f = g$, $\Phi = 0$ and $\lambda = 0$. Assumption (1) of Corollary 1.1.22 implies that the alternative (b) in Theorem 1.1.21 cannot occur.

1.1.8   MINTY'S PRINCIPLE

Let X be a real Banach space and C a nonempty closed convex subset of X. If $A : X \to X^*$ is a monotone and hemicontinuous operator (see Section 3.3) then a well-known result due to Minty [269] establishes that for $u \in C$ the inequality

$$\langle Au, v - u \rangle \ge 0, \ \forall \ v \in C,$$

is equivalent to the following one

$$\langle Av, v - u \rangle \geq 0, \ \forall \, v \in C.$$

Let us now state this result in a more general form which will be used later.

Theorem 1.1.23 Let X be a real Banach space and let C be a nonempty closed convex subset of X. Let $A : X \to X^*$ be a monotone and hemicontinuous operator, $\varphi : X \to \mathbb{R} \cup \{+\infty\}$ a proper convex l.s.c. function and $H : X \times X \to \mathbb{R}$ a mapping which is positively homogeneous with respect to the second variable, i.e.

$$H(u, \lambda v) = \lambda H(u, v), \ \forall \, u, \, v \in X, \ \lambda > 0.$$

Then $u \in C$ solves

$$\langle Au, v - u \rangle + \varphi(v) - \varphi(u) + H(u, v - u) \geq 0, \ \forall \, v \in C, \qquad (1.1.51)$$

if and only if

$$\langle Av, v - u \rangle + \varphi(v) - \varphi(u) + H(u, v - u) \geq 0, \ \forall \, v \in C. \qquad (1.1.52)$$

Proof. If u satisfies (1.1.51) then

$$\langle Au, v - u \rangle + \varphi(v) - \varphi(u) + H(u, v - u) \geq 0, \ \forall \, v \in C, \qquad (1.1.53)$$

and since A is monotone, we obtain

$$\langle Av, v - u \rangle \geq \langle Au, v - u \rangle, \forall \, v \in X. \qquad (1.1.54)$$

Using (1.1.53) together with (1.1.54) we obtain that u solves (1.1.52). Let us now suppose that u solves (1.1.52), i.e.

$$\langle Av, v - u \rangle + \varphi(v) - \varphi(u) + H(u, v - u) \geq 0, \ \forall \, v \in C. \qquad (1.1.55)$$

Let $h \in C$ and $\theta \in (0, 1]$ be given. Set $v = (1 - \theta)u + \theta h \in C$ in (1.1.55) to get

$$\theta \langle A((1 - \theta)u + \theta h), h - u \rangle + \theta(\varphi(h) - \varphi(u)) + \theta H(u, h - u) \geq 0.$$

Dividing the last inequality by θ and taking the limit as $\theta \to 0^+$, we obtain, due to the hemicontinuity of A, that

$$\langle Au, h - u \rangle + \varphi(h) - \varphi(u) + H(u, h - u) \geq 0.$$

Since h was chosen arbitrarily in C, we may conclude that u satisfies (1.1.51). ∎

1.1.9   THE COMPLEMENTARITY PRINCIPLE

Let X be a real Banach space, $A : X \to X^*$ a possibly nonlinear operator and C a nonempty closed convex subset of X. Let us consider the variational inequality:

$$u \in C : \quad \langle Au, v - u \rangle \geq 0, \ \forall\, v \in C. \qquad (1.1.56)$$

If the set C is a nonempty closed convex cone then the inequality (1.1.56) is equivalent to the complementarity system

$$u \in C, \qquad (1.1.57)$$

$$\langle Au, v \rangle \geq 0, \ \forall\, v \in C, \qquad (1.1.58)$$

$$\langle Au, u \rangle = 0. \qquad (1.1.59)$$

The proof of this important property is given below.

Theorem 1.1.24   (Complementarity Principle)
If C is a nonempty closed convex cone then (1.1.56) is equivalent to the system (1.1.57)-(1.1.59).

Proof.   Let u be a solution of (1.1.56). Set $v = 0$ in (1.1.56) to obtain

$$\langle Au, u \rangle \leq 0$$

and then set $v = 2u$ in (1.1.56) to get

$$\langle Au, u \rangle \geq 0.$$

Thus $\langle Au, u \rangle = 0$ and (1.1.59) is satisfied. Let $h \in C$ be given and set $v = u + h \in C$ in (1.1.56), one obtains

$$\langle Au, h \rangle \geq 0.$$

Since h is arbitrarily chosen in C, we conclude that (1.1.58) is satisfied. Conversely, if u is a solution of (1.1.57)-(1.1.59) then the difference between (1.1.58) and (1.1.59) leads to

$$\langle Au, v \rangle - \langle Au, u \rangle \geq 0, \ \forall\, v \in C,$$

that is u solves (1.1.56).   ∎

Recall that the dual cone of C is denoted by C^* and defined by the formula (see Section 1.1.2)

$$C^* := \{ w \in X^* : \langle w, v \rangle \geq 0, \ \forall\, v \in C \},$$

and the relation (1.1.58) can also be written as $Au \in C^*$. Sometimes it is considered the polar cone with respect to C defined to be $-C^*$

Remark 1.1.25 The natural connection between variational inequalities involving cone constraints and complementarity problems will be used later in this book so as to obtain some existence theorems for variational inequalities (see Section 5.6) and bifurcation theorems for eigenvalue problems in variational inequalities (see Section 10.3 and 10.4). It is worth to note that the mathematical model (1.1.57)-(1.1.59) as well as some generalized complementarity models have their own applications in Optimization, Game Theory, Structural Mechanics, Economic Equilibrium. For some applications of the complementarity models in Structural Mechanics and Robotics, we refer the reader to the following articles : [7], [8], [173], [154], [225], [226], [236], [237], [246] [248], [249], [258], [259], [395]. The Complementarity Mathematical Theory has recently known important developments and, for further details, we refer the interested reader to the books of Isac [197], [202] and Murty [301] and to the surveys of Bershchanskii and Meerov [50] and Moré [277]. Note here also that some important generalizations of the complementarity model (1.1.57)-(1.1.59) are discussed and studied in [55], [56], [399], [166], [179], [198], [199], [200], [201].

1.1.10   THE ORDERED COMPLEMENTARITY PRINCIPLE

Let X be a real Banach space. We say that $K \subset X$ is a pointed convex cone if K is a convex cone and

$$K \cap (-K) = \{0\}. \qquad (1.1.60)$$

We denote by " $\overset{K}{\leq}$ " the ordering defined by K, that is

$$x \overset{K}{\leq} y \Leftrightarrow y - x \in K. \qquad (1.1.61)$$

Assume now that the ordered vector space $(X, \overset{K}{\leq})$ is a vector lattice, that is for every pair $(x, y) \in X \times X$, the supremum $\vee \{x, y\}$ and the infinum $\wedge \{x, y\}$ exist in X. We recall that for every $x, y, z \in X$, $\alpha, \beta > 0$ the following fundamental relations hold [229]:

$$\vee \{x, y\} + z = \vee \{x + z, y + z\} \qquad (1.1.62)$$

$$\wedge \{x, y\} + z = \wedge \{x + z, y + z\} \qquad (1.1.63)$$

$$x + y = \vee\{x, y\} + \wedge\{x, y\} \tag{1.1.64}$$

$$\vee\Big\{\vee\{x, y\}, z\Big\} = \vee\Big\{\vee\{x, y\}, \vee\{y, z\}\Big\} = \vee\{x, y, z\} \tag{1.1.65}$$

$$\wedge\{x, y\} = 0 \quad \Leftrightarrow \quad \wedge\{\alpha x, \beta y\} = 0. \tag{1.1.66}$$

Setting $x^{+} = \vee\{0, x\}, x^{-} = \vee\{0, -x\}$ and $\mid x \mid = \vee\{x, -x\}$, we have also

$$\mid x + y \mid \leq \mid x \mid + \mid y \mid \tag{1.1.67}$$

$$\mid x^{+} - y^{+} \mid \leq \mid x - y \mid \tag{1.1.68}$$

and

$$\mid x^{-} - y^{-} \mid \leq \mid x - y \mid . \tag{1.1.69}$$

Let $A : X \to X$ be an operator. The ordered complementarity problem consists to find $x \in X$ such that

$$\wedge\{x, Ax\} = 0. \tag{1.1.70}$$

If X is a real Hilbert lattice, i.e.

$$\wedge\{x, y\} = 0 \Leftrightarrow x, y \in K \text{ and } (x, y) = 0 \tag{1.1.71}$$

then problem (1.1.70) is equivalent to the complementarity system

$$x \in K \tag{1.1.72}$$

$$Ax \in K \tag{1.1.73}$$

$$(x, Ax) = 0. \tag{1.1.74}$$

If in addition, the scalar product $(.,.)$ satisfies the property

$$x \in K, y \in K \Rightarrow (x, y) \geq 0 \tag{1.1.75}$$

then (1.1.73) yields

$$(Ax, v) \geq 0, \ \forall v \in K,$$

i.e. $Ax \in K^{*}$ and the classical formulation (1.1.57)-(1.1.59) is recovered.

Let $\sigma > 0$ be given. We have

$$\begin{aligned}
\wedge\{x, Ax\} = 0 \quad &\Leftrightarrow \quad \wedge\{x, \sigma Ax\} = 0 \\
&\Leftrightarrow \quad x - \wedge\{x, \sigma Ax\} = x \\
&\Leftrightarrow \quad x + \vee\{-x, -\sigma Ax\} = x \\
&\Leftrightarrow \quad \vee\{0, x - \sigma Ax\} = x.
\end{aligned}$$

This means that the formulation (1.1.70) is also equivalent to the fixed point problem.

Find $x \in X$ such that

$$x = \vee\{0, x - \sigma A x\}, \tag{1.1.76}$$

the real number $\sigma > 0$ being fixed.

Note that if X is a real Hilbert lattice endowed with a scalar product satisfying the condition (1.1.75), we have

$$\vee\{0, z\} = P_K(z). \tag{1.1.77}$$

Indeed, it suffices to prove that (1.1.47) holds, that is

$$(P_K z - z, h - P_K z) \geq 0, \ \forall h \in K.$$

Here

$$
\begin{aligned}
(\vee\{0, z\} - z, \ h - \vee\{0, z\}) \ &= \ (\vee\{-z, 0\}, \ h - \vee\{0, z\}) \\
&= \ (z^-, h - z^+) \\
&= \ (z^-, h) - (z^-, z^+).
\end{aligned}
$$

It is clear from (1.1.71) that $(z^-, z^+) = 0$. Moreover, condition (1.1.75) yields $(z^-, h) \geq 0$ for any $h \in K$. It results that in this framework, the fixed point problem (1.1.70) is equivalent to the following one

$$x = P_K(x - \sigma A x). \tag{1.1.78}$$

The formulation (1.1.49) is here also recovered.

1.1.11   THE VARIATIONAL PRINCIPLE

Let C be a nonempty closed convex subset of a Banach space X. We first show that if $\Phi : X \to \mathbb{R}$ is Gâteaux-differentiable, then any solution of the variational principle

$$u \in C : \Phi(u) \leq \Phi(v), \ \forall v \in C \tag{1.1.79}$$

is a solution of the variational inequality

$$u \in C : \langle \Phi'(u), v - u \rangle \geq 0, \ \forall v \in C. \tag{1.1.80}$$

Theorem 1.1.26 Let u be a solution of problem (1.1.79). Then u satisfies the inequality (1.1.80).

Proof. Let $\lambda \in]0, 1], h \in C$ and set $v := u + \lambda(h - u)$ in (1.1.79) to get

$$\Phi(u) \leq \Phi(u + \lambda(h - u)).$$

Thus

$$\frac{\Phi(u + \lambda(h - u)) - \Phi(u)}{\lambda} \geq 0$$

and taking the limit as $\lambda \to 0+$, we obtain the inequality

$$\langle \Phi'(v), h - u \rangle \geq 0.$$

$\blacksquare$

Let us now show that if in addition $\Phi : X \to \mathbb{R}$ is convex then the inequality problem (1.1.80) is equivalent to the variational principle (1.1.79).

Theorem 1.1.27   If Φ is convex then formulations (1.1.79) and (1.1.80) are equivalent.

Proof.   Let u be a solution of problem (1.1.80). The function Φ is convex and thus for $\lambda \in]0, 1]$ and $v \in X$, we have

$$\Phi((1 - \lambda)u + \lambda v) \leq (1 - \lambda)\Phi(u) + \lambda\Phi(v)$$

and so

$$\frac{\Phi((u + \lambda(v - u)) - \Phi(u)}{\lambda} \leq \Phi(v) - \Phi(u).$$

Taking the limit as $\lambda \to 0+$, we obtain

$$\langle \Phi'(u), v - u \rangle \leq \Phi(v) - \Phi(u).$$

It results that for $v \in C$,

$$0 \leq \Phi(v) - \Phi(u),$$

that is u is a solution of problem (1.1.79). The converse has been proved in Theorem 1.1.26 $\blacksquare$

For example, if $\Phi : X \to \mathbb{R}$ is defined by

$$\Phi(v) = \int_0^1 \langle A(\theta v), v \rangle d\theta - \langle l, v \rangle + K \qquad (1.1.81)$$

where $K = \Phi(0)$, $A : X \to X^*$ is a monotone, continuous and potential operator and $l \in X^*$, then Problems (1.1.80) and (1.1.79) are equivalent. Here we have

$$\Phi' = A - l.$$

Corollary 1.1.28 Suppose that Φ is given by (1.1.81). Then Formulations (1.1.79) and (1.1.80) are equivalent.

Proof. Since A is a potential operator, there exists a Gâteaux-differentiable functional Ψ such that $A = \Psi'$. Then

$$\langle A(\theta v), v \rangle = \langle \Psi'(\theta v), v \rangle = \frac{d}{dr}\Psi(rv)\,|_{r=\theta},$$

and thus

$$\int_0^1 \langle A(\theta v), v \rangle d\theta = \Psi(v) - \Psi(0).$$

Setting

$$\Phi(v) = \Psi(v) - \langle l, v \rangle,$$

we obtain

$$\Phi(0) = \Psi(0),$$

and thus (1.1.81). Moreover Φ is Gâteaux-differentiable and

$$\Phi' = \Psi' - l = A - l.$$

The monotonicity of A implies that Φ is convex and the result follows from Theorem 1.1.26.   ∎

From Theorem 1.1.26 and Proposition 1.1.5, we may deduce an important result in quadratic programming theory.

Let $X = \mathbb{R}^n$ and $\Phi : X \to \mathbb{R}$ be defined by

$$\Phi(x) = \frac{1}{2}x^T Q x + c^T x, \tag{1.1.82}$$

with $Q \in \mathbb{R}^{n \times n}$ symmetric and $c \in \mathbb{R}^n$. We consider the set

$$C = \{x \in X : Ax \leq b\}, \tag{1.1.83}$$

with $A \in \mathbb{R}^{N \times n}$ and $b \in \mathbb{R}^N$.

Corollary 1.1.29 Suppose that Φ and C are given by (1.1.82) and (1.1.83) respectively. Then a vector $\overline{x} \in C$ satisfies the variational inequality (1.1.80) if and only if there exists a vector $\overline{\lambda} \in \mathbb{R}^N$ such that

$$c + Q\overline{x} + A^T\overline{\lambda} = 0, \quad \overline{\lambda} \geq 0, \quad \overline{\lambda}^T(A\overline{x} - b) = 0.$$

Proof. Let $\overline{x}$ satisfying (1.1.80). From (1.1.28), we see that

$$Q\overline{x} + c \in -N_C(\overline{x}).$$

Denoting by $A_i \in \mathbb{R}^{1 \times n}$ the i-th line of the matrix A, we may write

$$C = \{x \in X : A_i x \leq b_i, \ i \in \{1, \cdots, N\}\}$$

and thanks to Proposition 1.1.5, we may find real numbers $\lambda_\alpha \in E(\overline{x})$ such that

$$Q\overline{x} + c = - \sum_{\alpha \in E(\overline{x})} \lambda_\alpha A_\alpha^T.$$

Here

$$E(\overline{x}) = \{\alpha \in \{1, \cdots, N\} : A_\alpha \overline{x} = b_\alpha\}.$$

Defining now the vector $\overline{\lambda}$ by setting $\overline{\lambda}_i = \lambda_i$ for $i \in E(\overline{x})$ and $\overline{\lambda}_i = 0$ for $i \in \{1, \cdots, N\}\backslash E(\overline{x})$, we see that $\overline{\lambda} \geq 0$, $\overline{\lambda}^T(A\overline{x} - b) = 0$ and we may write

$$Q\overline{x} + c = - \sum_{\alpha=1}^{N} \overline{\lambda}_\alpha A_\alpha^T,$$

that is also

$$c + Q\overline{x} + A^T\overline{\lambda} = 0$$

Let us now check the converse. Let $\overline{x} \in \mathbb{R}^n$ and $\overline{\lambda} \in \mathbb{R}^N$ such that

$$\overline{\lambda} \geq 0, \ A\overline{x} \leq b, \ c + Q\overline{x} = -A^T\overline{\lambda}, \ \overline{\lambda}^T(A\overline{x} - b) = 0.$$

Remarking that $\overline{\lambda}_i = 0$ as soon as $(A\overline{x} - b)_i = A_i\overline{x} - b_i < 0$, we obtain that

$$Q\overline{x} + c = - \sum_{\alpha \in E(\overline{x})} \lambda_\alpha A_\alpha^T$$

and thus thanks to Proposition 1.1.5,

$$Q\overline{x} + c \in -N_C(\overline{x}).$$

∎

From Theorem 1.1.26 and Corollary 1.1.29 we see that if $\overline{x}$ solves the variational principle (1.1.79) with Φ and C as defined in (1.1.82) and (1.1.83) respectively, then there exists a vector $\overline{\lambda} \in \mathbb{R}^N$ such that

$$c + Q\overline{x} + A^T\overline{\lambda} = 0, \ \overline{\lambda} \geq 0, \ \overline{\lambda}^T(A\overline{x} - b) = 0.$$

The converse holds if A is positive semi-definite since in this case the functional Φ in (1.1.82) is convex and Theorem 1.1.27 can be applied.

1.2   UNILATERAL ANALYSIS IN $L_{LOC}(X; \mathbb{R})$

This Section is devoted to the calculus developed by Clarke [88]– [90] and Rockafellar [375], [376] for locally Lipschitz functions on Banach spaces. We restrict ourselves to those basic elements of Clarke's theory that are needed in the sequel. Precisely, we discuss generalized directional derivative and generalized gradient, Lebourg's mean value theorem [239], chain rule, generalized gradient of integral functions and of restrictions to submanifolds. Finally, it is given the definition of critical point in the sense of Chang [80] for a locally Lipschitz function and it is pointed out the relationship between the hemivariational inequalities and the generalized critical point problem.

1.2.1   GENERALIZED DIRECTIONAL DERIVATIVE AND CLARKE'S SUBGRADIENT

A function $f : U \to \mathbb{R}$ on an open subset of a real Banach space X is called locally Lipschitz if each point $u \in U$ possesses a neighborhood $N_u \subset U$ such that

$$| f(u_1) - f(u_2) | \le K\|u_1 - u_2\|, \ \forall \, u_1, \, u_2 \, \in \, N_u,$$

for a constant $K > 0$ depending on N_u.

Definition 1.2.1   The generalized directional derivative of the locally Lipschitz function $f : U \to \mathbb{R}$ at the point $u \in U$ in the direction $v \in X$ is defined by

$$f^0(u; v) := \limsup_{\substack{w \to u \\ t \downarrow 0}} \frac{1}{t}(f(w + tv) - f(w)).$$

The next two results establish the relationship between the generalized directional derivative $f^0(u; v)$ and the classical (one-sided) directional derivative

$$f'(u; v) := \lim_{t \downarrow 0} \frac{1}{t}(f(u + tv) - f(u)),$$

when this exists, in some important situations.

Proposition 1.2.2   If $f : U \to \mathbb{R}$ is continuously differentiable, then the equality below holds

$$f^0(u;v) = f'(u;v), \ \forall\, u \, \in\, U, \ \forall\, v \, \in\, X. \qquad (1.2.1)$$

Proof.   Fix $u \in U$, $v \in X$ and $w \in U$. The function $g(t) = f(w + tv)$, for all $t \geq 0$ sufficiently small, is continuously differentiable with the derivative

$$g'(t) = f'(w + tv; v).$$

Given any admissible $t > 0$ the mean value theorem yields some $s \in (0, t)$ such that

$$\frac{1}{t}(f(w + tv) - f(w)) = \frac{1}{t}(g(t) - g(0)) = g'(s) = f'(w + sv; v).$$

In the view of the continuity of the differential of f, letting $w \to u$ in X and $t \to 0$ in $\mathbb{R}$, the desired result follows.

Proposition 1.2.3   Let the open set U be convex in X and suppose that $f : U \to \mathbb{R}$ is a convex and continuous function. Then f is locally Lipschitz and formula (1.2.1) is valid.

Proof.   The convexity of $f : U \to \mathbb{R}$ ensures the existence of the one-sided directional derivative $f'(u; v)$. The continuity of f in conjunction with the convexity of f implies that f is locally Lipschitz. Let us fix $u \in U$, $v \in X$ and an arbitrary small number $\delta > 0$ such that the Lipschitz condition for f holds on an open ball centered at u with radius δ and the Lipschitz constant $K > 0$. Definition 1.2.1 enables us to write

$$
\begin{aligned}
f^0(u;v) &= \lim_{\varepsilon \downarrow 0} \ \sup_{\|w-u\|<\varepsilon\,\delta} \ \sup_{0<t<\varepsilon} \frac{1}{t}(f(w + tv) - f(w)) \\[2mm]
&= \lim_{\varepsilon \downarrow 0} \ \sup_{\|w-u\|<\varepsilon\,\delta} \frac{1}{\varepsilon}(f(w + \varepsilon\, v) - f(w)),
\end{aligned}
$$

the second equality being the consequence of the convexity of f. The local Lipschitzianess of f implies

$$\left| \frac{1}{\varepsilon}(f(w + \varepsilon\, v) - f(w)) - \frac{1}{\varepsilon}(f(u + \varepsilon\, v) - f(u)) \right| \leq 2\delta K$$

for $\|w - u\| < \varepsilon\,\delta$ with $\varepsilon \in (0, 1)$. We derive that

$$f^0(u;v) \leq \lim_{\varepsilon \downarrow 0} \frac{1}{\varepsilon}(f(u + \varepsilon\, v) - f(u)) + 2\delta K = f'(u;v) + 2\delta K.$$

Letting $\delta \to 0$ we arrive at

$$f^0(u,;v) \leq f'(u,v).$$

The reverse inequality is a direct consequence of Definition 1.2.1.   ∎

The concept in Definition 1.2.1 does not presuppose the existence of any limit.

Remark 1.2.4 i) A function $f : U \to \mathbb{R}$ on an open set U of a Banach space X is said to be regular at $u \in U$ if the one-sided directional derivative $f'(u;v)$ exists whenever $v \in X$ and the relation (1.2.1) is true. Therefore Propositions 1.2.2 and 1.2.3 provide relevant examples of regular functions.

ii) One says that $f : U \to \mathbb{R}$ is strictly differentiable at $u \in U$ if there exists an element of $\mathcal{L}(X; \mathbb{R})$ denoted by $D_s f(u)$ such that

$$\lim_{\substack{w \to u \\ t \downarrow 0}} \frac{1}{t}(f(w + tv) - f(w)) = \langle D_s f(u), v \rangle, \ \forall \, v \in X$$

and provided the convergence is uniform for v in compact sets. If f is strictly differentiable at x then f is locally Lipschitz near x [90]. Continuously differentiable functions and locally Lipschitz functions such that the limit

$$\lim_{\substack{w \to u \\ t \downarrow 0}} \frac{1}{t}(f(w + tv) - f(w)) = \langle \zeta, v \rangle$$

for some $\zeta \in \mathcal{L}(X; \mathbb{R})$ and for all $v \in X$ holds are examples of strictly differentiable functions [90].

The following result collects the essential properties of the generalized directional derivative.

Proposition 1.2.5   Let $f : U \to \mathbb{R}$ be a locally Lipschitz function on an open set U of a Banach space X. Then the following properties hold:
(a) For every $u \in U$ the function $f^0(u, \cdot) : X \to \mathbb{R}$ is positively homogeneous and subadditive, so convex, satisfies the inequality

$$|f^0(u;v)| \leq K\|v\|, \ \forall \, v \in X, \tag{1.2.2}$$

and is Lipschitz continuous on X with the Lipschitz constant K if $K > 0$ is a Lipschitz constant of f near u.

(b) $f^0(\cdot;\cdot): U \times X \to \mathbb{R}$ is upper semi-continuous.
(c) $f^0(u;-v) = (-f)^0(u;v)$, $\forall\, u \in U$, $\forall\, v \in X$.
(d) $-f^0(u;v) \le (-f)^0(u;v)$, $\forall\, u \in U$, $\forall\, v \in X$.

Proof. (a) The positive homogeneity of $f^0(u;\cdot)$ and inequality (1.2.2) follow readily from Definition 1.2.1. To check the subadditivity of $f^0(u;\cdot)$ let $v_1, v_2 \in X$ be fixed. Then we see that

$$
\begin{aligned}
f^0(u; v_1 + v_2) &= \limsup_{\substack{w \to u \\ t \downarrow 0}} \frac{1}{t}\left(f(w + t(v_1 + v_2)) - f(w)\right) \\[2mm]
&\le \limsup_{\substack{w \to u \\ t \downarrow 0}} \frac{1}{t}\left(f(w + tv_1 + tv_2) - f(w + tv_2)\right) \\[2mm]
&\quad + \limsup_{\substack{w \to u \\ t \downarrow 0}} \frac{1}{t}\left(f(w + tv_2) - f(w)\right) \\[2mm]
&= f^0(u; v_1) + f^0(u; v_2).
\end{aligned}
$$

For arbitrary $v_1, v_2 \in X$, using the Lipschitz constant K on a neighborhood of u, we obtain

$$
\begin{aligned}
f(w + tv_1) - f(w) &= f(w + tv_1) - f(w + tv_2) + f(w + tv_2) - f(w) \\
&\le K\|v_1 - v_2\|t + f(w + tv_2) - f(w)
\end{aligned}
$$

whenever w is close to u and $t > 0$ is small enough. Then we infer that

$$
f^0(u; v_1) \le K\|v_1 - v_2\| + f^0(u; v_2).
$$

We check in the same way that

$$
f^0(u; v_2) \le K\|v_1 - v_2\| + f^0(u; v_1).
$$

Assertion (a) is thus verified.

(b) In order to justify the upper semicontinuity of f^0 let $\{u_n\} \subset U$ and $\{v_n\} \subset X$ be sequences with $u_n \to u \in U$ and $v_n \to v \in X$ as $n \to \infty$. Take any $x_n \in U$ and $t_n > 0$ with

$$
\|x_n - u_n\| + t_n < \frac{1}{n}
$$

and

$$
f^0(u_n; v_n) \le t_n^{-1}(f(x_n + t_n v_n) - f(x_n)) + \frac{1}{n}.
$$

It turns out that

$$f^0(u_n; v_n) - \frac{1}{n} \leq t_n^{-1}(f(x_n + t_n v) - f(x_n))$$

$$+ t_n^{-1}(f(x_n + t_n v_n) - f(x_n + t_n v))$$

$$\leq t_n^{-1}(f(x_n + t_n v) - f(x_n)) + K\|v_n - v\|,$$

where $K > 0$ is the Lipschitz constant of f around u. Passing to the limit for $n \to \infty$ it results

$$\limsup_{n \to \infty} f^0(u_n; v_n) \leq \limsup_{n \to \infty} t_n^{-1}(f(x_n + t_n v) - f(x_n)) \leq f^0(u; v)$$

which ends the proof of (b).

(c) For given $u \in U$, $v \in X$, by Definition 1.2.1 we have

$$\begin{aligned} f^0(u; -v) &= \limsup_{\substack{w \to u \\ t \downarrow 0}} \frac{1}{t}(f(w - tv) - f(w)) \\[2mm] &= \limsup_{\substack{w \to u \\ t \downarrow 0}} \frac{1}{t}(f(w - tv)) - f(w - tv + tv) \\[2mm] &= \limsup_{\substack{w \to u \\ t \downarrow 0}} \frac{1}{t}(-f(w - tv + tv) - (-f(w - tv))) \\[2mm] &= (-f)^0(u; v). \end{aligned}$$

(d)

$$\begin{aligned} -f^0(u; v) &= -\limsup_{\substack{w \to u \\ t \downarrow 0}} \frac{1}{t}(f(w + tv) - f(w)) \\[2mm] &= \liminf_{\substack{w \to u \\ t \downarrow 0}} \frac{1}{t}(-f(w + tv) - (-f(w))) \\[2mm] &\leq (-f)^0(u; v). \end{aligned}$$

This completes the proof.   ∎

The definition below introduces the main concept in this Section.

Definition 1.2.6 Let $f : U \to \mathbb{R}$ be a locally Lipschitz function on the open set U of a Banach space X. The generalized gradient $\partial f(u)$ of f at a point $u \in U$ is the subset of the dual space X^* defined as follows

$$\partial f(u) = \{z \in X^* : \langle z, v \rangle \leq f^0(u; v), \ \forall v \in X\}.$$

Useful properties of the generalized gradient are given below.

Proposition 1.2.7   Let $f : U \to \mathbb{R}$ be as in Definition 1.2.6. Then the following assertions hold :

(i) For every $u \in U$, $\partial f(u)$ is a nonempty, convex and weak *-compact subset of X^* which is bounded by the Lipschitz constant $K > 0$ of f near u.

(ii) For every $u \in U$, $f^0(u; \cdot)$ is the support function of $\partial f(u)$, i.e.,

$$f^0(u; v) = \max\{\langle z, v \rangle : z \in \partial f(u)\}, \ \forall \, v \in X.$$

(iii) The set-valued map ∂f from U to X^* is weak*-closed, that is, if $\{u_n\} \subset U$ and $\{z_n\} \subset X^*$ are sequences such that $u_n \to u$ (strongly in X), $z_n \in \partial f(u_n)$ and $z_n \rightharpoonup^* z$ (weakly * in X^*) for $u \in U$, $z \in X^*$, then $z \in \partial f(u)$. In particular, if X is finite dimensional, ∂f is upper semi-continuous.

(iv) The set-valued map ∂f is weak*-upper semi-continuous from U to X^* in the sense that for any $u \in U$, $\varepsilon > 0$, $v \in X$ there is a $\delta > 0$ such that for each $z \in \partial f(x)$ with $\|x - u\| < \delta$ there is some $\zeta \in \partial f(u)$ such that $|\langle z - \zeta, v \rangle| < \varepsilon$.

Proof.   (i) Applying (a) in Proposition 1.2.5 and the Hahn-Banach Theorem 1.1.8 with $p(x) := f^0(u; x)$, $Y = \{0\}$ and $F \equiv 0$, (the linear function f of Theorem 1.1.8 is here denoted by F) there exists $z \in X^*$ satisfying

$$\langle z, v \rangle \leq f^0(u; v), \ \forall \, v \in X.$$

Hence $\partial f(u)$ is nonempty. The convexity of $\partial f(u)$ is clear from Definition 1.2.6. For $z \in \partial f(u)$ and $v \in X$, by Definition 1.2.6 relation (1.2.2) and (c) in Proposition 1.2.5 we see that

$$\begin{aligned}
-K\|v\| \ &\leq \ -|(-f)^0(u; v)| \leq -(-f)^0(u; v) = -f^0(u; -v) \\
&\leq \ \langle z, v \rangle \leq f^0(u; v) \leq K\|v\|.
\end{aligned}$$

Thus

$$|\langle z, v \rangle| \leq K\|v\|.$$

It follows

$$\|z\|_* \leq K, \ \forall \, z \in \partial f(u). \tag{1.2.3}$$

Since $\partial f(u)$ is weak*-closed (see part (iii)), the boundedness in (1.2.3) and Banach-Alaoglu-Bourbaki's Theorem (see, e.g. [67], p.42) ensures that $\partial f(u)$ is weak*-compact in X^*.

(ii) Suppose by condradiction that there exists $v \in X$ with

$$f^0(u; v) > \max\{\langle z, v \rangle : z \in \partial f(u)\}. \tag{1.2.4}$$

The Hahn-Banach Theorem enables us to find $\zeta \in X^*$ that fulfills the properties $\langle \zeta, v \rangle = f^0(u; v)$ and

$$\langle \zeta, x \rangle \leq f^0(u; x), \ \forall \, x \in X.$$

We may indeed apply Theorem 1.1.8 with $p(x) = f^0(u; x)$, $Y = \mathbb{R}v$ and $F(tv) = f^0(u; v)t$ (the linear function f of Theorem 1.1.8 is here denoted by F) since

$$F(tv) \leq p(tv), \ \forall \, t \in \mathbb{R}.$$

Indeed, if $t \geq 0$ then using (a) of Proposition 1.2.5, we get

$$F(tv) = p(tv).$$

If $t < 0$ then using (a), (c) and (d) of Proposition 1.2.5, we obtain

$$\begin{aligned}
p(tv) &= f^0(u; -t(-v)) \\
&= -tf^0(u; -v) \\
&= -t(-f)^0(u; v) \\
&\geq -t(-f^0(u; v)) \\
&= tf^0(u; v).
\end{aligned}$$

Therefore $\zeta \in \partial f(u)$ contradicts (1.2.4) which establishes the stated result.

(iii) Suppose that the sequences $\{u_n\} \subset U$ and $\{z_n\} \subset X^*$ satisfy the requirements in (iii). Fix some $v \in X$. In view of (b) in Proposition 1.2.5, by passing to the limit superior in the inequality

$$\langle z_n, v \rangle \leq f^0(u_n; v)$$

we obtain

$$\langle z, v \rangle \leq \limsup_{n \to \infty} f^0(u_n; v) \leq f^0(u; v).$$

Since $v \in X$ is arbitrary we conclude $z \in \partial f(u)$.

(iv) Arguing by contradiction we assume that the data $u \in U$, $\varepsilon > 0$ and $v \in X$ can be found such that there exist sequences $\{x_n\} \subset U$ and $\{z_n\} \subset X^*$ with $x_n \to u$ strongly in X, $z_n \in \partial f(x_n)$ and

$$|\langle z_n - \zeta, v \rangle| \geq \varepsilon, \ \forall \, \zeta \in \partial f(u). \tag{1.2.5}$$

From (i) we deduce the boundedness $\|z_n\|_* \leq K$ for n sufficiently large. So, along a subsequence, we have for some $z \in X^*$ that

$$z_n \rightharpoonup^* z \text{ in } X^*.$$

Then assertion (iii) implies $z \in \partial f(u)$. This clearly contradicts (1.2.5). The proof is complete. ∎

Let us here recall that one says that $f : U \to \mathbb{R}$ is said to have a derivative in the Gâteaux sense at $u \in U$ if there exists $f'(u) \in X^*$ such that for all $v \in X$,

$$\frac{f(u + \lambda v) - f(u)}{\lambda} \to \langle f'(u), v \rangle \text{ as } \lambda \to 0.$$

The element $f'(u) \in X$ corresponds by the Riesz theorem to $\nabla f(u)$ called the gradient of f at u if X is a Hilbert space. One says that $f : U \to \mathbb{R}$ is Fréchet differentiable at $u \in U$ if there exists $h \in X^*$ such that $f(u + v) = f(u) + h(v) + o(\|v\|)\|v\|$. Any Fréchet differentiable function is Gâteaux differentiable. Conversely if f is Gâteaux differentiable and the mapping $v \to f'(v)$ is continuous then it is Fréchet differentiable. We determine the generalized gradient for the situations described in Propositions 1.2.2 and 1.2.3.

Proposition 1.2.8   Let $f : U \to \mathbb{R}$ be a locally Lipschitz function on an open set U of a Banach space X. The following properties holds :
(i) If f is Gâteaux differentiable at $u \in U$, then its Gâteaux derivative $f'(u)$ belongs to $\partial f(u)$.
(ii) If f is continuously differentiable at $u \in U$, then $\partial f(u) = \{f'(u)\}$. More generally, f is strictly differentiable at $u \in U$ if and only if f is locally Lipschitz near u and $\partial f(u)$ reduces to a singleton which is necessarily the strict derivative of f at u. In particular, if X is finite dimensional, $\partial f(u)$ consists of a singleton for all $u \in U$ if and only if f is continuously differentiable on U.
(iii) If U is a convex set and $f : U \to \mathbb{R}$ is a convex function, then $\partial f(u)$, at any $u \in U$, coincides with the subdifferential of f at u in the sense of convex analysis (see Section 1.3).
(iv) If X is finite dimensional, then $\partial f(u)$ at any $u \in U$ is the convex hull in $X^* \equiv X$ of all points

$$z = \lim_{n \to \infty} f'(u_n), \tag{1.2.6}$$

with $u_n \to u$ as $n \to \infty$, $\{u_n\}$ contained in the complement of a Lebesgue measure zero set and f differentiable at each u_n.

Proof. (i) The Gâteaux derivative $f'(u)$ satisfies

$$\begin{aligned}
\langle f'(u), v \rangle &= f'(u; v) \\
&= \lim_{t \downarrow 0} \frac{1}{t}(f(u + tv) - f(u)) \\
&\leq f^0(u; v), \ \forall v \in V.
\end{aligned}$$

By Definition 1.2.6 this means that $f'(u) \in \partial f(u)$.

(ii) Proposition 1.2.2, specifically its proof, shows that if f is continuously differentiable at $u \in U$ then formula (1.2.1) holds. The linearity of the map $v \to \langle f'(u), v \rangle = f'(u; v)$ and Definition 1.2.6 imply $\partial f(u) = \{f'(u)\}$. For the second assertion of (ii) we refer to the careful analysis in Clarke [90]. Finally, assuming that X is finite dimensional and $\partial f(u)$ is a singleton for each $u \in U$, it follows that f is differentiable on U. Then, by the final remark in (iii) of Proposition 1.2.7, we deduce the continuity of the derivative f', so f is continuously differentiable.

(iii) This is the direct consequence of Proposition 1.2.3, assertion (ii) in Proposition 1.2.7 and the convexity of f (see Proposition 1.3.1).

(iv) By Rademacher's Theorem f is known to be Gâteaux differentiable a.e., so the involved sequences $\{u_n\}$ exist. In view of (iii) in Proposition 1.2.7 we see that each limit z (1.2.6) belongs to $\partial f(u)$. Hence the convex hull of the foregoing limits z forms a compact subset of $\partial f(u)$. Then the converse inclusion will be obtained from

$$f^0(u; v) \leq \limsup_{w \to u} \langle f'(w), v \rangle, \ \forall v \in X, \qquad (1.2.7)$$

where w in (1.2.7) runs in the admissible subset of U. Denote by α the right-hand side of (1.2.7) and fix a nonzero $v \in X$. There is $\delta > 0$ with

$$\langle f'(w), v \rangle \leq \alpha + \varepsilon, \quad \text{for a.e.} \quad \|w - u\| < \delta.$$

Then, using Fubini's Theorem, we get

$$f(w + tv) = f(w) + \int_0^t \langle f'(w + sv, v) \rangle ds$$

for all $\|w - u\| < \delta/2$ and all $0 \leq t < \delta/(2\|v\|)$. Clearly, this leads to

$$f^0(u; v) \leq \alpha + \varepsilon.$$

Letting $\varepsilon \to 0$ we arrive at (1.2.7). This ends the proof.   ∎

Remark 1.2.9 It is worth to point out that for a locally Lipschitz function $f : U \to \mathbb{R}$ on an open subset U of a Banach space X that is

(Gâteaux) differentiable at $u \in U$ without being continuously differentiable at u the generalized gradient $\partial f(u)$ does not generally reduce to $\{f'(u)\}$ (for comparison see (i) of Proposition 1.2.8). For instance, in the case of the locally Lipschitz function $f : \mathbb{R} \to \mathbb{R}$ given by $f(0) = 0$ and $f(x) = x^2 \sin(1/x)$ otherwise, one has $f'(0) = 0$ and $\partial f(0) = [-1, 1]$. In this respect we mention a notion of generalized derivative due to Michel and Penot [266] which is equal to the usual one-sided directional derivative under a weak condition. Namely, if $f : U \to \mathbb{R}$ is a function on an open subset of a Banach space X, the radial strict derivative of f at $u \in U$ in the direction $v \in X$ is

$$f^\square(u; v) = \sup_{w \in X} \limsup_{t \downarrow 0} \frac{1}{t}(f(u + tw + tv) - f(u + tw)).$$

The radial strict subdifferential of f at $u \in U$ is

$$\partial^\square f(u) = \{z \in X^* : \langle z, v \rangle \leq f^\square(u; v), \ \forall\, v \in X\}.$$

It is proved by Michel and Penot [266] that if f has a convex one-sided directional derivative $f'(u; \cdot)$ at $u \in U$ then $f^\square(u; \cdot) = f'(u; \cdot)$. If, moreover, f is Gâteaux differentiable at $u \in U$ then $\partial^\square f(u) = \{f'(u)\}$. We also refer the reader to the book of Dem'yanov, Stavroulakis, Polyakova and Panagiotopoulos [105] which develops the field of unilateral analysis through the concepts of quasidifferential and codifferential.

Employing the notion of generalized gradient in Definition 1.2.6 the corresponding calculus can be developed.

Proposition 1.2.10
(i) For each $\lambda \in \mathbb{R}$ and a locally Lipschitz function $f : U \to \mathbb{R}$ one has

$$\partial(\lambda f)(u) = \lambda \partial f(u), \ \forall\, u \in U.$$

(ii) Given the locally Lipschitz functions $f, g : U \to \mathbb{R}$ the relation below holds

$$\partial(f + g)(u) \subset \partial f(u) + \partial g(u), \ \forall\, u \in U.$$

Proof. (i) is obvious for $\lambda \geq 0$. Thus it suffices to justify it for $\lambda = -1$. This follows from (c) of Proposition 1.2.5. Indeed,

$$\begin{aligned}
\partial(-f)(u) &= \{w \in X^* : \langle w, v \rangle \leq (-f)^0(u; v), \ \forall\, v \in X\} \\
&= \{w \in X^* : \langle w, v \rangle \leq f^0(u; -v), \ \forall\, v \in X\}
\end{aligned}$$

$$\begin{aligned}
&= \{w \in X^* : -\langle w, v\rangle \leq f^0(u; v), \ \forall \, v \in X\} \\
&= \{-w \in X^* : \langle w, v\rangle \leq f^0(u; v), \ \forall \, v \in X\} \\
&= -\partial f(u).
\end{aligned}$$

(ii) It is equivalent to show that

$$(f + g)^0(u; v) \leq f^0(u; v) + g^0(u; v), \ \forall \, v \in X$$

which is a direct consequence of Definition 1.2.1.   ∎

Corollary 1.2.11   If $u \in U$ is a local minimum or maximum of the locally Lipschitz function $f : U \to \mathbb{R}$ on an open set of a Banach space X, then $0 \in \partial f(u)$.

Proof. Applying (i) of Proposition 1.2.10 for $\lambda = -1$ we see that it is sufficient to consider the case where $u \in U$ is a local minimum. Then we note that

$$f^0(u; v) \geq \limsup_{t \downarrow 0} \frac{1}{t}(f(u + tv) - f(u)) \geq 0$$

for all $v \in X$. By Definition 1.2.6 the result follows.   ∎

The result below will be a basic tool in the following.

Theorem 1.2.12   (Lebourg's Mean Value Theorem [239])
Let U be an open subset of a Banach space X, let x, y be two points of U such that the line segment

$$[x, y] = \{(1 - t)x + ty : 0 \leq t \leq 1\}$$

is contained in U and let $f : U \to \mathbb{R}$ be a locally Lipschitz function. Then there exists $u \in [x, y] \setminus \{x, y\}$ satisfying

$$f(y) - f(x) = \langle z, y - x\rangle$$

for some $z \in \partial f(u)$.

Proof. The function $g : [0, 1] \to \mathbb{R}$ given by

$$g(t) = f((1 - t)x + ty), \ \forall \, t \in [0, 1]$$

is locally Lipschitz. It verifies

$$g^0(t; s) \ = \ \limsup_{\substack{\tau \to t \\ \lambda \downarrow 0}} \frac{1}{\lambda}(g(\tau + \lambda s) - g(\tau))$$

$$\begin{aligned}
=\ & \limsup_{\substack{\tau \to t \\ \lambda \downarrow 0}} \frac{1}{\lambda}(f(x + (\tau + \lambda s)(y - x)) - f(x + \tau(y - x))) \\[2mm]
\leq\ & \limsup_{\substack{w \to (1-t)x+ty \\ \lambda \downarrow 0}} \frac{1}{\lambda}(f(w + \lambda s(y - x)) - f(w)) \\[2mm]
=\ & f^0((1 - t)x + ty; s(y - x)), \ \forall\, t \in (0,1), \ \forall\, s \in \mathbb{R}.
\end{aligned}$$

We introduce the function $h : [0,1] \to \mathbb{R}$ by

$$h(t) = g(t) + t(f(x) - f(y)), \ \forall\, t \in [0,1].$$

Notice that $h(0) = h(1) = f(x)$, so h admits as a local minimum or maximum point some $t \in (0,1)$. By Corollary 1.2.11 this implies

$$0 \in \partial h(t) = \partial g(t) + f(x) - f(y),$$

so, by the computation above and (ii) of Proposition 1.2.7

$$f(y) - f(x) \in \partial g(t) \subset \partial f((1 - t)x + ty)(y - x).$$

The proof is complete. ∎

1.2.2 SUBDIFFERENTIATION OF COMPOSITE MAPPINGS AND RESTRICTIONS

The next objective is to establish the chain rule for the generalized gradients. To this end we recall that a map $F : U \to Y$ from an open set of a Banach space X to a Banach space Y is strictly differentiable at a point $u \in U$ if there is a continuous linear map $F'(u) : X \to Y$ such that

$$\limsup_{\substack{w \to u \\ t \downarrow 0}} \frac{1}{t}(F(w + tv) - F(w)) = F'(u)v, \ \forall\, v \in X,$$

and the convergence is uniform for v in compact sets.

Theorem 1.2.13 (Chain Rules)
(i) Let $F : U \to Y$ be a continuously differentiable map from an open set U of X with X and Y Banach spaces and let $g : Y \to \mathbb{R}$ be a locally Lipschitz function. Then one has

$$(g \circ F)^0(u; v) \leq g^0(F(u); F'(u)v), \ \forall\, u \in U, v \in X \qquad (1.2.8)$$

and

$$\partial(g \circ F)(u) \subset \partial g(F(u)) \circ F'(u), \; \forall\, u \in U. \tag{1.2.9}$$

More generally, (1.2.9) is valid if F is strictly differentiable. If in addition either g is regular at $F(u)$ or $F'(u)$ is surjective, then

$$(g \circ F)^0(u; v) = g^0(F(u); F'(u)v), \; \forall\, u \in U, \, v \in X$$

and

$$\partial(g \circ F)(u) = \partial g(F(u)) \circ F'(u), \; \forall\, u \in U. \tag{1.2.10}$$

(ii) Let $f : U \to \mathbb{R}$ be a locally Lipschitz function on an open set U of a Banach space X and let $h : \mathbb{R} \to \mathbb{R}$ be a locally Lipschitz function. Then one has

$$\partial(h \circ f)(u) \subset \overline{co}(\partial h(f(u)) \cdot \partial f(u)), \; \forall\, u \in U, \tag{1.2.11}$$

where the notation $\overline{co}$ stands for weak*-closed convex hull. Furthermore, if either h is strictly differentiable (in particular, continuously differentiable) at $f(u)$ or if h is regular at $f(u)$ and f is strictly differentiable at u, in (1.2.11) the equality holds and the symbol $\overline{co}$ is superfluous.

Proof. (i) For simplicity we suppose that F is continuously differentiable. The case of strict differentiability for F is treated in Clarke [90]. Clearly, by (ii) of Proposition 1.2.7 it is sufficient to show that

$$(g \circ F)^0(u; v) \leq \max\{\langle z, F'(u)v \rangle : z \in \partial g(F(u))\}, \; \forall\, v \in X. \tag{1.2.12}$$

Accordingly, Theorem 1.2.12 allows us to write

$$g \circ F(w + tv) - g \circ F(w) = \langle \zeta, F(w + tv) - F(w) \rangle$$

for some $\zeta \in \partial g(y)$ and y belonging to the line segment that joins $F(w)$ and $F(w + tv)$ in Y. In turn, the usual Mean Value Theorem ensures

$$F(w + tv) - F(w) = tF'(x)v$$

for a point x in the (open) segment between w and $w + tv$. Then we get

$$\begin{aligned}
(g \circ F)^0(u; v) &= \limsup_{\substack{w \to u \\ t \downarrow 0}} \frac{1}{t}(g \circ F(w + tv) - g \circ F(w)) \\
&= \limsup_{\substack{w \to u \\ t \downarrow 0}} \langle \zeta, F'(x)v \rangle = \langle \zeta, F'(u)v \rangle \\
&\leq \max\{\langle z, F'(u)v \rangle : z \in \partial g(F(u))\}.
\end{aligned}$$

Suppose now that g is regular at $F(u)$ (see Remark 1.2.4). By (ii) of Proposition 1.2.7 and the regularity assumption we obtain for any $v \in X$ that

$$\max\{\langle z, F^{'}(u)v \rangle : z \in \partial g(F(u))\} = g^0(F(u); F^{'}(u)v)$$

$$\begin{aligned} &= g^{'}(F(u); F^{'}(u)v) \\ &= \lim_{t\downarrow 0} \frac{1}{t}(g(F(u) + tF^{'}(u)v)) - g(F(u))) \\ &= \lim_{t\downarrow 0} \frac{1}{t}(g \circ F(u + tv) - g \circ F(u)) \\ &\leq (g \circ F)^0(u; v). \end{aligned}$$

This yields the equality in (1.2.9).

The remaining situation is when $F^{'}(u) : X \to Y$ is surjective. In view of Grave's Theorem [180] the mapping F is locally open at $u \in U$, so it is permitted to write

$$\max\{\langle z, F^{'}(u)v \rangle : z \in \partial g(F(u))\} = g^0(F(u); F^{'}(u)v)$$

$$\begin{aligned} &= \limsup_{\substack{y \to F(u) \\ t\downarrow 0}} \frac{1}{t}(g(y + tF^{'}(u)v) - g(y)) \\ &= \limsup_{\substack{x \to u \\ t\downarrow 0}} \frac{1}{t}(g(F(x) + tF^{'}(u)v) - g(F(x))) \\ &= \limsup_{\substack{x \to u \\ t\downarrow 0}} \frac{1}{t}(g \circ F(x + tv) - g \circ F(x)) \\ &= (g \circ F)^0(u; v) \end{aligned}$$

for all $v \in X$. It follows that (1.2.9) is verified with equality as required.

(ii) By a double application of Theorem 1.2.12 we derive, with a fixed $v \in X$, that

$$h \circ f(w + tv) - h \circ f(w) = \alpha(f(w + tv) - f(w)) = \alpha\langle z, tv \rangle, \quad (1.2.13)$$

for all w near u in U and all sufficiently small $t > 0$, where $\alpha \in \mathbb{R}$ belongs to $\partial h(s)$ at some s in the interval $[f(w), f(w + tv)]$ and $z \in \partial f(x)$ with $x \in U$ in the segment $[w, w + tv]$. Then, making use of property (iii) in Proposition 1.2.7 it is possible to show that (1.2.13) implies

$$(h \circ f)^0(u; v) \leq \max\{a\langle \zeta, v \rangle : \zeta \in \partial f(u), \ a \in \partial h(f(u))\}.$$

Then formula (1.2.11) holds true. The proof of the assertions regarding the equality cases in (1.2.11) follows the same lines as in statement (i) and will be omitted. In the mentioned cases the symbol $\overline{co}$ is not necessary in (1.2.11) due to (ii) of Proposition 1.2.8, under the additional assumptions of strict differentiability. This completes the proof.   ∎

For a later use we point out a special situation of (i) in Theorem 1.2.13.

Note that in formula (1.2.9),

$$\partial(g \circ F)(u) \subset X^* \;=\; L(X;\mathbb{R}),$$
$$\partial g(F(u)) \subset Y^* \;=\; L(Y;\mathbb{R})$$

and

$$F'(u) \in L(X;Y).$$

Let $\xi_u \in \partial(g \circ F)(u)$ be given. Equality (1.2.9) means that there exists $\chi_u \in \partial g(F(u))$ such that

$$\langle \xi_u, x \rangle_{X^*,X} = \langle \chi_u, F'(u)x \rangle_{Y^*,Y}, \ \forall\, x \in X.$$

Corollary 1.2.14   (Chang [80], Theorem 2.2, or Clarke [90], p.47)
Let X and Y be Banach spaces such that X is continuously embedded in Y and is dense in Y. Let $g : Y \to \mathbb{R}$ be a locally Lipschitz function and let $i : X \to Y$ denote the embedding operator. The restriction operator $g_{|X} : X \to \mathbb{R}$ is defined by

$$g_{|X}(u) = g \circ i(u), \ \forall\, u \in X.$$

Then, for every point $u \in X$ one has the formula

$$g^0_{|X}(u;v) = g^0(i(u);i(v)), \ \forall\, v \in X$$

and

$$\begin{aligned}
\partial(g_{|X})(u) &= \partial g(i(u)) \circ i \\
&= \{z_{|X} : z \in \partial g(i(u))\} \\
&\equiv \partial g(u)
\end{aligned}$$

in the sense that every element z of $\partial(g_{|X})(u)$ admits a unique extension to an element of $\partial g(u)$.

Proof. We have

$$g^0(i(u);i'(u)v) \;=\; g^0(i(u);i(v))$$

$$= \limsup_{\substack{y \to i(u) \\ t \downarrow 0}} \frac{1}{t}(g(y + ti(v)) - g(y)).$$

The operator i maps every neighborhood of u to a set which is dense in a neighborhood of $i(u)$ and thus

$$\limsup_{\substack{y \to i(u) \\ t \downarrow 0}} \frac{1}{t}(g(y + ti(v)) - g(y)) = \limsup_{\substack{x \to u \\ t \downarrow 0}} \frac{1}{t}(g(i(x) + ti(v)) - g(i(x))).$$

Therefore

$$\begin{aligned} g^0(i(u); i'(u)v) &= \limsup_{\substack{x \to u \\ t \downarrow 0}} \frac{1}{t}(g(i(x + tv)) - g(i(x))) \\ &= (g \circ i)^0(u; v). \end{aligned}$$

$\blacksquare$

The theory of generalized gradients can be extended for locally Lipschitz functions on Banach manifolds. This can be done in a straightforward way by using local charts as shown for the first time in Motreanu and Pavel [291]. According to our strict goals we limit ourselves to the case of typical submanifolds of Banach spaces arising in applications.

Suppose that $r \geq 1$ is an integer and X is a Banach space. A set $M \subset X$ is said to be a C^r-submanifold of codimension k, if for each $x \in M$, there exist an open set U containing x and a function $F \in C^r(U; \mathbb{R}^k)$ such that $F'(x) : X \to \mathbb{R}^k$ is onto for all $x \in U$ (i.e. x is a regular point for F) and

$$M \cap U = \{x \in U : F(x) = 0\}.$$

If x_0 is a regular point for F, then the tangent space, $T_{x_0}(M)$, to M at x_0 is defined by

$$T_{x_0}(M) = \{x \in X : \langle F'(x_0), x \rangle = 0\}.$$

For example, let $h : X \to \mathbb{R}$ be C^1 and

$$M := \{v \in X : h(v) = c\}$$

where $c \in \mathbb{R}$ is fixed. If for every $v \in M$, $h'(v)$ is nonzero, then M is a C^1-manifold of codimension 1 and, for each $v \in M$, the tangent space $T_v M$ can be identified with the subspace

$$\{w \in X : \langle h'(v), w \rangle = 0\}.$$

This entails that if $v \in M$ then we have the splitting

$$X = T_v M \oplus span\{N(v)\},$$

where $N(v) \in X, \|N(v)\| = 1$ is the normal vector to M. That means also that if X is a Hilbert space then each $\varphi \in T_v M$ can be written as

$$\varphi = x - (x, N(v))N(v),$$

for some $x \in X$.

Definition 1.2.15 Let M be a C^1-submanifold of a Banach space X and let $f : X \to \mathbb{R}$ be a locally Lipschitz function. The generalized gradient of the restriction $f\mid_M$ of f to M at a point $u \in M$, denoted $\partial(f\mid_M)(u)$, is defined by

$$\partial(f\mid_M)(u) = \{z\mid_{T_u M} : z \in \partial f(u)\}.$$

Thus $\partial(f\mid_M)(u)$ is contained in the cotangent space $T_u^* M$ of M at u.

All the preceding results admit appropriate versions in the setting of Definition 1.2.15. As an illustration of Definition 1.2.15 we determine the generalized gradient in the case where M is a sphere in a Hilbert space.

Example 1.2.16 Let X be a Hilbert space with the scalar product $(\cdot, \cdot)$ and the associated norm $\|\cdot\|$ and let $f : X \to \mathbb{R}$ be a locally Lipschitz function. For any number $r > 0$ consider the sphere

$$S_r = \{v \in X : \|v\| = r\}$$

which is known to be a smooth submanifold of X. Then $\partial(f\mid_{S_r})(u)$, at every $u \in S_r$, can be (isometrically) identified with the next subset of X^* described by

$$\begin{aligned}
\partial(f\mid_{S_r})(u) &= \partial f(u) - \frac{1}{r^2}\langle\partial f(u), u\rangle(u, \cdot) \\
&= \{z - \frac{1}{r^2}\langle z, u\rangle(u, \cdot) : z \in \partial f(u)\}.
\end{aligned}$$

Indeed, in view of Definition 1.2.15 and the characterization of the tangent space $T_u S_r$ as

$$T_u S_r = \{v \in X : (v, u) = 0\},$$

to each $\zeta \in \partial(f \mid_{S_r})(u)$ we can associate a unique $z \in \partial f(u)$ such that

$$\zeta = z - \frac{1}{r^2}\langle z, u \rangle (u, \cdot) \quad \text{on } T_u S_r$$

and

$$\|\zeta\|_{T_u^* S_r} = \|z - \frac{1}{r^2}\langle z, u \rangle (u, \cdot)\|_{X^*}.$$

The situation presented in Example 1.2.16 can be extended as follows.

Example 1.2.17   On the Hilbert space H let $g : H \to \mathbb{R}$ be a C^1-function for which $0 \in \mathbb{R}$ is a regular value, i.e., $g'(x) \neq 0$ whenever $g(x) = 0$. It is well-known that

$$M = g^{-1}(0)$$

is a C^1-submanifold of H whose tangent space $T_u M$ at any $u \in M$ is expressed by

$$T_u M = g'(u)^{-1}(0) = \{x \in H : \langle g'(u), x \rangle = 0\}$$

(see, e.g., Abraham and Robbin [1]). Given a locally Lipschitz functional $f : H \to \mathbb{R}$, a reasoning similar to the one in Example 1.2.16 shows that, in the sense of Definition 1.2.15, the generalized gradient $\partial(f \mid_M)(u)$ at some $u \in M$ admits the following characterization

$$\partial(f \mid_M)(u) = \{z - \frac{\langle z, \nabla g(u) \rangle}{\|\nabla g(u)\|^2} g'(u) : z \in \partial f(u)\},$$

where we recall that $\nabla g(u)$ denotes the gradient of g at u, that is the element $\nabla g(u) \in H$ satisfying

$$\langle g'(u), v \rangle = (\nabla g(u), v), \ \forall\, v \in H.$$

1.2.3   SUBDIFFERENTIATION OF INTEGRAL FUNCTIONALS

First of all we discuss the subdifferentiation of the real-valued function in a form of a primitive $j : \mathbb{R} \to \mathbb{R}$,

$$j(t) = \int_0^t \beta(\tau)d\tau, \ \forall\, t \in \mathbb{R}, \tag{1.2.14}$$

for $\beta \in L^{\infty}_{loc}(\mathbb{R})$. It is clear that j in (1.2.14) is locally Lipschitz. Its generalized gradient is described below. For other related results we indicate Chang [80] and Clarke [90].

For any $\delta > 0$ and $t \in \mathbb{R}$, let

$$\underline{\beta}_{\delta}(t) = \operatorname*{ess\,inf}_{|\tau - t| < \delta} \beta(\tau)$$

and

$$\overline{\beta}_{\delta}(t) = \operatorname*{ess\,sup}_{|\tau - t| < \delta} \beta(\tau).$$

For t fixed, $\underline{\beta}_{\delta}$ is decreasing in δ while $\overline{\beta}_{\delta}$ is increasing in δ. Thus, the limits

$$\underline{\beta}(t) = \lim_{\delta \to 0+} \underline{\beta}_{\delta}(t)$$

and

$$\overline{\beta}(t) = \lim_{\delta \to 0+} \overline{\beta}_{\delta}(t)$$

exist. Moreover

$$\underline{\beta}(t) = \sup_{\delta > 0} \underline{\beta}_{\delta}(t)$$

and

$$\overline{\beta}(t) = \inf_{\delta > 0} \overline{\beta}_{\delta}(t).$$

If in addition the left limit $\beta(t-0)$ and the right limit $\beta(t+0)$ exist then

$$\underline{\beta}(t) = \min\{\beta(t - 0), \beta(t + 0)\}$$

and

$$\overline{\beta}(t) = \max\{\beta(t - 0), \beta(t + 0)\}.$$

Proposition 1.2.18 The function $\underline{\beta}$ (resp. $\overline{\beta}$) is lower semi-continuous (resp. upper semi-continuous).

Proof. Let $\alpha > 0$ and $0 < \varepsilon < \frac{\alpha}{2}$ be given. Let $\{t_n\} \subset \mathbb{R}$ be a sequence converging to $t \in \mathbb{R}$. We prove that

$$\underline{\beta}(t) \leq \liminf_{n \to \infty} \underline{\beta}(t_n).$$

Indeed, for n great enough, we have $|\, t_n - t \,| < \frac{\alpha}{2}$ and thus

$$\underline{\beta}_{\varepsilon}(t_n) = \operatorname*{ess\,inf}_{|\tau - t_n| < \varepsilon} \beta(\tau)$$

$$\geq \operatorname*{ess\,inf}_{|\tau - t_n| < \frac{\alpha}{2}} \beta(\tau)$$

$$\geq \operatorname*{ess\,inf}_{|\tau - t| < \alpha} \beta(\tau)$$

$$= \underline{\beta}_\alpha(t).$$

Thus

$$\underline{\beta}(t_n) = \lim_{\varepsilon \to 0+} \underline{\beta}_\varepsilon(t_n) \geq \underline{\beta}_\alpha(t).$$

It results that

$$\liminf_{n \to \infty} \underline{\beta}(t_n) \geq \underline{\beta}_\alpha(t)$$

and finally that

$$\liminf_{n \to \infty} \underline{\beta}(t_n) \geq \lim_{\alpha \to 0+} \underline{\beta}_\alpha(t) = \underline{\beta}(t).$$

This proves that $\underline{\beta}$ is l.s.c. Using a similar argument, we prove that $\overline{\beta}$ is u.s.c. $\blacksquare$

Proposition 1.2.18 entails that the functions $\underline{\beta}$ and $\overline{\beta}$ are μ -measurable for any positive measure μ on $\mathbb{R}$.

Proposition 1.2.19   Under the foregoing conditions the next relation holds

$$\partial j(t) = [\underline{\beta}(t), \overline{\beta}(t)], \ \forall\, t \in \mathbb{R},$$

In particular, if the left limit $\beta(t - 0)$ and right limit $\beta(t + 0)$ exist at some $t \in \mathbb{R}$, then

$$\partial j(t) = [\min\{\beta(t - 0), \beta(t + 0)\}, \max\{\beta(t - 0), \beta(t + 0)\}].$$

Proof.   It is clear that if $z \in \partial j(t)$ then

$$z \leq j^0(t; 1) = \limsup_{\substack{\tau \to t \\ \lambda \downarrow 0}} \frac{1}{\lambda}(j(\tau + \lambda) - j(\tau))$$

$$= \limsup_{\substack{\tau \to t \\ \lambda \downarrow 0}} \frac{1}{\lambda} \int_\tau^{\tau + \lambda} \beta(s)\,ds \leq \overline{\beta}(t).$$

This implies that

$$\max_{z \in \partial j(t)} z \leq \overline{\beta}(t).$$

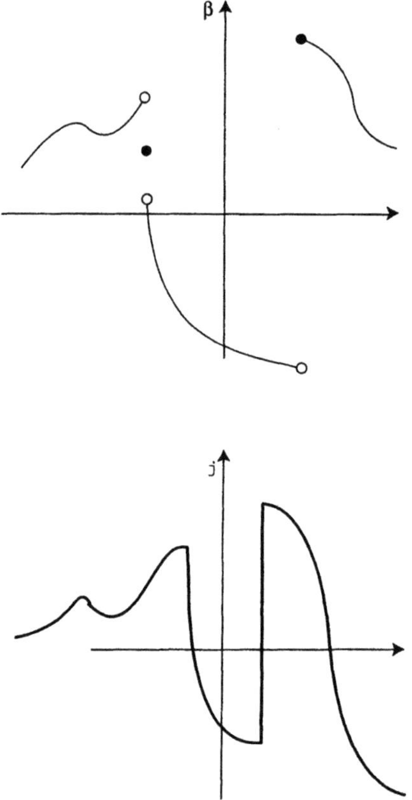

Figure 1.2.1.   Subdifferentiation of integral functionals.

Analogously one shows that

$$\underline{\beta}(t) \leq \min_{z \in \partial j(t)} z.$$

From the definition of $\overline{\beta}$ it is clear that there exists a sequence $t_n \to t$ with $\beta(t_n) \to \overline{\beta}(t)$. The function j in (1.2.14) is differentiable a.e. on $\mathbb{R}$ and we can choose $\{t_n\}$ such that $j'(t_n) = \beta(t_n)$. By (i) of Proposition 1.2.8 we know that $\beta(t_n) \in \partial j(t_n)$, so by virtue of (iii) in Proposition 1.2.7 we deduce that $\overline{\beta}(t) \in \partial j(t)$. In a similar way we get $\underline{\beta}(t) \in \partial j(t)$. The assertions above and the fact that $\partial j(t)$ is an interval yield the stated result.   ∎

Roughly speaking (Fig. 1.2.1) ∂j results from the generally discontinuous function by "filling in the gaps".

In the following we deal with the integral functionals on the spaces $L^p(T; \mathbb{R}^m)$ with integers $m \geq 1$, $+\infty > p \geq 1$ and a positive complete measure space $(T, \mathcal{T}, \mu)$ provided $\mu(T) < \infty$. The cases of interest for us will be $T = \Omega$ and $T = \partial\Omega$ for a bounded and open subset Ω of the Euclidean space $\mathbb{R}^n$ with the corresponding Lebesgue measure $\mathcal{L}_n$ of dimension n ($d\mathcal{L}_n$ is formally denoted by dx) and Hausdorff measure H_{n-1} (dH_{n-1} is formally denoted by ds) of dimension $n - 1$.

Let $j : T \times \mathbb{R}^m \to \mathbb{R}$ be a function such that $j(\cdot, y) : T \to \mathbb{R}$ is μ-measurable whenever $y \in \mathbb{R}^m$, $j(., e) \in L^1(T)$ for some $e \in L^p(T; \mathbb{R}^m)$ and satisfies either

$$|j(x, y_1) - j(x, y_2)| \leq k(x)|y_1 - y_2|, \ \forall \, x \in T, \ y_1, \ y_2 \in \mathbb{R}^m, \quad (1.2.15)$$

for a function $k \in L^q(T)$, with $1/p + 1/q = 1$, or, $j(x, \cdot) : \mathbb{R}^m \to \mathbb{R}$ is locally Lipschitz for all $x \in T$ and there is a constant $c > 0$ such that

$$|z| \leq c(1 + |y|^{p-1}), \ \forall \, x \in T, \ y \in \mathbb{R}^m, \ z \in \partial_y j(x, y). \quad (1.2.16)$$

The notation $|\cdot|$ used in (1.2.15), (1.2.16) stands for the Euclidean norm in $\mathbb{R}^m$, while $\partial_y j(x, y)$ in (1.2.16) means the generalized gradient of j with respect to the second variable $y \in \mathbb{R}^m$, i.e., $\partial j(x, \cdot)(y)$.

We are now in a position to handle the functional $J : L^p(T; \mathbb{R}^m) \to \mathbb{R}$ defined by

$$J(v) = \int_T j(x, v(x))d\mu, \ \forall \, v \in L^p(T; \mathbb{R}^m), \quad (1.2.17)$$

where the integral is computed with respect to the complete measure on T. The main subdifferentiation result regarding J in (1.2.17) is formulated below.

Theorem 1.2.20   Under either (1.2.15) or (1.2.16) the functional $J : L^p(T; \mathbb{R}^m) \to \mathbb{R}$ of (1.2.17) is Lipschitz continuous on bounded sets and satisfies

$$J^0(u; v) \leq \int_T j_y^0(x, u(x); v(x))d\mu, \ \forall \, u, \ v \in L^p(T; \mathbb{R}^m)$$

(where the subscript y indicates that the generalized derivative j^0 is taken with respect to $y \in \mathbb{R}^m$), and

$$\partial J(u) \subset \int_T \partial_y j(x, u(x))d\mu, \ \forall \, u \in L^p(T; \mathbb{R}^m) \quad (1.2.18)$$

in the sense that for every $z \in \partial J(u)$ and $x \in T$ there is $z(x) \in \mathbb{R}^m$ such that

$$z(t) \in \partial_y j(t, u(t)) \quad \text{for a.e. } t \in T,$$

$z(\cdot)\xi \in L^1(T)$ whenever $\xi \in \mathbb{R}^m$ and

$$\langle z, v \rangle = \int_T z(x).v(x)d\mu, \quad \forall\, v \in L^p(T; \mathbb{R}^m).$$

Moreover, if $j(x, \cdot)$ is regular at $u(x)$ for all $x \in T$, then J is regular at u and (1.2.18) holds with equality.

Proof. The first step in the proof is to check that J is Lipschitz continuous on bounded subsets of $L^p(T; \mathbb{R}^m)$. Suppose that (1.2.15) is verified. Then it is straightforward to establish that J is just Lipschitz continuous on $L^p(T; \mathbb{R}^m)$. Assume now that (1.2.16) holds. For a fixed number $r > 0$ and arbitrary elements $u, v \in L^p(T; \mathbb{R}^m)$ with $|\, u\, |_{0,p} \leq r$, $|\, v\, |_{0,p} \leq r$ we can write the estimate

$$|J(u) - J(v)| \leq \int_T |j(x, u(x)) - j(x, v(x))| d\mu$$

$$\leq c \int_T (1 + |u(x)|^{p-1} + |v(x)|^{p-1}) |u(x) - v(x)| dx$$

$$\leq c \left(\int_T (1 + |u(x)|^{p-1} + |v(x)|^{p-1})^{p/(p-1)} dx \right)^{\frac{p-1}{p}} |\, u - v\, |_{0,p}$$

$$\leq c_1 \left(\int_T (1 + |u(x)|^p + |v(x)|^p) dx \right)^{\frac{p-1}{p}} |\, u - v\, |_{0,p}$$

$$\leq c_2 |\, u - v\, |_{0,p},$$

with constants $c_1, c_2 > 0$ where c_2 depends on r. The inequalities above have been derived by using Theorem 1.2.12, assumption (1.2.16) and Hölder's inequality. The Lipschitzianess property on bounded sets for J is thus verified. The condition $j(., e) \in L^1(T)$ for some $e \in L^p(T; \mathbb{R}^m)$ has been used implicitly to ensure that $J(u) < +\infty$, $\forall\, u \in L^p(T; \mathbb{R}^m)$.

The map $x \to j_y^0(x, u(x); v(x))$ is measurable on T. Since $j(x, \cdot)$ is continuous, we may express $j_y^0(x, u(x); v(x))$ as the upper limit of

$$\frac{j(x, y + \lambda v(x)) - j(x, y)}{\lambda}$$

where $\lambda \downarrow 0$ taking rational values and $y \to u(x)$ taking values in a countable dense subset $\{y_i\}$ of $\mathbb{R}^m$. Thus $j_y^0(x, u(x); v(x))$ is measurable as the "countable limsup" of measurable functions of x.

The next task is to prove (1.2.18). To this end we are firstly concerned with the proof of the inequality

$$J^0(u; v) \leq \int_T j_y^0(x, u(x); v(x))dx, \ \forall \, u, \, v \ \in \ L^p(T; \mathbb{R}^m).   \quad (1.2.19)$$

Assuming (1.2.15), it is permitted to apply Fatou's Lemma that leads directly to (1.2.19). Suppose now that is assumption (1.2.16) which is satisfied. Then using Theorem 1.2.12, we obtain

$$\frac{j(x, u(x) + \lambda v(x)) - j(x, u(x))}{\lambda} = \langle \zeta_x, v(x) \rangle$$

for some $\zeta_x \in \partial_y j(x, u^*(x))$ and $u^*(x)$ lying in the interval $[u(x), u(x) + \lambda v(x)]$. We can now also use Fatou's lemma to obtain (1.2.19).

The final step, that we only sketch, is to pass from (1.2.19) to (1.2.18). Here the essential thing is to observe that, in view of (1.2.19), any $z \in \partial J(u)$ belongs to the subdifferential at $0 \in L^p(T; \mathbb{R}^m)$ (in the sense of convex analysis, see Section 1.3) of the convex function on $L^p(T; \mathbb{R}^m)$ mapping $v \in L^p(T; \mathbb{R}^m)$ to

$$\int_T j_y^0(x, u(x); v(x))d\mu \in \mathbb{R}.   \quad (1.2.20)$$

These properties and the subdifferentiation result in Ioffe and Levin [196] applied to (1.2.20) yields (1.2.18). Finally, we are concerned with the regularity assertion in the statement. Under either of hypotheses (1.2.15) or (1.2.16) we may apply Fatou's lemma to get, if the regularity of $j(x, \cdot)$ at $u(x)$ is imposed,

$$\liminf_{\lambda \downarrow 0} \frac{1}{\lambda}(J(u + \lambda v) - J(u)) \geq \int_T j_y'(x, u(x); v(x))d\mu$$

$$= \int_T j_y^0(x, u(x); v(x))d\mu, \ \forall \, v \ \in \ L^p(T; \mathbb{R}^m).$$

Combining with (1.2.19) it is seen that there exists $J'(u; v)$ and $J'(u; v) = J^0(u; v)$ whenever $v \in L^p(T; \mathbb{R}^m)$ which means the regularity of J at u. Moreover we deduced the equality

$$J^0(u; v) = \int_T j_y'(x, u(x); v(x))d\mu, \ \forall \, v \ \in \ L^p(T; \mathbb{R}^m).$$

Taking z belonging to the right-hand side of (1.2.18), the regularity assumption for $j(x, \cdot)$ implies through the formula above that

$$\langle z, v \rangle = \int_T z(x).v(x)d\mu \leq J^0(u; v), \ \forall \, v \ \in \ L^p(T; \mathbb{R}^m),$$

so $z \in \partial J(u)$. This completes the proof.   ∎

In the study of nonlinear boundary value problem an essential role is played by the potential constructed by means of Nemyckii operator (see, e.g. Zeidler [432]). In a nonsmooth setting this represents functionals J of the form (1.2.17) restricted to appropriate Sobolev spaces. As a consequence of Theorem 1.2.20 we point out a result due to Chang [80] that is very useful in such problems.

Corollary 1.2.21   Let $\beta \in L^{\infty}_{loc}(\mathbb{R})$ verify the growth condition

$$|\beta(t)| \leq c(1 + |t|^{p-1}), \ \forall\, t\, \in\, \mathbb{R}, \tag{1.2.21}$$

for constants $c > 0$ and $p \geq 1$. Then the functional $J : L^p(T) \to \mathbb{R}$, described by the Lebesgue integral

$$J(v) = \int_T \int_0^{v(x)} \beta(t)dtdx, \ \forall\, v\, \in\, L^p(T), \tag{1.2.22}$$

is Lipschitz continuous on bounded sets in $L^p(T)$ and satisfies at any $u \in L^p(T)$ the relation

$$\partial J(u)(x) \subset [\underline{\beta}(u(x)), \overline{\beta}(u(x))] \ \text{ for } \mu - \text{ a.e. } \ x \in T. \tag{1.2.23}$$

Proof.   Notice that with the notation j in (1.2.14) the functional J of (1.2.22) can be expressed in the form of (1.2.17). Since, by making use of Proposition 1.2.19, condition (1.2.16) is valid from (1.2.21) we may invoke Theorem 1.2.20. The Lipschitzianess of J on bounded subsets of $L^p(T)$ is established. Now, using (1.2.18) and Proposition 1.2.19, we obtain for each $u \in L^p(T)$ and $z \in \partial J(u)$ that

$$\langle z, v \rangle \ \leq \ \int_{v(x)<0} v(x)\underline{\beta}(u(x))d\mu$$
$$+ \int_{v(x)>0} v(x)\overline{\beta}(u(x))d\mu, \ \forall\, v\, \in\, L^p(T). \tag{1.2.24}$$

Note that the functions $\underline{\beta} \circ u$ and $\overline{\beta} \circ u$ are μ -measurable. Moreover

$$\begin{aligned}
|\underline{\beta}(t)| \ &= \ |\lim_{\delta \to 0+} \operatorname*{ess\,inf}_{|t-\tau|<\delta} \beta(\tau)| \\
&\leq \ |\lim_{\delta \to 0+} \operatorname*{ess\,inf}_{|t-\tau|<\delta} c(1+ |\tau|)^{p-1}| \\
&\leq \ c(1+ |t|^{p-1})
\end{aligned}$$

and thus

$$\underline{\beta} \circ u \in L^{\frac{p}{p-1}}(T).$$

Similarly, one checks that

$$\overline{\beta} \circ u \in L^{\frac{p}{p-1}}(T).$$

The integrals in (1.2.24) are thus well-defined.

We justify (1.2.23) arguing by contradiction. For instance, let us admit that on a set A in T of positive measure one has

$$z(x) < \underline{\beta}(u(x)), \ \forall\, x \in A.$$

Since (1.2.24) yields for $v = -\chi_A$ (the characteristic function of A) the following inequality

$$\int_A (z(x) - \underline{\beta}(u(x)))d\mu \geq 0,$$

we arrive at a contradiction. A similar reasoning shows that one cannot have the inequality

$$z(x) > \overline{\beta}(u(x))$$

on a set in T of positive measure. In this argument we employed implicitly the fact that $z \in L^{p/(p-1)}(T)$. The proof of (1.2.23) is achieved. ∎

Corollary 1.2.22   Let $J : L^p(T) \to \mathbb{R}$ be defined in (1.1.24) with $m = 1$ and suppose that either (1.1.22) or (1.1.23) is satisfied. Denoting for $j : T \times \mathbb{R} \to \mathbb{R}$,

$$\partial_y j(x,y) = [\underline{j}(x,y), \overline{j}(x,y)].$$

For a.e. $x \in T$ and $y \in \mathbb{R}$, we require that the functions $\underline{j}(x,y)$ and $\overline{j}(x,y)$ are $\mu \times \mathcal{L}_1$ -measurable. Then, for any $u \in L^p(T)$, the formula holds

$$\partial J(u)(x) \subset [\underline{j}(x,u(x)), \overline{j}(x,u(x))] \ \text{ for a.e. } \ x \in T, \qquad (1.2.25).$$

Proof. The proof is similar to the one of Corollary 1.2.21 with (1.2.24). For $u \in L^p(T)$ and $z \in \partial J(u)$, we obtain

$$\langle z, v \rangle \leq \int_{v(x)<0} \underline{j}(x,u(x))v(x)d\mu$$

$$+ \int_{v(x)>0} \overline{j}(x, u(x))v(x)d\mu, \ \ \forall \, v \, \in \, L^p(T).$$

Then, as in the proof of Corollary 1.2.21, we check that $\underline{j}(x, u(x)) \leq z(x) \leq \overline{j}(x, u(x))$, for a.e. $x \in T$. ∎

Remark 1.2.23 i) Corollary 1.2.21 can be easily generalized to give the following result. Let $J : L^p(T) \to \mathbb{R}$ be defined by

$$J(u) = \int_T \int_0^{u(x)} \beta(x, \tau)d\tau d\mu$$

with T and μ as specified above. Then if β is a $\mu \times \mathcal{L}_1-$ measurable function defined on $T \times \mathbb{R}$ satisfying the growth condition

$$|\beta(x, t)| \leq c(1 + |t|^{p-1})$$

then
$$\partial J(u) \subset [\underline{\beta}(x, u(x)), \overline{\beta}(x, u(x))] \text{ for a.e. } x \in T,$$

provided that the functions

$$\underline{\beta}(x, t) = \lim_{\delta \to 0+} \operatorname*{ess\,inf}_{|\tau - t| < \delta} \beta(x, \tau)$$

and

$$\overline{\beta}(x, t) = \lim_{\delta \to 0+} \operatorname*{ess\,sup}_{|\tau - t| < \delta} \beta(x, \tau).$$

are $\mu \times \mathcal{L}_1-$ measurable on $T \times \mathbb{R}$.

ii) If β is $\mu \times \mathcal{L}_1-$ measurable on $T \times \mathbb{R}$ and monotone non-decreasing in $t \in \mathbb{R}$ then $\underline{\beta}$ and $\overline{\beta}$ are $\mu \times \mathcal{L}_1-$ measurable.

Indeed, here

$$\begin{aligned}
\underline{\beta}(x, t) &= \min\{\beta(x, t - 0), \beta(x, t + 0)\} \\
&= \beta(x, t - 0)
\end{aligned}$$

and

$$\begin{aligned}
\overline{\beta}(x, t) &= \max\{\beta(x, t - 0), \beta(x, t + 0)\}. \\
&= \beta(x, t + 0)
\end{aligned}$$

Moreover,

$$\beta(x, t - 0) = \lim_{n \to \infty} \beta(x, t - \frac{1}{n})$$

and

$$\beta(x, t + 0) = \lim_{n \to \infty} \beta(x, t + \frac{1}{n}).$$

Thus $\beta(x, t - 0)$ and $\beta(x, t + 0)$ are $\mu \times \mathcal{L}_1$ -measurable on $T \times \mathbb{R}$. It results that $\underline{\beta}$ and $\overline{\beta}$ are $\mu \times \mathcal{L}_1$ -measurable too.

iii) A fundamental consequence of Theorem 1.2.20 is that if $j : T \times \mathbb{R}^m \to \mathbb{R}$ satisfies either (1.2.15) or (1.2.16) then (a) the functional J defined by (1.2.17) is Lipschitz continuous on bounded sets and (b) the relation (1.2.19) is satisfied, that is

$$\int_T j_y^0(x, u(x); v(x))d\mu \geq J^0(u; v), \ \forall \, u, v \ \in \ L^p(T; \mathbb{R}^m). \qquad (1.2.26)$$

Note that from the proof of Theorem 1.2.20, we see easily that the above mentioned properties (a) and (b) hold even if $\mu(T) = +\infty$ or $p = +\infty$. iv) As already noticed in the proof of Theorem 1.2.20, (1.2.26) implies that z in $\partial J(u)$ belongs to the subdifferential at $0 \in L^p(T; \mathbb{R}^m)$ (in the sense of convex analysis, see Section 1.3) of the convex function on $L^p(T; \mathbb{R}^m)$ mapping $v \in L^p(T; \mathbb{R}^m)$ to

$$\int_T j_y^0(x, u(x); v(x))d\mu.$$

The bilinear form $(u, v) \ \to \ \int_T u.vd\mu, u \ \in \ L^p(T; \mathbb{R}^m), v \ \in \ L^q(T; \mathbb{R}^m) \ (p^{-1} + q^{-1} = 1)$ can be used to set $L^p(T; \mathbb{R}^m)$ and $L^q(T; \mathbb{R}^m)$ in duality and from Theorem 1.3.21 we deduce that there exists $A \subset T$ such that $\mu(A) = 0$ and

$$j_y^0(x, u(x); h) \geq z(x).h, \ \forall \, x \ \in \ T \backslash A, \ \forall \, h \ \in \ \mathbb{R}^m,$$

i.e.

$$z(x) \in \partial_y j(x, u(x)), \ \forall \, x \ \in \ T \backslash A.$$

1.2.4 HEMIVARIATIONAL INEQUALITIES AND GENERALIZED CRITICAL POINT PROBLEMS

The critical point theory is nowdays one of the most important and fascinating tools in the modern nonlinear analysis. For our goals it is relevant in solving equations whose structure is variational, i.e., those equations that are obtained by vanishing the differential (possibly in a generalized sense) of a functional on a Banach space or a Banach manifold. Among the various generalizations of the classical notion of critical (or stationary) point in the smooth analysis we recall a significant concept of critical point due to Chang [80] that is very useful in the unilateral analysis.

Definition 1.2.24 Let X be a real Banach space. An element $u \in X$ is said to be a critical point of a locally Lipschitz function $I : X \to \mathbb{R}$ on a Banach space if

$$0 \in \partial I(u).$$

An alternate formulation of the foregoing condition is that

$$I^0(u; v) \geq 0, \ \forall \, v \ \in \ X.$$

Example 1.2.25 With X as above and I continuously differentiable on X, Definition 1.2.24 coincides with the usual concept of critical point for I, i.e., $I'(u) = 0$. A local minimum (or maximum) of a locally Lipschitz function $I : X \to \mathbb{R}$ is a critical point of I in the sense of Definition 1.2.24 as shown in Corollary 1.2.11. If X is a Banach space and $I : X \to \mathbb{R}$ is locally Lipschitz and convex then the converse is true, and thus, every critical point of I is a global minimum of I because $\partial I(u)$ is equal to the subdifferential of I at $u \in X$ (cf. (iii) of Proposition 1.2.8).

Definition 1.2.26 Let M be a C^1-submanifold of a real Banach space X. Let $I : X \to \mathbb{R}$ be a locally Lipschitz function. We say that $u \in M$ is a critical point of $I_{|M}$ if

$$0 \in \partial I_{|M}(u),$$

that is there exists $z \in \partial I(u)$ such that

$$\langle z, \theta \rangle = 0, \ \forall \, \theta \ \in \ T_u M.$$

Example 1.2.27 Let X be a real Hilbert space with the scalar product $(.,.)$ and the associated norm $\|.\|$. An element u of the sphere S_r is a critical point of $I_{|S_r}$ if and only if there exists $z \in \partial I(u)$ such that

$$\langle z, \theta \rangle - \frac{1}{r^2} \langle z, u \rangle \langle u, \theta \rangle = 0, \ \forall \, \theta \in X.$$

We briefly discuss now the notion of hemivariational inequality that has been introduced by P. D. Panagiotopoulos [329]. The following notion refer only to a simple but illustrative model. More general hemivariational inequalities will be discussed later in this book.

Let V be a real Banach space that is densely and continuously embedded in $L^p(\Omega; \mathbb{R}^m)$, with $+\infty > p \geq 1$ and an integer $m \geq 1$, for a bounded open set Ω in $\mathbb{R}^N$. Let $a : V \times V \to \mathbb{R}$ be a continuous, symmetric, bilinear form on V, let $f \in V^*$ and let $j : \Omega \times \mathbb{R}^m \to \mathbb{R}$ be a function measurable in the first variable such that $j(., e) \in L^1(T)$ for some $e \in L^p(T; \mathbb{R}^m)$ and satisfying the requirement (1.2.15) or (1.2.16). A typical hemivariational inequality model results in the following problem: find $u \in V$ such that

$$a(u, v) + \langle f, v \rangle + \int_\Omega j_y^0(x, u(x); v(x))dx \geq 0, \ \forall \, v \in V. \qquad (1.2.27)$$

Here, as previously, the notation j_y^0 designates the generalized directional derivative of j with respect to the second variable $y \in \mathbb{R}^m$.

The result presented in this Section reveals the relationship between the concept of critical point introduced in Definition 1.2.24 and the solution of the hemivariational inequality in (1.2.27).

Proposition 1.2.28 Let $I : V \to \mathbb{R}$ be the locally Lipschitz functional defined by

$$I(v) = \frac{1}{2}a(v, v) + \langle f, v \rangle + J_{|V}(v), \ \forall \, v \in V, \qquad (1.2.28)$$

with

$$J(v) = \int_\Omega j(x, v(x))dx, \ \forall \, v \in L^p(\Omega; \mathbb{R}^m).$$

Then any critical point $u \in V$ of I is a solution of (1.2.27). If in addition the bilinear form a is coercive and the function j satisfies (1.2.15) then there exists a constant $\sigma > 0$ such that for each $R > 0$:

$$\mathcal{L}_N\{x \in \Omega : u(x) > R\} \leq \frac{\sigma}{R^p}.$$

Proof. The functional I is locally Lipschitz and it makes sense to consider the critical points of I in the framework of Definition 1.2.24.

Assume that $u \in V$ is a critical point of I. Then Definition 1.2.6 and assertion (ii) of Proposition 1.2.8 and Proposition 1.2.10 imply

$$0 \in a(u, \cdot) + f + \partial J \mid_V (u).$$

Using Corollary 1.2.14 and relation (1.2.18) in Theorem 1.2.20 we infer that there exists $z \in L^{p/(p-1)}(\Omega; \mathbb{R}^m)$ such that

$$\alpha(u, v) + \langle f, v \rangle + \int_\Omega z(x) \cdot v(x) dx = 0, \ \forall \, v \, \in \, V,$$

and

$$z(x) \in \partial_y j(x, u(x)) \text{ for a.e. } x \in \Omega.$$

It is now clear that by means of (ii) in Proposition 1.2.7 it follows that u is a solution to (1.2.27).

Set $v = -u$ in (1.2.27) to obtain

$$a(u, u) \leq -\langle f, u \rangle + \int_\Omega j_y^0(x, u(x); -u(x)) dx.$$

Let $\alpha > 0$ be the constant of coercivity of a and $c_0 > 0$ the constant of continuity of the embedding $V \hookrightarrow L^p(\Omega; \mathbb{R}^m)$. If j satisfies (1.2.15) then we get

$$\alpha \|u\|_V^2 \leq \max\{c_0, 1\}(\|f\|_{V^*} \|u\|_V + \mid k \mid_{0,q} \|u\|_V).$$

Moreover, we have

$$\|u\|_V \ \geq \ \frac{1}{c_0}(\int_\Omega |u|^p)^{\frac{1}{p}} \geq \frac{1}{c_0}(\int_{u>R} |u|^p)^{\frac{1}{p}}$$

$$\geq \ \frac{1}{c_0}(\int_{u>R} R^p)^{\frac{1}{p}} = \frac{1}{c_0} R(\mathcal{L}_N(\{u > R\}))^{\frac{1}{p}}.$$

Thus

$$\mathcal{L}_N\{u > R\}^{\frac{1}{p}} \leq \frac{c_0 \max\{c_0, 1\}}{R\alpha}(\|f\|_{V^*} + \mid k \mid_{0,q}).$$

$\blacksquare$

Let us here mention that critical point methods for inequality problems are studied in Chapter 4.

Remark 1.2.29 The calculus with generalized gradients for locally Lipschitz functions has been introduced and developed by Clarke [88]-

[90] and Rockafellar [375], [376]. The notion of generalized gradient of a locally Lipschitz function plays the same crucial role as the concept of derivative or differential in the case of a smooth function. It extends and unifies the fundamental notions of Fréchet differential for a continuously differentiable function. For geometric different aspects initially motivated from optimization and related with the theory of generalized gradients we refer also the reader to the books of Aubin and Frankovska [29] and the article of Ekeland [125]. Related concepts and results are discussed in [40], [41], [127]. Set-valued analysis is intensively studied in [28], [29] and [48]. Among the basic calculus results discovered for generalized gradients there are the chain rules of Clarke [89], Lebourg's mean-value theorem [239] and the subdifferentiation of integral functionals theory developed by Aubin and Clarke [26] and Chang [80]. This material has been collected in Sections 1.2.1, 1.2.2 and 1.2.3 so as to specify and discuss the main devices of what we call in this book "unilateral analysis" in $L_{loc}(X; \mathbb{R})$. The concept of critical point of a locally Lipschitz function defined on a Banach space (Definition 1.2.24) has been introduced by Chang [80]. This works also on Banach manifolds and the possibility to define the generalized gradients on Banach manifolds was observed by Motreanu and Pavel [291]. See also the more recent work of Ribarska, Tsachev and Krastanov [372]. The concept of critical point has been used by Chang so as to define a concept of weak solution of elliptic partial differential equations involving discontinuous non–linearities. The concept of solution in an appropriate weak sense was considered before Chang by Rauch [370] and this goes back in the case of differential equations to Filippov [132]. Note that the theory of differential equations involving discontinuities has been the wealth of intensive researches in the Soviet Union and we refer the interested reader to the book of Filippov [133] for further details and references. Further applications of the approach of Rauch and Chang in the field of partial differential equations involving nonlinear discontinuities can be found in the following articles: [12], [13], [14], [57], [58], [79], [183], [242], [257], [270], [294], [292], [295], [297], [402], [403]. The relationship between the concept of critical point of Chang and the hemivariational inequality model has been later unmasked by Motreanu and Panagiotopoulos [293] through a unilateral eigenvalue problem. There is a close relationship between the theory of hemivariational inequalities and the study of semilinear elliptic equations involving discontinuities. Most problems of partial differential equations involving discontinuous non–linearities lead to hemivariational inequalities which are the corresponding weak (and well-defined) formulations of the aforementioned problems. On the other hand, the hemivariational inequality implies the classical relations of the

problem, e.g. in the sense of distributions and the boundary conditions, e.g. in the trace sense. The hemivariational inequality formulation constitutes however a more robust model expressing, in the case of concrete problems in Mechanics, the principle of virtual work or power (in the dynamic case) (see Chapter 2).

1.3   UNILATERAL ANALYSIS IN $\Gamma_0(X; I\!R \cup +\infty)$

This Section is devoted to the calculus for proper convex and l.s.c. functionals. We discuss the properties of the subdifferential of integral functionals. Finally it is given the definition of critical point in the sense of Szulkin [405] for the sum of a C^1- function and a proper convex l.s.c. function and we point out the relationship between this concept and the variational inequality problem. We rely primarily on the works of Aubin [25], Brézis [63], [64], [67], Burkill [73], Ioffe and Levin [196], Moreau [278], Panagiotopoulos [336], Rockafellar [373], [374], Schwartz [386], Szulkin [405] and Zeidler [432].

1.3.1   SUBDIFFERENTIAL OF CONVEX FUNCTIONALS

Let X be a real Banach space and let $f : X \to I\!R \cup \{+\infty\}$ be a proper ($f \not\equiv +\infty$) and convex functional. The subdifferential of f at x is the possibly empty closed convex subset of X^* defined by

$$\partial f(x) = \{w \in X^* : f(v) - f(x) \geq \langle w, v - x \rangle, \forall\, v \in X\}. \qquad (1.3.1)$$

Note that in this book we use the same notation to denote the subdifferential of a convex functional and the subdifferential of a locally Lipschitz functional. Let us recall that the link between the two subdifferentials is established in part (iii) of Proposition 1.2.8. The reader should at this point have in mind that if f is proper convex and l.s.c. then ∂f is defined by (1.3.1) while if f is locally Lipschitz then ∂f must be understood in the sense of Definition 1.2.6.

The set $\partial f(x)$ describes the differential properties of f by means of the supporting hyperplanes to the epigraph of f at $(x, f(x))$. An element w of $\partial f(x)$ is called a subgradient of f at x. The special case $f = \Psi_C$, where C is a nonempty convex subset of X and Ψ_C denotes the indicator function of C, i.e.

$$\Psi_C(x) = \begin{cases} 0 & \text{if } x \in C \\ +\infty & \text{if } x \notin C \end{cases} \tag{1.3.2}$$

is very important. Then, for all $x \in C$,

$$\partial\Psi_C(x) = \{w \in X^* : \langle w, v - x \rangle \leq 0, \ \forall v \in C\}.$$

This set forms a closed convex cone called the normal cone of C at x and denoted by $N_C(x)$ (see also Section 1.1.2). Thus

$$N_C(x) = \partial\Psi_C(x), \forall x \in C.$$

The next results establish the relationship between the subdifferential and the one-sided directional Gâteaux-differential (see Section 1.2.1). Note that if f is convex then for any $x, v \in X$ the mapping

$$\theta \to \frac{f(x + \theta v) - f(x)}{\theta}$$

is nondecreasing on $(0, +\infty)$. Indeed, if $0 < \theta_1 \leq \theta_2$ then

$$\begin{aligned} f(x + \theta_1 v) - f(x) &= f(\frac{\theta_1}{\theta_2}(x + \theta_2 v) + (1 - \frac{\theta_1}{\theta_2})x) - f(x) \\ &\leq \frac{\theta_1}{\theta_2}f(x + \theta_2 v) + (1 - \frac{\theta_1}{\theta_2})f(x) - f(x) \end{aligned}$$

and thus

$$\frac{f(x + \theta_1 v) - f(x)}{\theta_1} \leq \frac{f(x + \theta_2 v) - f(x)}{\theta_2}.$$

In particular, for any $\theta \in (0, 1]$, we obtain that

$$\frac{f(x + \theta v) - f(x)}{\theta} \leq f(x + v) - f(x),$$

We have also

$$f(x) - f(x - v) \leq \frac{f(x + \theta v) - f(x)}{\theta},$$

for any $\theta \geq 0$. Indeed,

$$\begin{aligned} f(x) &= f(\frac{1}{1 + \theta}(x + \theta v) + \frac{\theta}{1 + \theta}(x - v)) \\ &\leq \frac{1}{1 + \theta}f(x + \theta v) + \frac{\theta}{1 + \theta}f(x - v). \end{aligned}$$

Thus

$$\frac{\theta}{1+\theta}(f(x) - f(x - v)) \le \frac{1}{1+\theta}(f(x + \theta v) - f(x)).$$

Consequently, we obtain

$$
\begin{aligned}
f(x) - f(x - v) \;&\le\; \inf_{\theta > 0} \frac{f(x + \theta v) - f(x)}{\theta} \\
&=\; \lim_{\theta \to 0^+} \frac{f(x + \theta v) - f(x)}{\theta} \\
&=\; f'(x; v) \\
&\le\; f(x + v) - f(x).
\end{aligned}
$$

That means that a convex function whose domain has nonempty interior admits the directional derivative (from the right) $f'(x; v)$ at every point x interior to its domain and $f'(x; v) < +\infty$.

Proposition 1.3.1 Let $f : X \to (-\infty, +\infty]$ be a convex function whose domain has nonempty interior. Then for $x \in int\{D(f)\}$, one has

$$\partial f(x) = \{w \in X^* : f'(x; v) \ge \langle w, v \rangle, \forall\, v \in X\}.$$

Proof. Let $w \in \partial f(x)$ be given. For $0 < \theta < 1$, we have

$$
\begin{aligned}
\langle w, v \rangle \;&=\; \frac{1}{\theta}\,\langle w, \theta v + x - x \rangle \\
&\le\; \frac{1}{\theta}\,(f(x + \theta v) - f(x))
\end{aligned}
$$

Taking the limit as $\theta \to 0^+$, we obtain

$$\langle w, v \rangle \le f'(x; v). \tag{1.3.3}$$

Conversely, suppose that $w \in X^*$ satisfies (1.3.3). Then using the convexity of f, we obtain:

$$
\begin{aligned}
f'(x; v - x) \;&=\; \lim_{\theta \to 0^+} \frac{f(x + \theta(v - x)) - f(x)}{\theta} \\
&\le\; f(v) - f(x)
\end{aligned}
$$

and thus

$$\langle w, v - x \rangle \le f(v) - f(x),$$

which means that $w \in \partial f(x)$. ∎

Proposition 1.3.2   *[25]* Let $f : X \to (-\infty, +\infty]$ be a proper convex function. If f is finite and continuous at x then

$$f'(x; v) = \sup\{\langle w, v \rangle : w \in \partial f(x)\}$$

∎

The case in which ∂f has only one element is now discussed.

Proposition 1.3.3   Let $f : X \to \mathbb{R} \cup \{+\infty\}$ be convex, and suppose that the Gâteaux differential $f'(x)$ exists at x. Then $\partial f(x) = \{f'(x)\}$. Conversely, if f is finite and continuous at x and if $\partial f(x)$ has only one element, then f' exists at x and $\partial f(x) = \{f'(x)\}$.

Proof.   If $f'(x)$ exists at x then $f'(x; v) = \langle f'(x), v \rangle$ and since

$$f'(x; v - x) \le f(v) - f(x), \forall\, v \in X,$$

it results that $f'(x) \in \partial f(x)$. If $w \in \partial f(x)$ then

$$f'(x; v) \ge \langle w, v \rangle, \forall\, v \in X.$$

Thus

$$\langle f'(x), v \rangle \ge \langle w, v \rangle, \forall\, v \in X.$$

Setting $v = \pm h \in X$, we get

$$\langle f'(x) - w, h \rangle = 0, \forall\, h \in X,$$

and thus

$$f'(x) = w.$$

That means that $\partial f(x) = \{f'(x)\}$. Conversely, suppose that $\partial f(x) = \{w\}$. Then from Proposition 1.3.2, we obtain

$$f'(x; v) = \langle w, v \rangle, \forall\, v \in X.$$

The conclusion follows easily.   ∎

Proposition 1.3.4   Let $f : X \to (-\infty, +\infty]$ be a convex function and suppose that f is finite and continuous at $x_0 \in X$. Then

$$\partial f(x_0) \ne \emptyset.$$

Proof. Set $A = int\{epi\ f\}$ and $B = \{(x_0, f(x_0))\}$. The set A is convex, open and nonempty since the continuity of f at x_0 implies that $(x_0, f(x_0) + \varepsilon) \in int\{epi f\}$ for every $\varepsilon > 0$ sufficiently small. The set B is nonempty, convex and $A \cap B = \emptyset$. We may apply the Separation Theorem 1.1.9 to get the existence of $c \in \mathbb{R}, w \in X^*$ and $\alpha \in \mathbb{R}$ such that

$$\langle w, x \rangle + \alpha \lambda \leq c \leq \langle w, x_0 \rangle + \alpha f(x_0), \ \forall\ (x, \lambda) \in epi\ f. \qquad (1.3.4)$$

We claim that $\alpha \neq 0$. Indeed, if $\alpha = 0$ then

$$\langle w, x - x_0 \rangle \leq 0, \forall x \in D(f).$$

The continuity of f at x_0 entails that for each $\phi \in X$, there exists $\varepsilon > 0$ small enough so that $x_0 \pm \varepsilon \phi \in D(f)$. Thus

$$\langle w, \phi \rangle = 0, \forall\ \phi \in X$$

and $w = 0$, which is not possible. Thus $\alpha \neq 0$.
Set $(x, \lambda) = (x_0, f(x_0) + \varepsilon)$ with $\varepsilon > 0$ small in (1.3.4). We obtain

$$\langle w, x_0 \rangle + \alpha f(x_0) + \alpha \varepsilon \leq \langle w, x_0 \rangle + \alpha f(x_0)$$

and thus $\alpha \leq 0$. From our previous considerations we obtain that $\alpha < 0$ and if we set $(x, \lambda) = (z, f(z))$ (z arbitrarily chosen in X) in (1.3.4) then we obtain

$$f(z) - f(x_0) \geq \langle -\frac{w}{\alpha}, z - x_0 \rangle, \forall z \in X,$$

so that $-\frac{w}{\alpha} \in \partial f(x_0)$. $\blacksquare$

Remark 1.3.5 It is known that every proper convex functional f on a finite dimensional Banach space X is continuous on $int\{D(f)\}$. In this case, Proposition 1.3.4 ensures that $\partial f(x) \neq \emptyset$, $\forall x \in int\{D(f)\}$.

Let us end this Section by recalling some additional properties for l.s.c. proper convex functions. A proof of the results announced in the following theorem can be found in [25] and [358].

Theorem 1.3.6 Let X be a real Banach space and let $f : X \to (-\infty, +\infty]$ be a l.s.c. proper convex function. Then

(i) f is continuous at every $x \in int\{D(f)\}$,

(ii) $D(\partial f)$ is a dense subset of $D(f)$.

1.3.2 CONVEX SUBDIFFERENTIAL CALCULUS

In this Section some propositions concerning operations with subdifferentials are proved. Again, X is a real Banach space.

Proposition 1.3.7 Let $f : X \to (-\infty, +\infty]$ and $\lambda > 0$. Then for every $x \in D(\partial f)$

$$\partial(\lambda f)(x) = \lambda \partial f(x). \tag{1.3.5}$$

Proof. Property (1.3.5) results immediately from (1.3.1). ∎

Proposition 1.3.8 Let $f_1 : X \to (-\infty, +\infty]$ and $f_2 : X \to (-\infty, +\infty]$. Then for every $x \in D(\partial f_1) \cap D(\partial f_2)$, we have

$$\partial f_1(x) + \partial f_2(x) \subset \partial(f_1 + f_2)(x). \tag{1.3.6}$$

Proof. If $x' = x_1' + x_2'$ with $x_1' \in \partial f_1(x)$ and $x_2' \in \partial f_2(x)$, we have

$$(f_1 + f_2)(z) = f_1(z) + f_2(z) \geq f_1(x) + f_2(x) + \langle x_1', z - x \rangle$$
$$+ \langle x_2', z - x \rangle = (f_1 + f_2)(x) + \langle x', z - x \rangle, \forall \, z \in X$$

and hence $x' \in \partial(f_1 + f_2)(x)$. This proves the inclusion (1.3.6). ∎

This inclusion holds in the form of equality of sets, if certain additional conditions given in the following propositions hold.

Proposition 1.3.9 Assume that $f_1 : X \to (-\infty, +\infty]$ and $f_2 : X \to (-\infty, +\infty]$ are convex and that f_2' exists at the point x. Then for $x \in D(\partial f_1)$, we have

$$\partial(f_1 + f_2)(x) = \partial f_1(x) + f_2'(x). \tag{1.3.7}$$

Proof. It is sufficient to show that every $x' \in \partial(f_1 + f_2)(x)$ can be written in the form $x_1' + f_2'(x)$, where $x_1' \in \partial f_1(x)$ and $\partial f_2(x) = \{f_2'(x)\}$. If we replace z in the inequality

$$f_1(z) + f_2(z) - f_1(x) - f_2(x) \geq \langle x', z - x \rangle, \forall \, z \in X$$

by $x + \mu(z - x)$, $\mu > 0$, divide by μ and let $\mu \to 0^+$, we obtain the relation

$$f_1'(x; z - x) \geq \langle x' - f_2'(x), z - x \rangle, \forall\, z \in X$$

which, combined with Proposition 1.3.1, yields

$$x' - f_2'(x) \in \partial f_1(x).$$

$\blacksquare$

A more general result can be proved [25].

Proposition 1.3.10 Suppose that $f_1 : X \to (-\infty, +\infty]$ and $f_2 : X \to (-\infty, +\infty]$ are convex and l.s.c., and that a point $x_0 \in D(f_1) \cap D(f_2)$ exists at which f_1 is continuous. Then

$$\partial(f_1 + f_2)(x) = \partial f_1(x) + \partial f_2(x), \forall\, x \in X. \tag{1.3.8}$$

Let X, Y be real Banach spaces and let $f : Y \to \mathbb{R} \cup \{+\infty\}$ be a convex and l.s.c. function and $l : X \to Y$ a linear and continuous operator. We shall now compute the subdifferential of the composite functionnal

$$f \circ l : X \to \mathbb{R} \cup \{+\infty\}.$$

Let us recall that we denote by l^* the transpose operator of l.

Proposition 1.3.11 Assume that $f : Y \to (-\infty, +\infty]$ is convex and l.s.c. and that a point $y_0 = lx_0$ exists at which f is finite and continuous. Then

$$\partial(f \circ l)(x) = l^* \partial f(lx), \forall\, x \in X. \tag{1.3.9}$$

Proof. Assume that $y' \in \partial f(lx)$. Let us first note that the functional $f \circ l$ has the same properties as f. We have

$$f(y) - f(lx) \geq \langle y', y - lx \rangle_{Y^*, Y}, \forall\, y \in Y,$$

and thus

$$(f \circ l)(z) - (f \circ l)(x) \geq \langle y', l(z - x) \rangle_{Y^*, Y} = \langle l^* y', z - x \rangle_{X^*, X}, \forall\, z \in X$$

Accordingly,

$$l^* \partial f(lx) \subset \partial(f \circ l)(x), \forall\, x \in X.$$

Let us now show the reverse inclusion. Suppose that $x' \in \partial(f \circ l)(x)$. We set

$$A := int\{epi \, f\}$$

and

$$B = \{(lz, \langle x', z - x \rangle_{X^*,X} + (f \circ l)(x)) : z \in X\}.$$

The set A is convex, open and nonempty since $(y_0, f(y_0) + \varepsilon) \in int\{epif\}$ for every $\varepsilon > 0$. The set B is a closed affine hyperplane. We claim that $A \cap B = \emptyset$. Indeed, suppose that there exists $s_0 \in int\{D(f)\}, \lambda \in \mathbb{R}$ and $z_0 \in X$ such that $f(s_0) < \lambda, lz_0 = s_0$ and

$$\lambda = \langle x', z_0 - x \rangle_{X^*,X} + (f \circ l)(x). \tag{1.3.10}$$

Relation (1.3.10) and the fact that $x' \in \partial(f \circ l)(x)$ entail that

$$\lambda \leq (f \circ l)(z_0).$$

We obtain the contradiction $f(s_0) < f(lz_0) = f(s_0)$. We may apply Theorem 1.1.9 to get $c \in \mathbb{R}, w \in Y^*$ and $\alpha \in \mathbb{R}$ such that

$$\langle w, y \rangle_{Y^*,Y} + \alpha\lambda \leq c \leq \langle w, \zeta \rangle_{Y^*,Y} + \alpha\theta, \, \forall \, (y, \lambda) \in \bar{A}, \, (\zeta, \theta) \in B.$$

We claim that $\alpha \neq 0$. Indeed, if $\alpha = 0$ then

$$\langle w, lx_0 - y \rangle_{X^*,X}, \geq 0, \forall \, y \in D(f).$$

As in Proposition 1.3.4, we show that $w = 0$, which is impossible. As in Proposition 1.3.4, we show also that $\alpha \leq 0$ and thus $\alpha < 0$.

Choose $(y, f(y)) \in \bar{A}$. Then for every $z \in X$:

$$\langle w, y \rangle_{Y^*,Y} + \alpha f(y) \leq \langle w, lz \rangle_{Y^*,Y}$$

$$+\alpha \langle x', z - x \rangle_{X^*,X} + \alpha(f \circ l)(x). \tag{1.3.11}$$

For $z = 0$, we obtain

$$\langle w, y \rangle_{Y^*,Y} + \alpha f(y) \leq \alpha \langle x', -x \rangle_{X^*,X} + \alpha(f \circ l)(x)$$

and thus

$$f(y) - f(lx) \geq [\langle x', -x \rangle_{X^*,X} - \frac{1}{\alpha}\langle w, y \rangle_{Y^*,Y}], \forall \, y \in Y. \tag{1.3.12}$$

Choosing $(lx, f(lx)) \in \bar{A}$ in (1.3.11), we obtain

$$\langle w, lx - lz \rangle_{Y^*,Y} \leq \alpha \langle x', z - x \rangle_{X^*,X}, \forall \, z \in X,$$

which implies that $x' = -\frac{1}{\alpha}l^*(w)$. This together with (1.3.12) implies that

$$f(y) - f(lx) \geq -\frac{1}{\alpha}\langle w, y - lx \rangle_{Y^*,Y}, \forall \, y \in Y,$$

meaning that $-\frac{1}{\alpha}w \in \partial f(lx)$, or that $x' = l^*(-\frac{1}{\alpha}w) \in l^*\partial f(lx)$.   ■

1.3.3   MONOTONICITY PROPERTY OF THE CONVEX SUBDIFFERENTIAL

It is clear from the Definition 1.3.23 that $\partial f : X \to 2^{X^*}$ acts as a set-valued monotone operator, i.e.

$$\langle w - w', x - x' \rangle \geq 0, \ \forall \, x, x' \in X, \ w \in \partial f(x), \ w' \in \partial f(x').$$

Indeed, if $w \in \partial f(x)$ and $w' \in \partial f(x')$ then

$$\langle w, x - x' \rangle \geq f(x) - f(x')$$

and

$$-\langle w', x - x' \rangle \geq f(x') - f(x).$$

Adding these two last inequalities we obtain the result. If f maps $\mathbb{R}$ into $\mathbb{R}$ then the subgradients $w \in \partial f(x)$ are the slopes of the nonvertical lines through $(x, f(x))$, which have no point in common with $int\{epi \ f\}$.

From

$$f'_+(x) = f'(x; +1)$$

and

$$f'_-(x) = -f'(x; -1)$$

where $f'_+(x)$ and $f'_-(x)$ denotes the right and left derivatives at x, respectively, it follows that

$$f'_-(x) \leq w \leq f'_+(x).$$

If f is a convex l.s.c. proper functional on $\mathbb{R}$ then the right and left derivatives f'_+, f'_- may be extended, when $x \notin D(f)$, i.e. when $f(x) = +\infty$, by setting $f'_+ = f'_- = +\infty$. We may then write for $f : \mathbb{R} \to \mathbb{R} \cup \{+\infty\}$

$$\partial f(x) = \{w \in \mathbb{R} : f'_-(x) \leq w \leq f'_+(x)\}. \tag{1.3.13}$$

For the left and right derivatives of a convex functional on $\mathbb{R}$, the following result holds.

Proposition 1.3.12 Let f be a convex, l.s.c. proper functional on $\mathbb{R}$. Then f'_+ and f'_- are nondecreasing functions on $\mathbb{R}$ and have finite values on $int\{D(f)\}$, and for $x_1 < x < x_2$ the following inequalities hold

$$f'_+(x_1) \leq f'_-(x) \leq f'_+(x) \leq f'_-(x_2). \tag{1.3.14}$$

Proof. The left and right derivatives $f'_+(x)$ and $f'_-(x)$ exist for $x \in D(f)$ and $f'_-(x) \leq f'_+(x)$. The monotonicity of the difference quotients defining the right and left derivatives f'_+ and f'_- implies that these derivatives are finite in $int\{D(f)\}$. If $x_1 < x_2$ and $x_1, x_2 \notin D(f)$, the same inequality holds by definition. ∎

The following result ensures that for a proper l.s.c. convex function $f : X \to \mathbb{R} \cup \{+\infty\}$ defined on a reflexive Banach space, the subdifferential acts as a maximal monotone operator from X into 2^{X^*}.

Theorem 1.3.13 Let X be real reflexive Banach space and let $f : X \to \mathbb{R} \cup \{+\infty\}$ be a proper convex l.s.c. function. Then $\partial f : X \to 2^{X^*}$ is maximal monotone.

Proof. It suffices to show that for each $l \in X^*$, there exists $x \in D(f)$ such that

$$Jx + \partial f(x) \ni l.$$

This problem is equivalent to the variational inequality: Find $x \in X$ such that

$$\langle Jx - l, v - x \rangle + f(v) - f(x) \geq 0, \forall \, v \in X.$$

We can introduce an equivalent norm on X so that X and X^* are strictly convex (see e.g. [432]). With respect to this new norm, the operator J is monotone, coercive and hemicontinuous and we may apply Theorem 3.3.1 (see Chapter 3) to conclude. ∎

Proposition 1.3.14   Let $\beta : [a, b] \to \mathbb{R}(a < b, a \in \mathbb{R} \cup \{-\infty\}, b \in \mathbb{R} \cup \{+\infty\})$ be a nondecreasing function and let $x_0 \in (a, b)$ be given. Then the function $\varphi : \mathbb{R} \to \mathbb{R} \cup \{+\infty\}$ defined by

$$\varphi(x) = \begin{cases} \int_{x_0}^{x} \beta(t)dt & \text{if} \quad x \in [a, b] \\ \\ +\infty & \text{if not} \end{cases}$$

is proper, convex and s.c.i. Moreover,

$$\partial\varphi(x) = [\beta(x-), \beta(x+)], \, \forall \, x \in \,]a, b[.$$

Proof. The function φ is proper since $\varphi(x_0) = 0$.

Let us now check that φ is convex. Let $x, y \in D(\beta)$ with $x \leq y$ and set $z = (1 - \lambda)x + \lambda y$ for some $\lambda \in [0, 1]$.

Then

$$\varphi(z) - \varphi(x) = \int_x^z \beta(t)dt \leq (z - x)\beta(z)$$

and

$$\varphi(y) - \varphi(z) = \int_z^y \beta(t)dt \geq (y - z)\beta(z).$$

Thus

$$
\begin{aligned}
(1 - \lambda)(\varphi(z) - \varphi(x)) &+ \lambda(\varphi(z) - \varphi(y)) \\
&\leq \ [(z - x)(1 - \lambda) + (z - y)\lambda]\beta(z) \\
&= \ [z - (1 - \lambda)x - \lambda y]\beta(z) \\
&= \ 0.
\end{aligned}
$$

It results that

$$\varphi(z) \leq \lambda\varphi(y) + (1 - \lambda)\varphi(x).$$

Let us now verify that φ is l.s.c. Let $\{z_n\}$ be a sequence satisfying $z_n \to z$ and $\varphi(z_n) \leq c$ for some arbitrary $c \in \mathbb{R}$. The inequality $\varphi(z_n) \leq c$ yields $z_n \in [a, b]$ and thus $z \in [a, b]$. Moreover

$$\varphi(z) \ = \ \int_{x_0}^z \beta(r)dr = \lim_{z_n \to z} \int_{x_0}^{z_n} \beta(r)dr$$

and thus $\varphi(z) \leq c$.

For $z > x, z, x \in]a, b[$, we have

$$\frac{\varphi(z) - \varphi(x)}{z - x} = \frac{1}{z - x}\int_x^z \beta(t)dt \geq \beta(x)$$

so that

$$\varphi'_+(x) \geq \beta(x).$$

We check also that

$$\beta(x) \geq \varphi'_-(x).$$

For $y < x < z, y, x, z \in]a, b]$, we see by application of Proposition 1.3.12 that

$$\varphi'_+(y) \leq \varphi'_-(x) \leq \varphi'_+(x) \leq \varphi'_-(z).$$

Thus

$$\varphi'_+(x) \leq \varphi'_-(z) \leq \beta(z) \leq \varphi'_+(z).$$

and

$$\varphi'_+(x) \leq \lim_{z \to x+} \beta(z) \leq \lim_{z \to x+} \varphi'_+(z)$$

so that

$$\beta(x+) = \varphi'_+(x).$$

On the other hand

$$\varphi'_-(y) \leq \beta(y) \leq \varphi'_+(y) \leq \varphi'_-(x)$$

and thus

$$\lim_{y \to x-} \varphi'_-(y) \leq \lim_{y \to x-} \beta(y) \leq \varphi'_-(x)$$

so that

$$\beta(x-) = \varphi'_-(x)$$

$\blacksquare$

We end this Section with an important result concerning maximal monotone graph of $\mathbb{R}^2$.

Proposition 1.3.15 Let β be a maximal monotone graph of $\mathbb{R}^2$, then

$$\beta = \partial j$$

for some proper, convex and lower semi-continuous function $j : \mathbb{R} \to \mathbb{R} \cup \{+\infty\}$.

Proof. Let us first note that there exist $-\infty \leq a \leq b \leq +\infty$ such that $(a, b) \subset D(\beta) := \{x \in \mathbb{R} : \beta(x) \neq \emptyset\} \subset [a, b]$. We denote by $\beta^0 : D(\beta) \to \mathbb{R}$ the minimal section of β, i.e. $\beta^0(x) \in \beta(x)$ and $\mid \beta^0(x) \mid = \inf\{\mid w \mid ; w \in \beta(x)\}$. Let $x_0 \in (a, b)$ be given and set

$$j(x) = \begin{cases} \int_{x_0}^{x} \beta^0(s)ds & \text{if} \quad x \in [a, b] \\ +\infty & \text{if} \quad x \in \mathbb{R} \setminus [a, b] \end{cases}$$

Let $x \in D(\beta)$ be given. Then for every $y \in D(h), y > x$ and $w \in \beta(x)$ we have

$$\begin{aligned} j(x) - j(y) &= \int_{x_0}^{x} \beta^0(s)ds - \int_{x_0}^{y} \beta^0(s)ds \\ &= \int_{y}^{x} \beta^0(s)ds \leq (x - y)w. \end{aligned}$$

The function j is proper convex and l.s.c. (see Proposition 1.3.14) and thus

$$\beta(x) \subset \partial j(x), \ \forall \, x \, \in \, D(\beta).$$

We conclude that $\beta = \partial j$ since β is maximal monotone. ∎

Example 1.3.16   i) Let $f : \mathbb{R} \to \mathbb{R} \cup \{+\infty\}$ be given by

$$f(x) = \begin{cases} -g(x - a_1) & \text{if} \quad x \leq a_1 \\ 0 & \text{if} \quad a_1 \leq x \leq a_2 \\ \infty & \text{if} \quad x > a_2, \end{cases} \tag{1.3.15}$$

with $g, a_1, a_2 \in \mathbb{R}$. Then (fig. 1.3.1)

$$\partial f(x) = \begin{cases} -g & \text{if} \quad x < a_1 \\ [-g, 0] & \text{if} \quad x = a_1 \\ 0 & \text{if} \quad a_1 < x < a_2 \\ [0, \infty) & \text{if} \quad x = a_2 \\ \emptyset & \text{if} \quad x > a_2. \end{cases} \tag{1.3.16}$$

1.3.4   CONVEX INTEGRAL FUNCTIONALS

Let $(T, \mathcal{T}, \mu)$ be a positive complete measure space such that $0 < \mu(T) < +\infty$. Let X be a Banach space and $\varphi : X \to \mathbb{R} \cup \{+\infty\}$ a proper convex and l.s.c. function. We have the following result on the integrability of φ.

Proposition 1.3.17   For each $u \in L^1(T; X)$ the function $\varphi(u)$ is measurable. Moreover, the function $\varphi(u)$ is integrable if and only if there exists $h \in L^1(T)$ such that $\varphi(u) \leq h, \ \mu.$ -a.e. on T.

Proof.   The function $\varphi(u)$ is measurable if and only if (see Corollary 5.2.20. in [386]) the set

$$A := \{x \in T : \varphi(u(x)) > \lambda\}$$

is measurable for all $\lambda \in \mathbb{R}$. The function $\varphi : X \to \mathbb{R} \cup \{+\infty\}$ being l.s.c., the set

$$\varphi^{-1}((\lambda, +\infty)) = \{z \in X : \varphi(z) > \lambda\}$$

is open. Thus

$$A = \{x \in T : u(x) \in \varphi^{-1}(\lambda, +\infty)\}$$

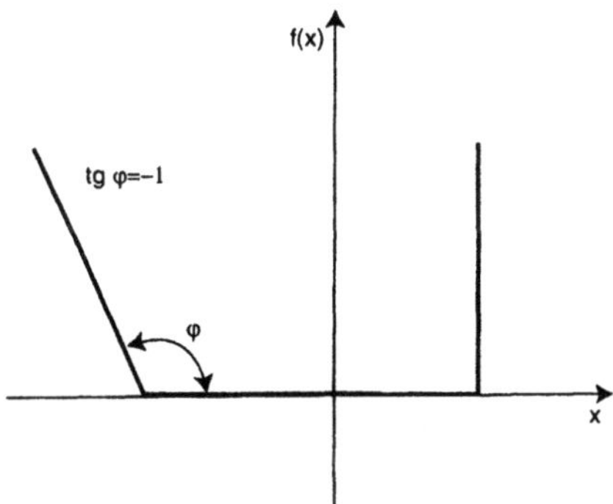

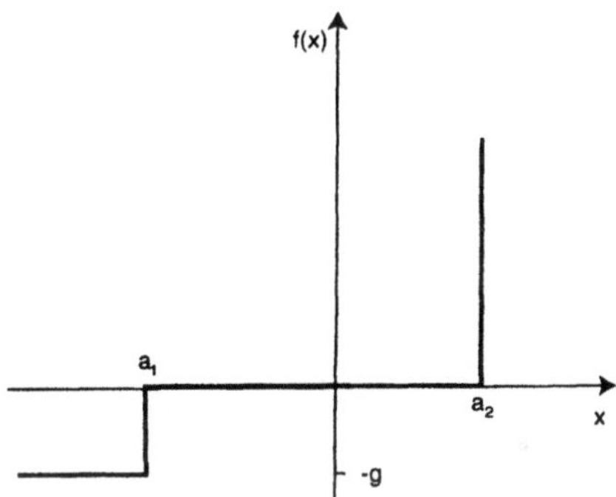

Figure 1.3.1.   The graphs of (1.3.15) and (1.3.16).

is measurable. From Theorem 1.1.11, there exist $\alpha \in X^*$ and $\beta \in \mathbb{R}$ such that

$$\varphi(x) \geq \langle \alpha, x \rangle + \beta, \forall\, x \in X.$$

Thus

$$\mid \varphi(u(x)) \mid \leq \max\{h(x), \|\alpha\|_* \|u(x)\| + \mid \beta \mid\}, \ \mu - \text{a.e. on } T.$$

This together with the measurability of $\varphi(u)$ entails that

$$\int_T \mid \varphi(u(x)) \mid d\mu < +\infty.$$

The converse is obvious since we may choose $h = \varphi(u)$. ∎

Let us now discuss the properties of the important integral functional model defined for $u \in L^1(T; X)$ by

$$\Phi(u) = \begin{cases} \int_T \varphi(u)d\mu & \text{if} \qquad \varphi(u(.)) \in L^1(T) \\ +\infty & \text{if not.} \end{cases} \tag{1.3.17}$$

The following result is known as Jensen inequality.

Proposition 1.3.18 Let $\Phi : L^1(T) \to \mathbb{R} \cup \{+\infty\}$ be defined by (1.3.17). If $u \in L^1(T; X)$ and $\overline{u} := \frac{1}{\mu(T)} \int_T u(x)d\mu$ then:

$$\Phi(u) \geq \mu(T)\varphi(\overline{u}). \tag{1.3.18}$$

Proof. If $\varphi(u) \notin L^1(T)$ then $\Phi(u) = +\infty$ and the result is clear. Let us now suppose that $\varphi(u) \in L^1(T)$. Then $\Phi(u) = \int_T \varphi(u(x))d\mu$. We have

$$\varphi(\alpha) \geq \varphi(\beta) + \langle w, \alpha - \beta \rangle, \ \forall \, \alpha, \, \beta \in X$$

provided that $w \in \partial\varphi(\beta)$. Choose

$$\begin{aligned} \alpha &= u(x) \\ \beta &= \overline{u} \end{aligned}$$

and fix

$$w \in \partial\varphi(\beta) \neq \emptyset.$$

We obtain

$$\varphi(u(x)) \geq \varphi(\overline{u}) + \langle w, u(x) - \overline{u} \rangle.$$

Thus

$$\begin{aligned} \int_T \varphi(u(x))d\mu &\geq \mu(T)\varphi(\overline{u}) + \langle w, [\int_T u(x)d\mu - \overline{u}\mu(T)]\rangle \\ &= \mu(T)\varphi(\overline{u}). \end{aligned}$$

$\blacksquare$

Proposition 1.3.19 The mapping $\Phi : L^1(X; T) \to \mathbb{R} \cup \{+\infty\}$ defined by (1.3.17) is proper convex and l.s.c.

Proof. The function Φ is clearly convex. It is proper since $x \to \varphi(x_0) = c < +\infty$ for some $x_0 \in X$ and therefore $\Phi(u_0) = c\mu(T) < +\infty$ for

$u_0 \in L^1(T; X)$ defined by $u_0(x) \equiv x_0, \forall\, x \in T$. It remains to prove that

$$\{u \in L^1(T; X) : \int_T \varphi(u)d\mu \leq \lambda\}$$

is closed for arbitrary $\lambda \in \mathbb{R}$.

Let $\{u_n\} \subset L^1(T; X)$ such that $\int_T \varphi(u_n)d\mu \leq \lambda$ and $u_n \to u$ in $L^1(T; X)$. There exists a subsequence $\{u_{n_k}\}$ such that $u_{n_k}(x) \to u(x)$ a.e. on T. From Theorem 1.1.11, there exist $\alpha \in X^*$ and $\beta \in \mathbb{R}$ such that

$$\varphi(x) \geq \langle \alpha, x \rangle + \beta, \forall\, x \in X.$$

Set

$$f_{n_k} = \varphi(u_{n_k}) - \langle \alpha, u_{n_k} \rangle - \beta.$$

We have $f_{n_k} \geq 0$ and thus

$$\int_T \liminf f_{n_k} d\mu \leq \liminf \int_T f_{n_k} d\mu.$$

It results that

$$\int_T \varphi(u)d\mu - \int_T (\langle \alpha, u \rangle + \beta)d\mu \leq \lambda - \int_T (\langle \alpha, u \rangle + \beta)d\mu$$

and thus

$$\int_T \varphi(u)d\mu \leq \lambda.$$

■

The main subdifferentiation results regarding Φ are now formulated.

Theorem 1.3.20   Let $u \in L^p(T; X) \cap D(\Phi)$ and $w \in L^q(T; X^*)(1 \leq p \leq +\infty, \frac{1}{p} + \frac{1}{q} = 1)$. The following two conditions are equivalent:

$$\Phi(v) - \Phi(u) \geq \int_T \langle w, v - u \rangle d\mu, \;\forall\, v \in L^p(T; X) \cap D(\Phi). \qquad (1.3.19)$$

For all $v \in L^p(T; X) \cap D(\Phi)$, we have

$$\varphi(v(x)) - \varphi(u(x)) \geq \langle w(x), v(x) - u(x) \rangle, \;\mu-\text{a.e. } x \in T. \qquad (1.3.20)$$

Proof.   It is clear that (1.3.20) entails (1.3.19). Suppose now that (1.3.19) is satisfied. Let E be a μ-measurable subset of T and set

$$h := \begin{cases} v & \text{on} \quad E \\ u & \text{on} \quad T \backslash E \end{cases}$$

in (1.3.19). It is clear that $h \in L^p(T; X)$. Moreover

$$\int_T |\, \varphi(h)\, |\, d\mu \;=\; \int_E |\, \varphi(v)\, |\, d\mu + \int_{T\backslash E} |\, \varphi(u)\, |\, d\mu$$

$$\leq \int_T |\, \varphi(v)\, |\, d\mu + \int_T |\, \varphi(u)\, |\, d\mu < +\infty$$

since v and u belong to $D(\Phi)$. We have thus

$$\int_E (\varphi(v) - \varphi(u) - \langle w, v - u \rangle) d\mu \geq 0.$$

The result being true for any μ -measurable subset of T, we may conclude. ∎

Theorem 1.3.21   Let $u \in L^p(T; X) \cap D(\Phi)$ and $w \in L^q(T; X^*)$. If X is separable then

$$\Phi(v) - \Phi(u) \geq \int_T \langle w, v - u \rangle d\mu, \; \forall\, v \; \in \; L^p(T; X) \; \cap D(\Phi) \qquad (1.3.21)$$

if and only if there exists $A \subset T$ such that $\mu(A) = 0$ and

$$w(x) \in \partial\varphi(u(x)), \; \forall\, x \in \; T\backslash A. \qquad (1.3.22)$$

Proof.   If (1.3.22) is satisfied then for any $x \in T\backslash A$, we have

$$\varphi(h) - \varphi(u(x)) \geq \langle w(x), h - u(x) \rangle, \forall\, h \; \in \; X. \qquad (1.3.23)$$

So, if $v \in L^p(T; X)$ then setting $h = v(x)$ in (1.3.23) and integrating over T, we obtain

$$\int_T \varphi(h) d\mu - \int_T \varphi(u) d\mu \geq \int_T \langle w(x), h - u(x) \rangle d\mu,$$

for all $h \in D(\Phi)$. The right integral is well-defined since we have supposed that $w \in L^q(T; X^*)$ and we may conclude that (1.3.21) is satisfied. Let us now prove the converse. Suppose that X is separable and let $w \in L^q(T; X^*)$ satisfying (1.3.21). From Theorem 1.3.20 we know that for all $v \in L^p(T; X) \cap D(\Phi)$, we have

$$\varphi(v(x)) - \varphi(u(x)) \geq \langle w(x), v(x) - u(x) \rangle, \mu - \text{a.e. } x \in T. \qquad (1.3.24)$$

The graph of φ is separable for the topology induced by $X \times \mathbb{R}$. Thus there exists a sequence $\{v_n\} \subset D(\varphi)$ such that for all $v \in D(\varphi)$ there exists a subsequence $\{v_{n_k}\}$ depending on v satisfying $v_{n_k} \to v$ and $\varphi(v_{n_k}) \to \varphi(v)$. Using (1.3.24) for the constant mapping $v_n \in L^p(T; X) \cap D(\Phi)$, we obtain for each $n \in \mathbb{N}$ the relation

$$\varphi(v_n) - \varphi(u(x)) \geq \langle w(x), v_n - u(x) \rangle, \mu - \text{a.e. } x \in T.$$

Thus there exists $A \subset T$ such that $\mu(A) = 0$ and

$$\varphi(v_n) - \varphi(u(x)) \geq \langle w(x), v_n - u(x) \rangle, \ \forall \, n \in \mathbb{N}, \ \forall \, x \in T \backslash A. \quad (1.3.25)$$

Indeed, for each $n \in \mathbb{N}$ there exists $A_n \subset T$ such that

$$\varphi(v_n) - \varphi(u(x)) \geq \langle w(x), v_n - u(x) \rangle, \ \forall \, x \in T \backslash A_n.$$

The choice $A = \cup_{n \in \mathbb{N}} A_n$ is convenient. Considering (1.3.25) for the sequence $\{v_{n_k}\}$ and taking the limit as $n_k \to +\infty$, we obtain the result. ∎

Remark 1.3.22 If $\varphi(x, z)$ is a function defined on $T \times X$ with values in $\mathbb{R}$ such that the functions $\varphi_x : X \to \mathbb{R}, z \to \varphi_x(z) := \varphi(x, z)$ are convex and continuous for almost $x \in T$ and the functions $\varphi_z : T \to \mathbb{R}, x \to \varphi_z(x) := \varphi(x, z)$ are μ -measurable for each $z \in X$ then the function

$$x \to \varphi(x, u(x))$$

is measurable on T, for every $u \in L^1(T; X)$ (see e.g. [432]).

We may consider the functional

$$\Phi(u) = \begin{cases} \int_T \varphi(x, u(x)) d\mu & \text{if } \varphi(., u(.)) \in L^1(T; X) \\ +\infty & \text{if not.} \end{cases}$$

If there exists $u_0 \in X$ such that $\varphi(., u_0(.)) \in L^1(T; X)$ then Φ is proper. As in Proposition 1.3.19, we can show that Φ is convex and l.s.c. The analogue of Theorems 1.3.20 and 1.3.21 follow. In particular, if X is separable and if $u \in L^p(T; X) \cap D(\Phi)$ then

$$\Phi(v) - \Phi(u) \geq \int_T \langle w, v - u \rangle d\mu, \ \forall \, v \in L^p(T; X) \cap D(\Phi)$$

if and only if $w \in L^p(T; X^*)$ and there exists $A \subset T$ such that $\mu(A) = 0$ and

$$\varphi(x, y) - \varphi(x, u(x)) \geq \langle w, y - u(x) \rangle, \ \forall \, x \in T \backslash A, \ y \in X.$$

1.3.5 VARIATIONAL INEQUALITIES AND GENERALIZED CRITICAL POINT PROBLEMS

In this Section we indicate a second notion of critical point in the nonsmooth setting that concerns possibly discontinuous functions.

Definition 1.3.23 (Szulkin [405]) Let $I : X \to \mathbb{R} \cup \{+\infty\}$ be a functional on the Banach space X satisfying the hypothesis $I = E + F$ with $E \in C^1(X;\mathbb{R})$ and $F : X \to \mathbb{R} \cup \{+\infty\}$ convex, proper and lower semi-continuous. A point $u \in X$ is called a critical point of I if $-E'(u) \in \partial F(u)$, where we recall that $E'(u)$ and $\partial F(u)$ stand for the Fréchet differential of E at u and the subdifferential of F at u in the sense of convex analysis, respectively. Consequently, this means that

$$\langle E'(u), v - u \rangle + F(v) - F(u) \geq 0, \ \forall \, v \, \in \, X. \tag{1.3.26}$$

Example 1.3.24 Given $I \to \mathbb{R} \cup \{+\infty\}$ as in Definition 1.3.23, a local minimum $u \in X$ is a critical point of I in the sense of Definition 1.3.23. Indeed, if $u \in X$ minimizes locally I, for all small $t > 0$ the differentiability of E and the convexity of F yield

$$\begin{aligned}
0 \ &\leq \ I((1-t)u + tv) - I(u) \\
&\leq \ E(u + t(v - u)) - E(u) + t(F(v) - F(u)), \forall \, v \, \in \, X,
\end{aligned}$$

and thus (1.3.26) by dividing by t and then letting $t \to 0$. A very useful situation in Definition 1.3.23 which is relevant especially for solving variational inequalities is when F is equal to the indicator function of a nonempty closed convex set K in X. Then (1.3.26) reduces to

$$\langle E'(u), v - u \rangle \geq 0, \forall \, v \, \in \, K.$$

Various problems in Mechanics (see Chapter 2) involve both convex energy functionals and locally Lipschitz functionals and the material developed in Section 1.2 and 1.3 must be used together. In this sense a concept of critical point recovering the one of Definition 1.2.24 as well as the one of Definition 1.3.23 is discussed in Chapter 4.

Remark 1.3.25   The concept of critical point in Definition 1.3.23 has been introduced by Szulkin [405]. Related results are discussed in [126], [158], [159] and [170].

1.4   ASYMPTOTIC UNILATERAL ANALYSIS

A general inequality problem may invoke several kinds of data among which a closed convex set, a proper convex and l.s.c. functional, a possibly nonlinear operator and an integral of the generalized derivative of a locally Lipschitz function. In this Section we discuss the mathematical tools we can use to describe the asymptotic behavior of these data, that are respectively the recession cone, the recession function, the recession mapping of Brézis and Nirenberg and the asymptotic potential mapping. Finally, we show how this material can be used to derive necessary conditions for the existence of possibly noncoercive inequality problems. Later in Chapter 3 we will use the same material to derive sufficient conditions for the existence of solutions of possibly noncoercive inequality problems. We rely herein on the works of Baiocchi, Buttazzo, Gastaldi and Tomarelli [33], Brézis and Nirenberg [66], Goeleven and Théra [162], Goeleven [165] and Rockafellar [374].

1.4.1   THE RECESSION CONE OF A CLOSED CONVEX SET

If C is a nonempty closed convex set of a real Banach space X then its behavior at infinity can be described in terms of what is called the recession directions, i.e. the directions which recede from C. The set of recession directions is usually denoted by C_∞, that is

$$C_\infty = \cap_{\lambda>0} \frac{1}{\lambda}(C - x_0), \qquad (1.4.1)$$

where x_0 is any element of C. Let us firstly note the following fundamental property.

Proposition 1.4.1   Let C be a nonempty closed convex set of X. Then

$$\begin{aligned}
C_\infty \;=\; & \{d \in X : \exists\{\lambda_n\} \subset \mathbb{R}_+ \backslash \{0\}, \exists\{d_n\} \subset X, \\
& \text{with } \lambda_n \to +\infty, d_n \overset{\sigma}{\to} d \text{ and } \lambda_n d_n \in C, \forall\, n \in \mathbb{N}\},
\end{aligned}$$

where $\overset{\sigma}{\to}$ denotes either the weak ($\rightharpoonup$) or the strong ($\to$) convergence in X.

Proof. If $d \in C_\infty$ then for each $\lambda > 0$, there exists $x(\lambda) \in C$ such that

$$d = \frac{x(\lambda)}{\lambda} - \frac{x_0}{\lambda}.$$

Let $\{\lambda_n\}$ be a sequence of positive real numbers such that $\lambda_n \to +\infty$ and set $d_n := \frac{x_n}{\lambda_n}$ with $x_n := x(\lambda_n)$. It is clear that $\lambda_n d_n \in C$ and $d_n = d + \frac{x_0}{\lambda_n} \overset{\sigma}{\to} d$ as $n \to +\infty$. Conversely, if $d \in X$ is such that there exist $\{\lambda_n\} \subset \mathbb{R}_+ \setminus \{0\}, \{d_n\} \subset X$ with $\lambda_n \to +\infty, d_n \overset{\sigma}{\to} d$ and $\lambda_n d_n \in C$, then for a given $\alpha > 0$ and a given $x_0 \in C$, we have

$$\frac{\alpha}{\lambda_n} \lambda_n d_n + (1 - \frac{\alpha}{\lambda_n}) x_0 \in C$$

provided that n is great enough to have $\frac{\alpha}{\lambda_n} \leq 1$. We obtain

$$\alpha d_n + (1 - \frac{\alpha}{\lambda_n}) x_0 \overset{\sigma}{\to} \alpha d + x_0 \in C \text{ as } n \to +\infty,$$

since C is σ-closed. The parameter $\alpha > 0$ has been chosen arbitrarily and thus

$$\alpha d + x_0 \in C, \ \forall \, \alpha > 0.$$

That means that $d \in C_\infty$.   ∎

Further properties are now given.

Proposition 1.4.2 Let C be a nonempty closed convex set of X. Then C_∞ is a closed convex cone, i.e.

$$0 \in C_\infty, \tag{1.4.2}$$

$$\alpha C_\infty \subset C_\infty, \ \forall \, \alpha > 0, \tag{1.4.3}$$

and

$$C_\infty + C_\infty \subset C_\infty. \tag{1.4.4}$$

Proof. The set C_∞ is closed as the intersection of a family of closed sets. Let us now check formulae (1.4.2), (1.4.3) and (1.4.4).

i) Let $\lambda > 0$ be given. If $x_0 \in C$ then $0 \in \frac{1}{\lambda}(C - x_0)$ and it results that $0 \in \cap_{\lambda > 0} \frac{1}{\lambda}(C - x_0)$.

ii) If $z \in \alpha C_\infty (\alpha > 0)$ then $z = \alpha d$ for some $d \in C_\infty$ and thus there exists $\lambda_n \to +\infty$, $d_n \to \frac{z}{\alpha}$ such that $\lambda_n d_n \in C$. If we set $\mu_n = \frac{\lambda_n}{\alpha}$ and $z_n = \alpha d_n$ then we see that $\mu_n z_n \in C, \mu_n \to +\infty$ and $z_n \to z$. This yields $z \in C_\infty$.

iii) If $z \in C_\infty + C_\infty$ then $z = z^1 + z^2$ and there exist $\lambda_n^1 \to +\infty, \lambda_n^2 \to +\infty, z_n^1 \to z^1, z_n^2 \to z^2$ such that $\lambda_n^1 z_n^1 \in C$ and $\lambda_n^2 z_n^2 \in C$. Set $\mu_n = \frac{\lambda_n^1 \lambda_n^2}{\lambda_n^1 + \lambda_n^2}$ and $z_n = z_n^1 + z_n^2$. We have $\mu_n \to +\infty$ and $\frac{\lambda_n^1}{\lambda_n^1 + \lambda_n^2} \leq 1$ and $\frac{\lambda_n^2}{\lambda_n^1 + \lambda_n^2} \leq 1$ so that

$$\mu_n z_n = \frac{\lambda_n^2}{\lambda_n^1 + \lambda_n^2} \lambda_n^1 z_n^1 + \frac{\lambda_n^1}{\lambda_n^1 + \lambda_n^2} \lambda_n^2 z_n^2 \in C.$$

Moreover $z_n \to z$ and thus $z \in C_\infty$. ∎

Proposition 1.4.3 Let C be a nonempty closed convex subset of X. Then

(i) if $x \in C, e \in C_\infty$ then $x + e \in C$,

(ii) if C is bounded then $C_\infty = \{0\}$,

(iii) $(C - z_0)_\infty = C_\infty$, for all $z_0 \in X$.

Proof. i) This property is a consequence of the relation

$$C_\infty = \cap_{\lambda > 0} \frac{1}{\lambda}(C - x) \subset C - x.$$

ii) Suppose by contradiction that $d \neq 0 \in C_\infty$. Then there exist sequences $\lambda_n \to +\infty$ and $d_n \to d$ such that $\lambda_n d_n \in C$. However, $\lambda_n \|d_n\| \to +\infty$, which is a contradiction since C is bounded.

iii) If $z \in (C - z_0)_\infty$ then there exists $\lambda_n \to +\infty, z_n \to z$ such that $\lambda_n z_n \in C - z_0$. Set $w_n = z_n + \frac{z_0}{\lambda_n}$. We have $\lambda_n w_n = \lambda_n z_n + z_0 \in C$ and $w_n \to z$. That means that $z \in C_\infty$. Conversely, if $z \in C_\infty$ then there exist $\lambda_n \to +\infty, z_n \to z$ such that $\lambda_n z_n \in C$. If we set $w_n = z_n - \frac{z_0}{\lambda_n}$ then we have $\lambda_n w_n = \lambda_n z_n - z_0 \in C - z_0$ and $w_n \to z$, so that $z \in (C - z_0)_\infty$. ∎

Example 1.4.4 For example of recession cones of closed convex sets, for

$$A = \{(x_1, x_2) \in \mathbb{R}^2 : x_1 > 0, x_1 x_2 \geq 1\}$$

one has (see fig. 1.4.1a)

$$A_\infty = \{(x_1, x_2) \in \mathbb{R}^2 : x_1 \geq 0, x_2 \geq 0\}$$

and for

$$B = \{(x_1, x_2) \in \mathbb{R}^2 : 2 \leq x_1, -x_1 + 4 \leq x_2 \leq x_2 + 4\}$$

one has (see fig 1.4.1b)

$$B_\infty = \{(x_1, x_2) \in \mathbb{R}^2 : 0 \leq x_1, \mid x_2 \mid \leq x_1\}.$$

1.4.2   THE RECESSION FUNCTION OF A PROPER CONVEX AND L.S.C. FUNCTION

Let $\Phi : X \to (-\infty, +\infty]$ be a proper, convex and l.s.c. function. The recession function Φ_∞ of Φ is defined by

$$\Phi_\infty(x) = \lim_{\lambda \to +\infty} \frac{1}{\lambda} \Phi(x_0 + \lambda x), \tag{1.4.5}$$

for any element $x_0 \in D(\Phi)$. Note that

$$\Phi_\infty(x) = \lim_{\lambda \to +\infty} \frac{1}{\lambda}(\Phi(x_0 + \lambda x) - \Phi(x_0))$$

$$= \sup_{\lambda > 0} \frac{1}{\lambda}(\Phi(x_0 + \lambda x) - \Phi(x_0)). \tag{1.4.6}$$

Let us now prove the following property.

Proposition 1.4.5   Let X be a real Banach space and $\Phi \in \Gamma_0(X; \mathbb{R} \cup \{+\infty\})$. We have

$$\Phi_\infty(x) = \liminf_{\substack{t \to +\infty \\ v \xrightarrow{\sigma} x}} \frac{\Phi(tv)}{t}$$

$$= \inf\{\liminf \frac{\Phi(t_n x_n)}{t_n} : \{t_n\} \subset \mathbb{R}_+ \setminus \{0\}, \{x_n\} \subset X,$$

$$t_n \to +\infty, x_n \xrightarrow{\sigma} x\}, \tag{1.4.7}$$

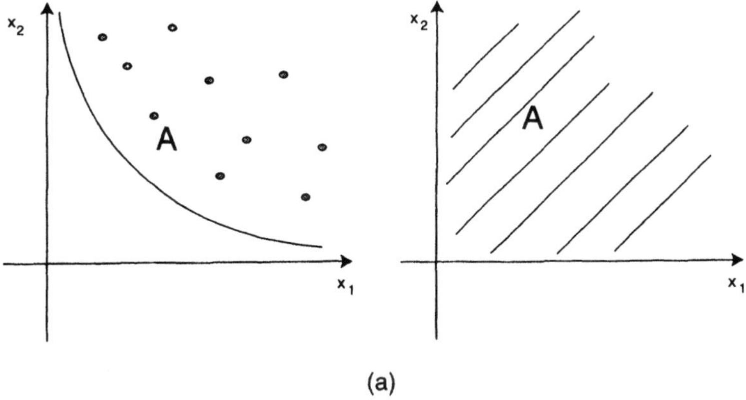

(a)

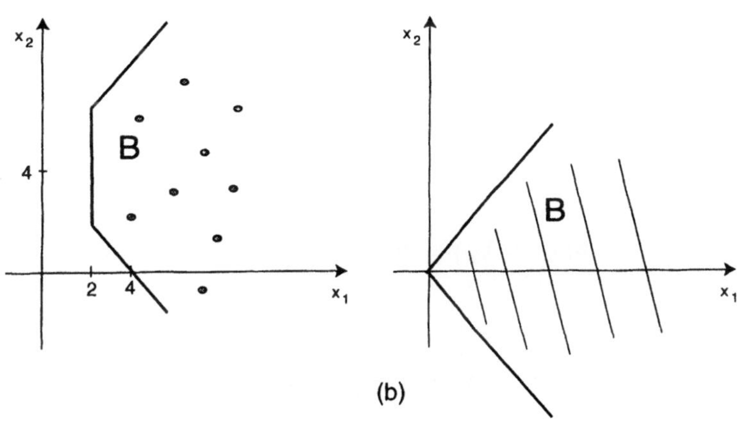

(b)

Figure 1.4.1. Recession cones.

where $\xrightarrow{\sigma}$ denotes either the weak or the strong convergence in X.

Proof. Set

$$\Lambda_\infty(x) := \liminf_{\substack{t \to +\infty \\ v \to x}} \frac{\Phi(tx)}{t}.$$

Let $x \in X$ and $x_0 \in D(\Phi)$ be given and consider a sequence $\{t_n\} \subset \mathbb{R}_+ \setminus \{0\}$ such that $t_n \to +\infty$. We have

$$\Lambda_\infty(x) \leq \liminf_{n \to \infty} \frac{\Phi(t_n z_n)}{t_n}$$

with $z_n = x + \frac{x_0}{t_n} \xrightarrow{\sigma} x$. Therefore

$$\Lambda_\infty(x) \le \liminf_{n \to +\infty} \frac{\Phi(t_n x + x_0)}{t_n}$$

$$= \Phi_\infty(x). \tag{1.4.8}$$

Let $\{\lambda_n\} \subset \mathbb{R}_+ \setminus \{0\}$, $\{x_n\} \subset X$ be sequences such that $\lambda_n \to +\infty$ and $x_n \xrightarrow{\sigma} x$. For all $\lambda > 0$, we have thanks to the convexity and σ-l.s.c. of Φ :

$$\begin{aligned}
\Phi(x_0 + \lambda x) &\le \liminf_{n \to \infty} \Phi((1 - \frac{\lambda}{\lambda_n})x_0 + \frac{\lambda}{\lambda_n}\lambda_n x_n) \\
&\le \liminf_{n \to \infty} \{(1 - \frac{\lambda}{\lambda_n})\Phi(x_0) + \frac{\lambda}{\lambda_n}\Phi(\lambda_n x_n)\} \\
&= \Phi(x_0) + \lambda \liminf_{n \to \infty} \frac{\Phi(\lambda_n x_n)}{\lambda_n}.
\end{aligned}$$

Thus

$$\frac{\Phi(x_0 + \lambda x) - \Phi(x_0)}{\lambda} \le \liminf_{n \to +\infty} \frac{\Phi(\lambda_n x_n)}{\lambda_n}. \tag{1.4.9}$$

The inequality (1.4.9) holds true for any sequence $\{\lambda_n\} \subset \mathbb{R}_+ \setminus \{0\}$, $\{x_n\} \subset X$ such that $\lambda_n \to +\infty$ and $x_n \xrightarrow{\sigma} x$. Thus

$$\frac{\Phi(x_0 + \lambda x) - \Phi(x_0)}{\lambda} \le \Lambda_\infty(x),$$

and taking now the limit as $\lambda \to +\infty$, we get

$$\Phi_\infty(x) \le \Lambda_\infty(x). \tag{1.4.10}$$

The result follows from (1.4.8) and (1.4.10). ∎

In the following result, we show that Φ_∞ turns out to be proper, convex, l.s.c. and positively homogeneous of order 1.

Proposition 1.4.6   If $\Phi \in \Gamma_0(X; \mathbb{R} \cup \{+\infty\})$ then $\Phi_\infty \in \Gamma_0(X; \mathbb{R} \cup \{+\infty\})$ and

$$\Phi_\infty(\alpha x) = \alpha \Phi_\infty(x), \ \forall \, \alpha \ge 0, \ \forall \, x \in X.$$

Proof. It is clear from (1.4.5) that $\Phi_\infty(0) = 0$ and thus Φ_∞ is proper. Let us now prove the convexity of Φ_∞. Let $\lambda \in [0, 1]$ and $x_1, x_2 \in X$. We have

$$\Phi_\infty(\lambda x_1 + (1 - \lambda)x_2)$$

$$= \lim_{t \to +\infty} \frac{1}{t}[\Phi(\lambda x_0 + (1-\lambda)x_0 + \lambda t x_1 + (1-\lambda)t x_2)$$
$$-\lambda\Phi(x_0) - (1-\lambda)\Phi(x_0)]$$
$$\leq \lim_{t \to +\infty} \frac{\lambda(\Phi(x_0 + tx_1) - \Phi(x_0))}{t}$$
$$+ \lim_{t \to +\infty} \frac{(1-\lambda)(\Phi(x_0 + tx_2) - \Phi(x_0))}{t}$$
$$= \lambda\Phi_\infty(x_1) + (1-\lambda)\Phi_\infty(x_2).$$

Let us now check that Φ_∞ is positively homogeneous of degree 1. We have

$$\Phi_\infty(\alpha x) = \lim_{\lambda \to +\infty} \frac{\Phi(x_0 + \lambda \alpha x)}{\lambda}$$

and setting $t = \alpha\lambda$, we get

$$\Phi_\infty(\alpha x) = \lim_{t \to +\infty} \alpha \frac{\Phi(x_0 + tx)}{t}$$
$$= \alpha\Phi_\infty(x).$$

It remains to prove that Φ_∞ is σ -l.s.c. Let $\{x_n\}$ be a sequence such that $x_n \overset{\sigma}{\to} x$ and

$$\Phi_\infty(x_n) \leq c,$$

with $c \in \mathbb{R}$. That is also

$$\liminf_{n \to \infty} \sup_{\lambda > 0} \frac{\Phi(x_0 + \lambda x_n) - \Phi(x_0)}{\lambda} \leq c,$$

where x_0 is chosen in $D(\Phi)$. We know that for each $\lambda > 0$,

$$\frac{\Phi(x_0 + \lambda x) - \Phi(x_0)}{\lambda} \leq \liminf_{n \to +\infty} \frac{\Phi(x_0 + \lambda x_n) - \Phi(x_0)}{\lambda}$$

Thus

$$\Phi_\infty(x) \leq \sup_{\lambda > 0} \liminf_{n \to +\infty} \frac{\Phi(x_0 + \lambda x_n) - \Phi(x_0)}{\lambda}$$
$$\leq \liminf_{n \to +\infty} \sup_{\lambda > 0} \frac{\Phi(x_0 + \lambda x_n) - \Phi(x_0)}{\lambda}$$
$$\leq c.$$

It results that the set

$$\{x \in X : \Phi(x) \leq c\}$$

is σ-closed for any $c \in \mathbb{R} \cup \{+\infty\}$ (the case $c = +\infty$ is trivial) and from Proposition 1.1.2, we conclude that Φ is σ-l.s.c. ∎

Let us now list further useful properties.

Proposition 1.4.7 Let X be a real Banach space and let Φ be a proper convex and l.s.c. function. Then

(i)
$$\Phi_\infty(e) \geq \Phi(x + e) - \Phi(x), \ \forall \ x \in D(\Phi), e \in X. \qquad (1.4.11)$$

(ii)   If C denotes a nonempty closed convex subset of X then

$$(\Psi_C)_\infty \equiv \Psi_{C_\infty}. \qquad (1.4.12)$$

(iii) If $\Phi_1, \Phi_2 \in \Gamma_0(X; \mathbb{R} \cup \{+\infty\})$ then

$$(\Phi_1 + \Phi_2)_\infty(x) \geq (\Phi_1)_\infty(x) + (\Phi_2)_\infty(x), \forall \ x \ \in \ X. \qquad (1.4.13)$$

(iv)   If $\Phi_1 \in \Gamma_0(X; \mathbb{R} \cup \{+\infty\})$ and Φ_2 is σ-continuous and positively homogeneous of degree 1 then

$$(\Phi_1 + \Phi_2)_\infty(x) = (\Phi_1)_\infty(x) + \Phi_2(x), \forall \ x \ \in \ X.$$

(v)   If $\Phi \in \Gamma_0(X; \mathbb{R} \cup \{+\infty\})$ is bounded from below then

$$\Phi_\infty(x) \geq 0, \forall \ x \ \in \ X.$$

Proof. i) Property (1.4.11) follows from (1.4.6). Indeed,

$$\begin{aligned}
\Phi_\infty(e) \ &= \ \sup_{\lambda > 0} \frac{\Phi(x + \lambda e) - \Phi(x)}{\lambda} \\
&\geq \ \Phi(x + e) - \Phi(x).
\end{aligned}$$

ii) Let $x_0 \in C$ be given. If $x \in C_\infty$ then $x_0 + \lambda x \in C_\infty$, $\forall \ \lambda \ > \ 0$ and thus $\frac{1}{\lambda}\Psi_C(x_0 + \lambda x) = 0, \forall \lambda > 0$. We deduce that

$$(\Psi_C)_\infty(x) = 0 \text{ if } x \in C_\infty. \qquad (1.4.14)$$

If $x \notin C_\infty$ then there exists $\lambda > 0$ such that

$$x_0 + \lambda x \notin C$$

and thus

$$\frac{1}{\lambda}(\Psi_C(x_0 + \lambda x) - \Psi_C(x_0)) = +\infty.$$

It results that

$$(\Psi_C)_\infty(x) = \sup_{\lambda>0} \frac{1}{\lambda}(\Psi_C(x_0 + \lambda x) - \Psi_C(x_0)) = +\infty.$$

Thus

$$(\Psi_C)_\infty(x) = +\infty \text{ if } x \notin C_\infty. \qquad (1.4.15)$$

From (1.4.14) and (1.4.15) we deduce that

$$(\Psi_C)_\infty \equiv \Psi_{C_\infty}.$$

iii) Property (1.4.13) follows directly from Proposition 1.4.5. Indeed

$$\liminf_{\substack{v \xrightarrow{\sigma} x \\ t \to +\infty}} \frac{\Phi_1(tv)}{t} + \liminf_{\substack{v \xrightarrow{\sigma} x \\ t \to +\infty}} \frac{\Phi_2(tv)}{t} \leq \liminf_{\substack{v \xrightarrow{\sigma} x \\ t \to +\infty}} \frac{(\Phi_1 + \Phi_2)(tv)}{t}.$$

iv) If Φ_2 is σ -continuous and positively homogeneous of order 1 then

$$\liminf_{\substack{v \xrightarrow{\sigma} x \\ t \to +\infty}} \frac{(\Phi_1 + \Phi_2)(tv)}{t} = (\Phi_1)_\infty(x) + \lim_{v \xrightarrow{\sigma} x} \Phi_2(v),$$

and we obtain the result.

v) If $\Phi(x) \geq c, \ \forall \ x \in X$, then $\frac{\Phi(tv)}{t} \geq ct^{-1}, \ \forall \ t > 0, \ v \in X$ and thus $\Phi_\infty(x) \geq \lim_{t \to \infty} ct^{-1} = 0.$ ∎

1.4.3   ASYMPTOTIC BEHAVIOR OF CONVEX ENERGY FUNCTIONALS AND CONVEX CONSTRAINTS

Let $(T, \mathcal{T}, \mu)$ be a positive complete measure space such that $0 < \mu(T) < +\infty$. Let X be a real Banach space and let $\gamma : X \to L^1(T; \mathbb{R}^m)$ $(m \in \mathbb{N} \setminus \{0\})$ be a linear and continuous mapping. Let $\varphi : \mathbb{R}^m \to \mathbb{R} \cup \{+\infty\}$ be a proper, convex and l.s.c. function. We set for $u \in L^1(T; \mathbb{R}^m)$

$$\Phi(u) = \begin{cases} \int_T \varphi(u(x))d\mu & \text{if } \varphi(u) \in L^1(T) \\ +\infty & \text{elsewhere} \end{cases}$$

and

$$\Psi(u) = \Phi(\gamma(u)), \forall \ u \in X.$$

Proposition 1.4.8 If $\gamma(X) \cap D(\Phi) \neq \emptyset$ then

$$\Psi_\infty(u) = \Phi_\infty(\gamma(u)), \forall \ u \in X$$

and

$$\Phi_\infty(u) = \begin{cases} \int_T \varphi_\infty(u(x))d\mu & \text{if } \varphi_\infty(u) \in L^1(T) \\[2mm] +\infty & \text{elsewhere.} \end{cases}$$

Proof. We may use Proposition 1.3.19 to ensure that

$$\Phi : L^1(X; \mathbb{R}^m) \to \mathbb{R} \cup \{+\infty\}$$

is proper, convex and l.s.c. It results that the composite mapping $\Phi \circ \gamma$ is convex and l.s.c. since γ is linear and continuous. The condition $\gamma(X) \cap D(\Phi) \neq \emptyset$ ensures that Ψ is proper. Let $u_0 \in X$ such that $\gamma(u_0) \in D(\Phi)$ be given. We have

$$\begin{aligned} \Psi_\infty(u) &= \lim_{\lambda \to +\infty} \frac{1}{\lambda} \Phi \circ \gamma(u_0 + \lambda u) \\ &= \lim_{\lambda \to +\infty} \frac{1}{\lambda} \Phi(\gamma(u_0) + \lambda \gamma(u)) \\ &= \Phi_\infty(\gamma(u)). \end{aligned}$$

Let us now compute $\Phi_\infty(u)$. Let $z_0 \in D(\Phi)$ be given. Let $\{\lambda_n\} \subset \mathbb{R}_+ \setminus \{0\}$ be a sequence such that $\lambda_n \to +\infty$ as $n \to +\infty$ and let

$$g_n(u(x)) := \frac{1}{\lambda_n}[\varphi(z_0(x) + \lambda_n u(x)) - \varphi(z_0(x))],$$

non-decreasing with respect to n, for almost all $x \in T$. If $\varphi_\infty(u) \in L^1(T)$ then from Proposition 1.3.17 and definition of φ_∞, we deduce that $g_n(u) \in L^1(T)$, $\forall \, n \in \mathbb{N}$. Using now Lebesgue's convergence theorem for non-decreasing sequence, we obtain

$$\int_T \sup_n g_n(u(x))d\mu = \sup_n \int_T g_n(u(x))d\mu,$$

that is

$$\int_T \varphi_\infty(u(x))d\mu = \Phi_\infty(u).$$

Suppose now that $\varphi_\infty(u) \notin L^1(T)$. Then there exists a sequence $\{\lambda_n\} \subset \mathbb{R}_+ \setminus \{0\}$ such that $\lambda_n \to +\infty$ and

$$\frac{\Phi(\lambda_n u + z_0) - \Phi(z_0)}{\lambda_n} \to +\infty. \tag{1.4.16}$$

Indeed, suppose by contradiction that there exist $c \in \mathbb{R}$ such that

$$\frac{\Phi(\lambda_n u + z_0) - \Phi(z_0)}{\lambda_n} \leq c, \ \forall \{\lambda_n\} \subset \mathbb{R}_+, \ \lambda_n \to +\infty.$$

Then we obtain that

$$\sup_n \int_T g_n(u(x))d\mu \leq c. \tag{1.4.17}$$

Since $\{g_n\}$ is nondecreasing, the condition (1.4.17) is a necessary and sufficient condition for the $\sup_n g_n(u(x))$ to be integrable. We obtain that

$$\varphi_\infty(u(x)) \in L^1(T)$$

and a contradiction. Thus, from (1.4.16), we deduce that

$$\begin{aligned} \Phi_\infty(u) &= \lim_{\lambda \to +\infty} \frac{\Phi(\lambda u + z_0) - \Phi(z_0)}{\lambda} \\ &= \lim_{\lambda_n \to +\infty} \frac{\Phi(\lambda_n u + z_0) - \Psi(z_0)}{\lambda_n} \\ &= +\infty. \end{aligned}$$

$\blacksquare$

Let us now suppose that $Q(x)$ is a nonempty closed convex subset of $\mathbb{R}^m$, for a.e. $x \in T$ and set

$$C = \{u \in X : \gamma u(x) \in Q(x), \ \text{a.e.} \ x \in T\}.$$

Proposition 1.4.9 Suppose that $C \neq \emptyset$. Then C is convex, closed and

$$C_\infty = \{u \in X : \gamma u(x) \in Q_\infty(x), \ \text{a.e.} \ x \in T\}.$$

Proof. The convexity of C is a direct consequence of the convexity of Q and the linearity of γ. Let us now check that C is closed. If $u_n \to u, u_n \in C$, then $\gamma(u_n) \to \gamma(u)$ in $L^1(T; \mathbb{R}^m)$ and along a subsequence, we may assume that

$$\gamma u_n(x) \to \gamma u(x), \ \text{a.e.} \ x \in T.$$

Thus

$$\gamma u(x) \in Q(x), \ \text{a.e.} \ x \in T,$$

since $Q(x)$ is closed, for all $x \in T$. We claim that

$$C_\infty \subset \{u \in X : \gamma u(x) \in Q_\infty(x), \text{ a.e. } x \in T\}.$$

Indeed, if $e \in C_\infty$ then

$$\lambda e + u \in C, \ \forall \, \lambda > 0, \ u \in C.$$

That means that

$$\lambda \gamma e(x) + \gamma u(x) \in Q(x), \text{ a.e. } x \in T, \ \forall \, \lambda > 0.$$

Moreover $\gamma u(x) \in Q(x)$ and

$$\gamma e(x) \in \frac{1}{\lambda}[Q(x) - \gamma u(x)], a.e. x \in T, \ \forall \, \lambda > 0.$$

Therefore

$$\gamma e(x) \in \cap_{\lambda > 0} \frac{1}{\lambda}[Q(x) - \gamma u(x)] = Q_\infty(x), \text{ a.e. } x \in T.$$

It remains to prove that

$$\{u \in X : \gamma u(x) \in Q_\infty(x), \text{ a.e. } x \in T\} \subset C_\infty.$$

If $\gamma u(x) \in Q_\infty(x)$ for a.e. $x \in T$ and if $u_0 \in C \neq \emptyset$ then $\gamma u_0(x) + \lambda \gamma u(x) \in Q(x), a.e. x \in T, \ \forall \, \lambda > 0$. Thus

$$u_0 + \lambda u \in C, \ \forall \, \lambda > 0.$$

It results that

$$u \in \cap_{\lambda > 0} \frac{1}{\lambda}[C - u_0] = C_\infty.$$

$\blacksquare$

Remark 1.4.10   i) An usual set of constraints encountered in the applications is given by

$$C = \{u \in X : x + \gamma u(x) \in \Lambda, \text{ a.e. } x \in T\},$$

where Λ is a given nonempty closed convex subset of $\mathbb{R}^3$ and T is a $\mu-$measurable subset of Λ. In this case C is nonempty since $0 \in C$ and from Proposition 1.4.9 (set $Q(x) = \Lambda - x$) and Proposition 1.4.3, iii), we deduce that

$$C_\infty = \{u \in X : \gamma u(x) \in \Lambda_\infty, \text{ a.e. } x \in T\}.$$

ii) Let us consider the set

$$C = \{u \in X : f(x, \gamma u(x)) \leq 0, \text{ a.e. } x \in T\},$$

where $f : T \times \mathbb{R}^m \to \mathbb{R}$ is a function convex in the second variable and such that $f(x, 0) \equiv 0$. Let us first note that C can be written equivalently as follows:

$$C = \{u \in X : \gamma u(x) \in Q(x), \text{ a.e. } x \in T\},$$

with

$$Q(x) = \{z \in \mathbb{R}^m : f(x, z) \leq 0\}.$$

The set $Q(x)$ is clearly nonempty $(0 \in Q(x))$, $\forall x \in T$, closed and convex. From Proposition 1.4.9, we deduce that

$$C_\infty = \{u \in X : \gamma u(x) \in Q_\infty(x), \text{ a.e. } x \in T\}.$$

We claim that

$$Q_\infty(x) = \{z \in \mathbb{R}^m : f_\infty(x, z) \leq 0\}.$$

Indeed, if $e \in Q_\infty(x)$ then for some $z_0 \in Q(x)$,

$$f(x, z_0 + \lambda e) \leq 0, \ \forall \lambda > 0.$$

It results that

$$f_\infty(x, e) \leq 0.$$

Conversely, if $f_\infty(x, e) \leq 0$ then for all $\lambda > 0$ and $z_0 \in Q(x)$ we have

$$\frac{f(x, z_0 + \lambda e) - f(x, z_0)}{\lambda} \leq 0.$$

For $z_0 = 0$, this yields

$$f(x, \lambda e) \leq 0, \ \forall \lambda > 0.$$

Thus

$$\lambda e \in Q(x), \ \forall \lambda > 0$$

and therefore (since $0 \in Q(x)$)

$$e \in \cap_{\lambda > 0} \frac{1}{\lambda} Q(x) = Q_\infty(x).$$

1.4.4   THE RECESSION MAPPING OF BRÉZIS AND NIRENBERG

Let X be a real Banach space and $A : X \to X^*$ a possibly nonlinear operator. The recession mapping of Brézis and Nirenberg $\underline{r}_A : X \to [-\infty, +\infty]$ is defined by

$$\underline{r}_A(u) = \liminf_{\substack{t \to +\infty \\ v \to u}} \langle A(tv), v \rangle. \tag{1.4.18}$$

Let us now restrict our investigations to the class $\Lambda(X)$ of operators $A : X \to X^*$ such that $\underline{r}_A(u) > -\infty$.

Proposition 1.4.11   Let $A \in \Lambda(X)$ be given. The following properties hold.

(i) $\underline{r}_A$ is positively homogeneous of order 1 and l.s.c.

(ii)   If B denotes another operator in $\Lambda(X)$, then

$$\underline{r}_{A+B}(u) \geq \underline{r}_A(u) + \underline{r}_B(u), \forall\, u\, \in\, X.$$

(iii)   If $A = \Phi'$ for some convex and continuously differentiable functional Φ, then

$$\underline{r}_A(u) = \Phi_\infty(u), \forall\, u\, \in\, X.$$

Proof.   i) Let $\lambda > 0$ be given. We have

$$\underline{r}_A(\lambda u) = \liminf_{\substack{t \to +\infty \\ v \to u}} \langle A(\lambda t v), \lambda v \rangle.$$

Setting $\alpha = \lambda t$, we get

$$\begin{aligned}
\underline{r}_A(\lambda u) &= \liminf_{\substack{\alpha \to +\infty \\ v \to u}} \lambda \langle A(\alpha v), v \rangle \\
&= \lambda \underline{r}_A(u).
\end{aligned}$$

To prove the l.s.c. of $\underline{r}_A$, it suffices to check that for any $c \in (-\infty, +\infty]$ the set

$$\{u \in X : \underline{r}_A(u) \leq c\}$$

is closed. Let $u_n \to u$ in X satisfying

$$\underline{r}_A(u_n) \leq c.$$

If $c = +\infty$ then the result is trivial. Suppose now that $c \in (-\infty, +\infty)$. For every $n \in \mathbb{N}$, there exists $t_n > n$ and $u_n' \in X$ such that

$$\|u_n - u_n'\| < \frac{1}{n}$$

and

$$\langle A(t_n u_n'), u_n' \rangle \le c + \frac{1}{n}.$$

Taking the limit as $n \to +\infty$, we have $t_n \to +\infty, u_n' \to u$ and thus

$$\underline{r}_A(u) \le c.$$

ii) This property is a direct consequence of the definition of $\underline{r}_{A+B}$.

iii) We have

$$\Phi(tu) - \Phi(x) \ge \langle Ax, tu - x \rangle, \ \forall \, u, \, x \, \in \, X, \, \forall \, t \, > \, 0.$$

Thus

$$\lim_{t \to +\infty} \frac{\Phi(tu)}{t} \ge \lim_{t \to +\infty} \frac{\Phi(x)}{t} + \lim_{t \to +\infty} \frac{\langle Ax, tu - x \rangle}{t}$$

and therefore

$$\Phi_\infty(u) \ge \langle Ax, u \rangle, \ \forall \, u, x \, \in \, X.$$

Thus

$$\Phi_\infty(u) \ge \langle A(tu), u \rangle, \ \forall \, u \, \in \, X, \, \forall \, t \, > \, 0$$

so that

$$\Phi_\infty(u) \ge \liminf_{t \to +\infty} \langle A(tu), u \rangle \ge \underline{r}_A(u).$$

On the other hand

$$\Phi(0) - \Phi(tu) \ge -\langle A(tu), tu \rangle, \forall \, u \, \in \, X, \, \forall \, t \, > \, 0$$

and thus

$$\frac{\Phi(tu)}{t} \le \frac{\Phi(0)}{t} + \langle A(tu), u \rangle, \forall \, u \, \in \, X, \, \forall \, t \, > \, 0.$$

It results that

$$\Phi_\infty(u) \le \underline{r}_A(u).$$

$\blacksquare$

More generally, if u_0 denotes a given element of X, we set

$$\underline{r}_{A,u_0}(u) = \liminf_{\substack{v \to u \\ t \to +\infty}} \frac{\langle A(tv), tv - u_0 \rangle}{t}. \tag{1.4.19}$$

It is clear that

$$\underline{r}_{A,0} \equiv \underline{r}_A.$$

The following result is trivial.

Proposition 1.4.12 Let $A : X \to X^*$ be a w^*-continuous and positively homogeneous operator. Then

$$\underline{r}_{A,u_0}(x) = \underline{r}_A(x) - \langle Ax, u_0 \rangle.$$

Let $(T, \mathcal{T}, \mu)$ be a positive complete measure space such that $\mu(T) > 0$. Let $h : T \times \mathbb{R} \to \mathbb{R}$ be a Caratheodory function, i.e. $h(.,y) : T \to \mathbb{R}$ measurable for all $y \in \mathbb{R}$ and $h(x,.) : \mathbb{R} \to \mathbb{R}$ continuous for almost all $x \in T$. Let $1 \le p < +\infty$ and $q = \frac{p}{p-1}$. Assume for a.e. $x \in T$ and all $u \in \mathbb{R}$ that

$$| h(x,u) | \le a \mid u \mid^{\frac{p}{q}} + b(x), \quad a \in \mathbb{R}, b \in L^q(T), \tag{1.4.20}$$

and

$$u.h(x,u) \ge -c(x) \mid u \mid - d(x), c \in L^q(T), d \in L^1(T). \tag{1.4.21}$$

Set

$$h_+(x) = \liminf_{u \to +\infty} h(x,u)$$

and

$$h_-(x) = \limsup_{u \to -\infty} h(x,u).$$

From assumption (1.4.21) we obtain

$$-c(x) \le h_+(x) \le +\infty$$

and

$$-\infty \le h_-(x) \le c(x).$$

Proposition 1.4.13 Set $(Au)(x) = h(x, u(x))$ for $u \in L^p(T)$. Then

$$\underline{r}_A(u) \geq \int_T h_+(x)u^+(x)d\mu - \int_T h_-(x)u^-(x)d\mu, \ \forall \ u \ \in \ L^p(T). \quad (1.4.22)$$

Proof. Let $\{v_n\} \subset L^p(T), \{t_n\} \subset \mathbb{R}_+ \setminus \{0\}$ be such that $t_n \to +\infty, v_n \to u$ in $L^p(T)$. Extracting a subsequence, we may assume that $v_n(x) \to u(x)$ a.e. $x \in T$ and $\mid v_n \mid \leq l, \ \forall \ n \ \in \ \mathbb{N}$, for some fixed function $l \in L^p(T)$. Set

$$T_+ = \{x \in T : u(x) > 0\},$$
$$T_- = \{x \in T : u(x) < 0\}$$

and

$$T_0 = \{x \in T : u(x) = 0\}.$$

We have

$$\int_T h(x, t_n v_n)v_n d\mu \ = \ \int_{T_+} h(x, t_n v_n)v_n d\mu + \int_{T_-} h(x, t_n v_n)v_n d\mu$$
$$+ \int_{T_0} h(x, t_n v_n)v_n d\mu.$$

Moreover

$$\liminf_{n \to +\infty} h(x, t_n v_n)v_n \ \geq \ h_+(x)u \ \text{ on } T_+$$
$$\liminf_{n \to +\infty} h(x, t_n v_n)v_n \ \geq \ h_-(x)u \ \text{ on } T_-$$

and the relation

$$h(x, t_n v_n)v_n \geq -c(x) \mid v_n \mid - \frac{d(x)}{t_n}$$

yields

$$\liminf_{n \to +\infty} h(x, t_n v_n)v_n \geq 0 \ \text{ on } T_0.$$

From assumption (1.4.20), we deduce that

$$h(x, t_n v_n)v_n \in L^1(T),$$

and we may apply Fatou's lemma to get

$$\liminf_{n \to +\infty} \int_T h(x, t_n v_n)v_n d\mu \geq \int_{T_+} h_+(x)u(x)d\mu + \int_{T_-} h_-(x)u(x)d\mu$$

and thus (1.4.22). $\blacksquare$

1.4.5   THE ASYMPTOTIC POTENTIAL MAPPING

Let $(T, \mathcal{T}, \mu)$ be a positive complete measure space T and suppose that $0 < \mu(T) < +\infty$. Let $j : T \times \mathbb{R}^m \to \mathbb{R}$ $(m \geq 1, m \in \mathbb{N})$ be a function of the type considered in Section 1.2.3 (see (1.1.22) and (1.1.23)). We introduce the recession mapping

$$\underline{\gamma}_j(u) = \liminf_{\substack{v \to u \\ t \to +\infty}} \int_T \min_{z \in \partial_y j(x,tv)} \langle z, v \rangle d\mu \tag{1.4.23}$$

defined on $L^p(T; \mathbb{R}^m)(+\infty > p \geq 1)$. We call this mapping the asymptotic potential mapping. Note that

$$\min_{z \in \partial_y j(x,tv)} \langle z, v \rangle = - \max_{z \in \partial_y j(x,tv)} \langle z, -v \rangle$$
$$= -j_y^0(x, tv; -v)$$

and thus the integral

$$\int_T \min_{z \in \partial_y j(x,tv)} \langle z, v \rangle d\mu$$

is well defined. Note also that if j satisfies condition (1.1.22) then

$$\int_T \min_{z \in \partial_y j(x,tv)} \langle z, v \rangle d\mu \leq \int_T k(x) \mid v(x) \mid d\mu$$
$$\leq \mid k \mid_{0,q} \mid v \mid_{0,p}$$

and thus

$$\underline{\gamma}_j(u) \leq \mid k \mid_{0,q} \mid u \mid_{0,p}$$

in this case.

Let $m = 1$ and denoting for $j : T \times \mathbb{R} \to \mathbb{R}$,

$$\partial_y j(x, y) = [\underline{j}(x, y), \overline{j}(x, y)],$$

for a.e. $x \in T$, we assume that $\underline{j}$ and $\overline{j}$ are $\mu \times \mathcal{L}_1$-measurable functions. Assume also for a.e. $x \in \Omega$ and $u \in \mathbb{R}$:

$$u\underline{j}(x, u) \geq -\underline{c}(x) \mid u \mid -\underline{d}(x), \underline{c} \in L^q(T), \underline{d} \in L^1(T); \tag{1.4.24}$$

$$u\overline{j}(x, u) \geq -\overline{c}(x) \mid u \mid -\overline{d}(x), \overline{c} \in L^q(T), \overline{d} \in L^1(T); \tag{1.4.25}$$

$$\mid \underline{j}(x, u) \mid \leq \underline{a} \mid u \mid^{\frac{p}{q}} +\underline{b}(x), \underline{a} \in \mathbb{R}, \underline{b} \in L^q(T); \tag{1.4.26}$$

and

$$| \, \overline{j}(x,u) \, | \le \bar{A} \, | \, u \, |^{\frac{p}{q}} + \bar{b}(x), \, \bar{A} \in \mathbb{R}, \bar{b} \in L^q(T); \qquad (1.4.27)$$

Set

$$\underline{j}_+(x) = \liminf_{u \to +\infty} \underline{j}(x,u)$$

and

$$\overline{j}_-(x) = \limsup_{u \to -\infty} \overline{j}(x,u).$$

From condition (1.4.24) and (1.4.25), we deduce that

$$-\underline{c}(x) \le \underline{j}_+(x) \le +\infty, \text{ a.e. } x \in T$$

and

$$-\infty \le \overline{j}_-(x) \le \overline{c}(x), \text{ a.e. } x \in T.$$

We obtain the following lower estimate of $\underline{\gamma}_j$.

Proposition 1.4.14   We have:

$$\underline{\gamma}_j(u) \ge \int_T \underline{j}_+(x)u^+(x)d\mu - \int_T \overline{j}_-(x)u^-(x)d\mu, \, \forall \, u \, \in \, L^p(T). \quad (1.4.28)$$

Proof.   Let $v_n \in L^p(T), t_n \to +\infty$ be such that $v_n \to u \in L^p(T)$. Along a subsequence, we may assume that $v_n(x) \to u(x)$ a.e. $x \in T$ and $| \, v_n \, | \le h, \forall \in \mathbb{N}$, for some fixed function $h \in L^p(T)$.

If $z_n \in \partial_y j(x, t_n v_n)$ then

$$z_n v_n \ge \underline{j}(x, t_n v_n)v_n \text{ if } v_n \ge 0$$

and

$$z_n v_n \ge \overline{j}(x, t_n v_n)v_n \text{ if } v_n \le 0.$$

We have

$$\int_T \min_{z_n \in \partial_y j(x, t_n v_n)} z_n v_n d\mu \ge \int_T \underline{j}(x, t_n v_n)v_n^+ d\mu + \int_T \overline{j}(x, t_n v_n)(-v_n^-)d\mu$$

$$= \int_T \underline{j}(x, t_n v_n^+)v_n^+ d\mu + \int_T \overline{j}(x, -t_n v_n^-)(-v_n^-)d\mu$$

$$= \int_{T \setminus T_0} \underline{j}(x, t_n v_n^+)v_n^+ d\mu + \int_{T \setminus T_0} \overline{j}(x, -t_n v_n^-)(-v_n^-)d\mu$$

$$+ \int_{T_0} \underline{j}(x, t_n v_n^+) v_n^+ d\mu + \int_{T_0} \overline{j}(x, -t_n v_n^-)(-v_n^-) d\mu$$

where

$$T_0 = \{x \in T : u(x) = 0\}.$$

We have

$$\int_{T_0} \underline{j}(x, t_n v_n) v_n^+ d\mu = \int_{T_0} \underline{j}(x, t_n v_n^+) v_n^+ d\mu$$

$$\geq \int_{T_0} -\underline{c}(x) \mid v_n^+ \mid -\underline{d}(x) t_n^{-1} d\mu$$

and

$$\int_{T_0} -\overline{j}(x, t_n v_n) v_n^- d\mu = \int_{T_0} -\overline{j}(x, -t_n v_n^-) v_n^- d\mu$$

$$\geq \int_{T_0} -\overline{c}(x) \mid -v_n^- \mid -\overline{d}(x) t_n^{-1} d\mu.$$

Using assumptions (1.4.26) and (1.4.27) together with Fatou's lemma, we obtain

$$\liminf_{n \to +\infty} \int_T \min_{z_n \in \partial_y j(x, t_n v_n)} z_n v_n d\mu \geq$$

$$\int_{T \setminus T_0} \liminf_{n \to +\infty} \underline{j}(x, t_n v_n^+) v_n^+ d\mu + \int_{T \setminus T_0} \liminf_{n \to +\infty} \overline{j}(x, -t_n v_n^-)(-v_n^-) d\mu$$

$$+ \liminf_{n \to +\infty} \left(\int_{T_0} \underline{j}(x, t_n v_n^+) v_n^+ d\mu + \int_{T_0} \overline{j}(x, -t_n v_n^-)(-v_n^-) d\mu \right)$$

$$\geq \int_{T \setminus T_0} \underline{j}_+(x) u^+ d\mu - \int_{T \setminus T_0} \overline{j}_-(x) u^- d\mu$$

$$= \int_T \underline{j}_+(x) u^+ d\mu - \int_T \overline{j}_-(x) u^- d\mu \qquad (1.4.29)$$

and thus (1.4.28).   ∎

Remark 1.4.15 A particular form of the asymptotic potential mapping has been introduced by Goeleven and Théra in [162]. The concept is really studied for the first time in this book.

1.4.6   ASYMPTOTIC PROPERTIES OF THE SOLUTIONS OF INEQUALITY PROBLEMS

Let us first consider the variational inequality

$$u \in C : \langle Au - f, v - u \rangle + \Phi(v) - \Phi(u) \geq 0, \forall\, v \in C \qquad (1.4.30)$$

where C is a nonempty closed convex subset of a real Banach space X, $A : X \to X^*$ is an operator, $f \in X^*$ and $\Phi \in \Gamma_0(X; \mathbb{R} \cup \{+\infty\})$. A list of necessary conditions for the existence of a solution of (1.4.30) is now given.

Proposition 1.4.16 Suppose that the solutions set of (1.4.30) is nonempty. Then

(i)   For any solution $u \in C$ of (1.4.30), one has

$$\langle Au - f, e \rangle + \Phi_\infty(e) \geq 0, \ \forall\, e \in C_\infty. \qquad (1.4.31)$$

(ii)   Suppose that $A : X \to X^*$ is bounded and linear then

$$\langle f, e \rangle \leq \Phi_\infty(e), \ \forall\, e \in C_\infty \cap Ker A^*. \qquad (1.4.32)$$

(iii) Suppose that $A : X \to X^*$ is monotone and sublinear, i.e. $\frac{\|Av\|_*}{\|v\|} \to 0$ as $\|v\| \to +\infty$, then

$$\underline{r}_A(e) + \Phi_\infty(e) \geq \langle f, e \rangle, \ \forall\, e \in C_\infty. \qquad (1.4.33)$$

(iv)   Suppose that $A : X \to X^*$ can be written as

$$A = A_1 + A_2,$$

where A_1 is a monotone and sublinear operator and A_2 satisfies

$$\langle A_2 z - A_2 v, z \rangle \geq H(v), \ \forall\, z, v \in X,$$

where $H(v)$ is independent of z, then

$$\underline{r}_{A_1}(e) + \underline{r}_{A_2}(e) + \Phi_\infty(e) \geq \langle f, e \rangle, \ \forall\, e \in C_\infty. \qquad (1.4.34)$$

Proof. Let $u \in C$ be a solution of (1.4.30). Note that necessarily $u \in D(\Phi)$.

i) We have

$$\langle Au - f, v - u\rangle + \Phi(v) - \Phi(u) \geq 0, \forall\, v \in C.$$

If $e \in C_\infty$ then $u + e \in C$ and thus

$$\langle Au - f, e\rangle + \Phi(u + e) - \Phi(u) \geq 0.$$

Using (1.4.11), we get

$$\langle Au - f, e\rangle + \Phi_\infty(e) \geq 0.$$

ii) If A is bounded linear then (1.4.31) can be written as

$$\langle A^* e, u\rangle + \Phi_\infty(e) \geq \langle f, e\rangle, \ \forall\, e \in C_\infty.$$

Thus

$$\Phi_\infty(e) \geq \langle f, e\rangle, \ \forall\, e \in C_\infty \cap Ker A^*.$$

iii) If A is monotone and sublinear then we claim that

$$\underline{r}_A(u) \geq \sup_{\varphi \in R(A)} \langle \varphi, u\rangle, \forall u \in X.$$

Indeed, from the monotonicity of A, we obtain

$$\langle A(tv) - Ax, v - \frac{x}{t}\rangle \geq 0, \ \forall\, x,\, v \in X,\, \forall\, t > 0.$$

Thus for $v \neq 0$

$$\begin{aligned}
\langle A(tv), v\rangle \ &\geq \ \langle A(tv), \frac{x}{t}\rangle + \langle Ax, v\rangle - \langle Ax, \frac{x}{t}\rangle \\
&\geq \ -\frac{\|A(tv)\|_*}{t}\|x\| + \langle Ax, v\rangle - \langle Ax, \frac{x}{t}\rangle \\
&= \ -\frac{\|A(tv)\|_*}{\|tv\|}\|x\|\,\|v\| + \langle Ax, v\rangle - \langle Ax, \frac{x}{t}\rangle,
\end{aligned}$$

and thus

$$\underline{r}_A(u) \geq \langle Ax, u\rangle, \forall\, x \in X, \forall\, u \in X.$$

Thus

$$\underline{r}_A(u) \geq \sup_{\varphi \in R(A)} \langle \varphi, u\rangle, \forall\, u \in X.$$

From (1.4.31) we have

$$\langle Au, e\rangle + \Phi_\infty(e) \geq \langle f, e\rangle$$

and since

$$\underline{r}_A(e) \geq \sup_{\varphi \in R(A)} \langle \varphi, e \rangle \geq \langle Au, e \rangle,$$

we obtain

$$\underline{r}_A(e) + \Phi_\infty(e) \geq \langle f, e \rangle.$$

iv) We have

$$\langle A_1 u, e \rangle + \langle A_2 u, e \rangle + \Phi_\infty(e) \geq \langle f, e \rangle, \ \forall\, e \in C_\infty.$$

By part iii), we know that

$$\underline{r}_{A_1}(e) \geq \sup_{\varphi \in R(A_1)} \langle \varphi, e \rangle.$$

We claim that the same property holds for A_2. Indeed,

$$\langle A_2(tv) - A_2(x), v \rangle \geq \frac{H(x)}{t}, \ \forall\, x, \ v \ \in \ X, \ \forall\, t \ > \ 0.$$

Thus

$$\underline{r}_{A_2}(u) \geq \langle A_2(x), u \rangle, \ \forall\, x, \ u \ \in \ X.$$

It results that

$$\underline{r}_{A_2}(u) \geq \sup_{\varphi \in R(A_2)} \langle \varphi, u \rangle.$$

Therefore

$$\begin{aligned}
\underline{r}_{A_2}(e) + \underline{r}_{A_1}(e) + \Phi_\infty(e) \ &\geq \ \langle A_2 u, e \rangle + \langle A_1 u, e \rangle + \Phi_\infty(e) \\
&\geq \ \langle f, e \rangle.
\end{aligned}$$

$$\blacksquare$$

From the proof of the previous result, we see that if either A is monotone and sublinear or if $A = A_1 + A_2$ satisfies

$$\langle A_2 z - A_2 v, z \rangle \geq H(v), \ \forall\, z, \ v \ \in \ X,$$

with $H(v)$ independent of z, then

$$\underline{r}_A(u) \geq \sup_{\varphi \in R(A)} \langle \varphi, u \rangle, \forall\, u \in X. \tag{1.4.35}$$

Let $(T, \mathcal{T}, \mu)$ be as usually a positive complete measure space and let us now consider the variational-hemivariational inequality problem

$$u \in C : \langle Au - f, v - u \rangle + \int_T j^0(x, \gamma u(x); \gamma v(x) - \gamma u(x))d\mu$$

$$+\Phi(v) - \Phi(u) \geq 0, \forall\, v \in C \tag{1.4.36}$$

where $\gamma : X \subset L^p(T)(p \geq 1)$ is a linear and continuous mapping and $j : T \times \mathbb{R} \to \mathbb{R}$ is a function measurable in the first variable and locally Lipschitz in the second one. Suppose also that $j(., e) \in L^1(T)$ for some $e \in L^p(T)$. Denoting for $j : T \times \mathbb{R} \to \mathbb{R}$,

$$\partial_y j(x, y) = [\underline{j}(x, y), \overline{j}(x, y)],$$

we assume that

$$\overset{\smallsmile}{\overline{j}}(x) := \sup_{y \in \mathbb{R}} \overline{j}(x, y) \in L^q(T)$$

and

$$\overset{\wedge}{\underline{j}}(x) := \inf_{y \in \mathbb{R}} \underline{j}(x, y) \in L^q(T),$$

where $q^{-1} = 1 - p^{-1}$. Then

$$\int_T j_y^0(x, \gamma u(x); \gamma e(x))d\mu \leq \int_T \overset{\smallsmile}{\overline{j}}(x)(\gamma e)^+(x)d\mu$$

$$- \int_T \overset{\wedge}{\underline{j}}(x)(\gamma e)^-(x)d\mu, \,\forall\, u, e \in X \tag{1.4.37}$$

Indeed, one has

$$\underline{j}(x, u(x)) \leq z(x) \leq \overline{j}(x, u(x))$$

provided that $z(x) \in \partial_y j(x, u(x))$. Thus

$$j_y^0(x, \gamma u(x); \gamma e(x)) = \max_{z \in \partial_y j(x, \gamma u(x))} z\gamma e(x)$$

$$\leq \begin{cases} \overline{j}(x, \gamma u(x))\gamma e(x) & \text{if} \quad \gamma e(x) > 0 \\[2mm] \underline{j}(x, \gamma u(x))\gamma e(x) & \text{if} \quad \gamma e(x) < 0. \end{cases}$$

It results that

$$j_y^0(x, \gamma u(x); \gamma e(x)) \leq \begin{cases} \overset{\smallsmile}{\overline{j}}(x)\gamma e(x) & \text{if} \quad \gamma e(x) > 0 \\[2mm] \overset{\wedge}{\underline{j}}(x)\gamma e(x) & \text{if} \quad \gamma e(x) < 0. \end{cases}$$

We have

$$\int_T j_y^0(x, \gamma u(x); \gamma e(x))d\mu = \int_{T_+} j^0(x, \gamma u(x); \gamma e(x))d\mu$$

$$+ \int_{T_-} j^0(x, \gamma u(x); \gamma e(x))d\mu,$$

where

$$T_+ = \{x \in T : \gamma e(x) > 0\}$$

and

$$T_- = \{x \in T : \gamma e(x) < 0\}.$$

We obtain

$$\int_T j_y^0(x, \gamma u(x); \gamma e(x))d\mu \leq \int_{T_+} \breve{\bar{j}}(x)\gamma e(x)d\mu + \int_{T_-} \hat{\underline{j}}(x)\gamma e(x)d\mu$$

$$= \int_T \breve{\bar{j}}(x)(\gamma e)^+(x)d\mu - \int_T \hat{\underline{j}}(x)(\gamma e)^-(x)d\mu.$$

It results that inequality (1.4.37) is satisfied.

If u denotes a solution of (1.4.36) then

$$\langle Au, e\rangle + \int_T j_y^0(x, \gamma u(x); \gamma e(x))d\mu + \Phi_\infty(e) \geq \langle f, e\rangle, \ \forall\, e \in C_\infty \quad (1.4.38)$$

thanks to Proposition 1.4.3 i) and Proposition 1.4.7 i). Therefore, we may use inequality (1.4.37) together with (1.4.38) in order to get as in Proposition 1.4.16 necessary conditions for the existence of a solution of problem (1.4.37). A general result is given below.

Proposition 1.4.17 Suppose that the aforementioned conditions are satisfied and assume that the solutions set of (1.4.36) is nonempty. i) Let $A : X \to X^*$ be an operator satisfying

$$A = A_1 + A_2,$$

with A_1 monotone and sublinear and A_2 satisfying

$$\langle A_2 z - A_2 v, z\rangle \geq H(v), \ \forall\, z, v \in X, \quad (1.4.39)$$

with $H(v)$ independent of z. Then

$$\underline{r}_{A_1}(e) + \underline{r}_{A_2}(e) + \int_T \breve{\bar{j}}(x)(\gamma e)^+(x)d\mu - \int_T \hat{\underline{j}}(x)(\gamma e)^-(x)d\mu$$

$$+\Phi_\infty(e) \geq \langle f, e \rangle, \ \forall \, e \in C_\infty. \qquad (1.4.40)$$

ii) If A_1 is bounded and linear, and A_2 satisfies (1.4.39)-(1.4.40) then

$$\underline{r}_{A_2}(e) + \int_T \check{\bar{j}}(x)(\gamma e)^+(x)d\mu - \int_T \hat{\underline{j}}(x)(\gamma e)^-(x)d\mu$$

$$+\Phi_\infty(e) \geq \langle f, e \rangle, \ \forall \, e \in C_\infty \cap Ker \, A_1^*. \qquad (1.4.41)$$

Proof. The proof of i) follows from (1.4.37), (1.4.38) and the property (1.4.35) satisfied by the operators A_1 and A_2 considered. Part ii) follows from (1.4.37), (1.4.38), the property (1.4.35) satisfied by A_2 and the fact that

$$\langle A_1 u, e \rangle = \langle A_1^* e, u \rangle = 0$$

for $e \in Ker \, A_1^*$. ∎

Chapter 2

UNILATERAL MECHANICS

In this Chapter we explain the origins of Unilateral Mechanics and of the Inequality Problems. To do this we use the two notions of convex and of nonconvex superpotentials. We consider boundary conditions resulting from convex or nonconvex, nonsmooth energy functions using the concept of subdifferential or of generalized gradient studied in Chapter 1.

In virtually every mechanical systems there is a situation in which unilateral effects occur. Contact problems in elasticity [17], [18], [20], [19], [43], [44], [54], [83], [84], [130]-[131], [135], [136]-[137], [139]-[140], [156], [174], [189], [208], [213], [214], [228], [220], [224], [248], [275], [284], [285], [333], [336], [343], [344], [350], [351], [353], [357], [388], [389], [399], [407], friction problems [16], [36], [60], [61], [76], [91], [98], [102], [116], [117], [120]-[122], [147], [148], [177], [178], [187], [204]-[205], [212], [225]-[226], [236], [237], [246], [249] [260], [261], [267], [274], [280], [282], [283], [299], [314], [318]-[319], [325], [336], [391]-[392], [406], adhesive grasping problems in robotics [7], [8], [9], [169], [317], [353], [398], fracture and crack problems [37], [93], [94], [118], [100], [112]-[113], [114], [115], [138], [141], [272], [353], [400], [401], [409], [410], [412], debonding and delamination effects [10], [38], [104], [134], [304], [307], [338], [340], [347], [348], [352], [353], [384], [393], [396], [397], [411], [427], [430], effects of connections, contacts, etc in structural mechanics [153], [256] [258], [259], [271], [273], [315], [321], [324], [326], [327], [328], [339], [363], [364], [381], [420], [433], hysteresis problems [353], [423], interface problems in composite material structures [23], [34]-[35], [78], [175], [209], [211], [276], [322], [323], [335], [353], [379],[394], [429], unilateral buckling problems [51], [85]-[86], [109], [110], [216], [217], [232]-[233], [316], [336], impact and shocks [81], [142], [230], [240], [282], [345], [354], [355], cavitation phenomena in fluids [160], [194], [320], phase change problems [377], [378], kinematical constraints and frictions in rigid bodies systems [103], [171], [206], [207], [255], [262]-[263], [281], [286], [356], [434], semipermeable media problems [156], [337], analysis of material behavior problems [95], [96], [101], [149], [184], [185], [195], [221], [222], [234], [247], [243], [244], [305], [308], [311], [330], [331], [332], [334], [336], [341], [342], [346], [349], [353], [408], [419], [426], [428] are important examples of unilateral problems. More generally, a great variety of free boundary value problems are also discussed in the books [31], [105], [121], [143], [156], [165], [189], [190], [191], [220], [223], [251], [252], [311], [336], [353], [377], [378], and [417]. In this book, we limit our investigations to some chosen applications, that are contact and friction problems in linear elasticity, adhesive contact problems, laminated von Kármán plates, unilateral buckling of a plate, unilateral bending of a beam, loading and unloading problems, obstacle problem for a membrane, semipermeable media problems, nonsmooth oscillator models and frictionless contact problems in structural mechanics.

The theory developed in this Chapter rely on the superpotential approaches introduced by Moreau [278], [279] and Panagiotopoulos [329], [331]. In preparing this chapter we have primarily followed the books of Panagiotopoulos [336], [353] and Naniewicz and Panagiotopoulos [355].

2.1   MATHEMATICAL FORMALISM

We first recall some notions from continuum mechanics (see e.g. [21], [22], [45], [150]-[152], [155], [182], [238], [264], [298], [385], [413], [418] and [431]) which will be used in the sequel. Let Ω be an open subset of $\mathbb{R}^3$ with boundary Γ. The set Ω is occupied by a deformable body and is referred to a fixed orthogonal Cartesian coordinate system $\{O; X_1; X_2; X_3\}$. Thus a one-to-one correspondence between the material particles X of the body and the point $\{X_1, X_2, X_3\}$ (material coordinates of X) is established. Henceforth we will refer to the body simply as body Ω. With respect to another orthogonal Cartesian coordinate system $\{\bar{0}; x_1; x_2; x_3\}$ we may consider the coordinate transformation:

$$x_k = x_k(X_1, X_2, X_3), \quad k = 1, 2, 3. \tag{2.1.1}$$

Any deformation process may generally be described by means of the trajectory of each material particle X, i.e. by

$$x_i = \chi_i(X_k, t), \quad i, k = 1, 2, 3, \tag{2.1.2}$$

where $t \in [0, T]$ is the time variable. The point $x = \chi(X, t)$ is the position occupied by X at time t and let $x \equiv X$ for $t = 0$. The coordinates x_k, $k = 1, 2, 3$, are called the spatial coordinates of X. A mapping $\chi = \{\chi_i\} : \Omega \times [0, T] \to \Omega_t \subset \mathbb{R}^3$ is called "motion" of Ω. We denote by Ω_t the subset of $\mathbb{R}^3$ occupied by Ω at time t. We shall assume that χ and χ^{-1} exist and are appropriately regular functions. Let $A = A(X, t)$ be a function describing a quantity A. We call it the material (or Lagrangian) description of A, whereas $A = A(x, t)$ is the spatial (or Eulerian) description. We define the local or spatial derivative $\partial A(x, t)/\partial t$ of A, and its material derivative $\partial A(X, t)/\partial t$, which we will denote simply by dA/dt. Between material and local derivatives there holds the relation

$$\frac{dA}{dt} = \frac{\partial A(x, t)}{\partial t} + \frac{\partial A}{\partial x_i}\frac{\partial x_i(X, t)}{\partial t}, \quad i = 1, 2, 3. \tag{2.1.3}$$

The velocity $v = v(X, t)$ is obtained by differentiating (2.1.2) with respect to t keeping X unchanged, i.e.

$$v_i(X, t) = \frac{\partial \chi_i(X, t)}{\partial t}, \tag{2.1.4}$$

whereas the acceleration $\gamma = \gamma(X, t)$ is given by

$$\gamma_i = \frac{\partial v_i(X, t)}{\partial t} = \frac{dv_i}{dt} = \frac{\partial v_i(x, t)}{\partial t} + \frac{\partial v_i}{\partial x_j}v_j. \tag{2.1.5}$$

Inverting (2.1.2) implies $v = v(x, t)$. In the spatial description we consider the velocity gradient $L = \{L_{ij}\} = \{v_{i,j}\}$. Its symmetric and anti-symmetric parts are called the rate of deformation and spin. Thus

$$D = \frac{1}{2}(L + L^T) = \text{sym}\,(\nabla v) \tag{2.1.6a}$$

and

$$W = \frac{1}{2}(L - L^T) = \text{asym}\,(\nabla v). \tag{2.1.6b}$$

The instantaneous position x of a material particle is related to the initial position X by means of the displacement vector u, i.e. $x = X + u$. We call $F = \{F_{i,j}\} = \{\partial x_i/\partial X_j\}$ the deformation gradient. Then $F^T F$ is the right Cauchy-Green tensor and

$$E = \frac{F^T F - I}{2} \tag{2.1.7}$$

is the Green strain tensor. Hereafter, for a vector a_i the derivative $\partial a_i/\partial X_j$ will be denoted by $a_{i,\bar{j}}$. The deformation gradient rate takes the form

$$\frac{\partial F_{ik}(X, t)}{\partial t} = v_{i,j}x_{j,\bar{k}} \quad \text{or} \quad \frac{dF}{dt} = LF. \tag{2.1.8}$$

Thus the rate $\dot{E}$ of the Green strain tensor reads

$$
\begin{aligned}
\dot{E} &= \frac{dE}{dt} \\
&= \frac{1}{2}\left(\frac{dF^T}{dt}F + F^T\frac{dF}{dt}\right) \\
&= \frac{1}{2}(F^T L^T F + F^T L F) \\
&= F^T D F.
\end{aligned}
\tag{2.1.9}
$$

After some manipulations we obtain that

$$
E_{i,j} = \frac{1}{2}(u_{i,\bar{j}} + u_{j,\bar{i}} + u_{k,\bar{i}}u_{k,\bar{j}})
\tag{2.1.10}
$$

and

$$
\frac{dE_{ij}}{dt} = \frac{1}{2}(v_{i,\bar{j}} + v_{j,\bar{i}} + v_{k,\bar{i}}u_{k,\bar{j}} + u_{k,\bar{i}}v_{k,\bar{j}}).
\tag{2.1.11}
$$

The Cauchy stress tensor $\sigma = \sigma(x, t)$ is used in the case of spatial description. This last tensor has a direct physical interpretation in describing the stress state. When it acts on the unit vector n in the Lagrangian configuration, it gives the contact force of the unit area element with the normal vector n. From the tensor σ the (second) Piola-Kirchhoff stress tensor $\Sigma = \Sigma(X, t)$ is defined by

$$
\Sigma = (\det F)F^{-1}\sigma(F^T)^{-1}.
\tag{2.1.12}
$$

If the displacement gradients are small enough we may write that

$$
E_{ij} \simeq \varepsilon_{ij} = \frac{1}{2}(u_{i,j} + u_{j,i}).
\tag{2.1.13}
$$

The tensor $\varepsilon = \{\varepsilon_{ij}\}$ is called the (small) strain tensor. The replacing of E by ε is the geometric or kinematic linearisation.

Let $\Omega_t \subset \mathbb{R}^3$ be an open subset of $\mathbb{R}^3$ occupied by a body Ω at time t. The set Ω_t is referred to an orthogonal Cartesian coordinate system attached to the body. Let P be a point of Ω_t having coordinates $\{x_{p1}, x_{p2}, x_{p3}\}$. A velocity field such that

$$
v(x_p, t) = v_0(t) + \omega(t)\wedge x_p, \ \forall \ x_p = \{x_{p_i}\} \in \Omega_t, \quad i = 1, 2, 3,
\tag{2.1.14}
$$

is called a rigidifying velocity field, because it corresponds to a rigid motion of the body. By v_0 we denote the velocity of the origin of the coordinate system. It is readily seen that such a velocity field results from a motion described by the equation

$$
x = Q(t)x_p + x_0(t)
\tag{2.1.15}
$$

where $Q^T Q = Q Q^T = I$ and det $Q = 1$. We have also the relation

$$\omega_k = \frac{1}{2}\varepsilon_{kij}\Omega_{ji}$$

where $\{\Omega_{ij}\}$ is a skew-symmetric tensor given by $\{\Omega_{ij}\} = (dQ/dt)Q^T$, and where $\varepsilon_{ijk} = 0$ if any two indices are alike, $\varepsilon_{ijk} = \varepsilon_{123} = 1$ if (i,j,k) is an even permutation of $(1,2,3)$ and $\varepsilon_{ijk} = -1$ if (i,j,k) is an odd permutation.

Let x_i and $\bar{x}_i$, $i = 1,2,3$, be the coordinates of the same material point with respect to two orthogonal Cartesian coordinate systems which rotate and translate each with respect to the other arbitrarily. Then a relation of the form

$$\bar{x} = \tilde{Q}(t)x + \tilde{x}_0(t) \tag{2.1.16}$$

holds, where $\tilde{Q}$ has the same properties as Q and $\tilde{x}_0$ is an arbitrary vector.

A tensor field is said to be frame-indifferent, or objective, if it transforms in the well-known tensorial manner, whenever (2.1.16) is considered as a classical coordinate transformation. A mapping between tensor fields is objective or frame-indifferent if all dependent and independent variables transform in the previous manner.

2.2   PRINCIPLE OF VIRTUAL POWER

We now consider a mechanical system Σ corresponding to the body Ω and we assume that all the admissible velocity fields, which may occur in a time interval at which the observation takes place, are known. The admissibility is understood with respect to the kinematical or geometrical constraints imposed to the body. Let us denote by U, the space of velocity fields assumed to be a real Banach space, and by U_{ad} the subset of admissible velocity fields. The forces f acting on Σ are supposed to be linear continuous functionals on the space U that is $f \in U^*$. We denote by $\langle .,. \rangle$ the duality product between U and U^* and in the terminology of Mechanics U is the space of virtual velocities and $\langle f, v \rangle$, $v \in U$, $f \in U^*$ is the virtual power produced by f. In order to create a framework for the study of continuous systems we place this system in an inertial frame of reference without any kinematical constraints and we define the real Banach space U of virtual velocities. Let us denote by U_0 the subspace of rigidifying velocity fields and let us apply the following postulate [151]:

$[P_1]$: The virtual power Π_i of the internal forces of the body is zero for any rigidifying velocity field at any time.

As we have defined before the virtual power Π_i is a linear function of v, i.e. the value $\Pi_i(v)$ at v of the virtual power remains unchanged if v is replaced by $v + v_0$, $v_0 \in U_0$. Further, we write $\Pi_i(v)$ in the form $\langle f, v \rangle$, and $[P_1]$ is equivalent to the statement that $\Pi_i(v) = \langle f, v \rangle = 0$, $\forall\, v \in U_0$. We are thus led to consider the quotient space $\bar{U} = U/U_0$, which is called the space of objective virtual velocities. The internal forces are by definition the continuous linear functionals on $\bar{U}$. We denote their space by $\bar{F}$. Obviously, the elements of both $\bar{U}$ and $\bar{F}$ are objective quantities, as can easily be seen by considering the invariance of the duality mapping. For the complete formulation of a continuum theory, we have to choose in addition to the space U, the precise form of the linear mapping $v \to \langle f, v \rangle$ expressing the virtual power of the internal forces of the system. Hence in the framework of a local theory we may consider that the successive gradients of the vector field v, i.e. $v_{i,j}, v_{i,jk}$, etc. are involved in the mapping $v \to \langle v, f \rangle$, and thus we are led to first-, second-, etc. order gradient theories respectively. Thus we may consider as a space U the cartesian product space $U^{(0)} \times U^{(1)} \times \ldots \times U^{(m)}$ where $U^{(m)} = \{v_{i,jk} \ldots (m-\text{ spatial derivations })\}$. Let Ω be the body considered. Then in the first-order gradient theory, which is the most common, the power of the continuous system depends on the velocities v_i and the first gradients $v_{i,j}$. On the assumption that the power Π_i of the internal forces acting in Ω can be expressed in the integral form,

$$\Pi_i(v) = -\int_\Omega p_i(v)dx \tag{2.2.1}$$

we may show [151] that at any point of Ω one has

$$p_i(v) = t_{ij}D_{ij}, \quad D_{ij} = \frac{1}{2}(v_{i,j} + v_{j,i}). \tag{2.2.2}$$

Indeed, $p_i(v)$ may be written at any point of Ω in the general form

$$p_i(v) = t_{ij}D_{ij} + r_{ij}\Omega_{ij} + q_iv_i, \quad \Omega_{ij} = \frac{1}{2}(v_{i,j} - v_{j,i}). \tag{2.2.3}$$

Obviously, t_{ij} is a component of a symmetric tensor. If $q_i \neq 0$ in a neighborhood of a point $M \in \Omega_1 \subset \Omega$, then one can easily determine a subsystem containing M and a translational virtual velocity such that $p_i(v)$ is not zero on this neighborhood. But this contradicts $[P_1]$. Analogously, by means of a rotational virtual velocity we find that $r_{ij} = 0$. Here $t = \{t_{ij}\}$ is called the intrinsic stress tensor. The symmetric tensor $D = \{D_{ij}\}$ is the rate of deformation (or stretching) tensor. For the theory of the second gradient and the respective theories of materials with microstructure, the reader is referred to [151], [152].

Let us now give the main postulate of mechanics, which governs the motion of any body. It is the "principle of virtual power" and reads:

[P_2]:   At any time and for any field of kinematically admissible virtual velocities (i.e. elements of U_{ad}), the virtual power of all the internal and external forces impressed on the system is equal to the virtual power of the inertial (or d'Alembert) forces.

Further, we pay some attention to the virtual power of the external and inertial forces in Ω. The external forces are volume forces (e.g., gravity, electromagnetic forces, etc.), or boundary forces (e.g., contact forces) acting on the boundary Γ of Ω. In the first case they are defined as continuous linear functionals on the space U, and in the second as continuous linear functionals on a vector space U_Γ defined on the boundary Γ of Ω. The latter space is assumed to exist and is such that it includes the traces $v|_\Gamma$ of the elements v of U. The elements of both U and U_Γ are not necessarily objective. Finally, the virtual power of the inertial forces is a continuous linear functional on $U^{(0)}$ whose elements are not objective. If we denote by Π_v, Π_c and Π_α the three virtual powers of volume, contact and inertial forces, then [P_2] implies that at any time t_0,

$$\Pi_i + \Pi_v + \Pi_c = \Pi_\alpha \ \forall \, v \, \in \, U_{ad}. \tag{2.2.4}$$

Equation (2.2.4) may also be formulated for any subsystem of Ω occupying the domain $\Omega_1 \subset \Omega$ at time t. Assuming that $U_{ad} \equiv U$, i.e. that the body is not constrained, we can obtain the possible forms of the terms Π_v, Π_c and Π_α. In the context of a first-order gradient theory and with regard to Ω_1, we assume that $\Pi_i(v), \Pi_v(v)$ and $\Pi_\alpha(v)$ (resp. $\Pi_c(v|_\Gamma)$) can be expressed as integrals over Ω_1 (resp. over Γ_1) of $p_i(v), p_v(v)$ and $p_a(v)$ (resp. $p_c(v|_\Gamma)$). Then, as is readily verified, the most general forms of $p_v(v)$ and $p_c(v|_\Gamma)$ are respectively

$$p_v(v) = f_i v_i + b_{ij} D_{ij} + c_{ij}\Omega_{ij}, \quad b_{ij} = b_{ji}, \quad c_{ij} = -c_{ji} \tag{2.2.5}$$

and

$$p_c(v|_\Gamma) = S_i(v|_\Gamma)_i, \tag{2.2.6}$$

where $\{f_i\}, \{b_{ij}\}$ and $\{c_{ij}\}$ are respectively the volume force, the double symmetric force, and the couple tensor in Ω_1, and $\{S_i\}$ is the stress vector on the boundary Γ_1 of Ω_1. In a mechanical theory (i.e. without Maxwellian fields [264]) the density of the inertial force power is

$$p_\alpha(v) = \rho \frac{d\hat{v}_i}{dt} v_i, \tag{2.2.7}$$

where ρ is the density of the body and $d\hat{v}/dt$ denotes the material derivative of the real velocity $\hat{v}$. If all the quantities are sufficiently smooth we may apply the Green-Gauss theorem (see Section 2.10.4), and since Ω_1 is arbitrary we obtain the equations (cf. [150]-[152])

$$\sigma_{ij,j} + f_i = \rho\frac{d\hat{v}_i}{dt}, \quad \sigma_{ij} - \sigma_{ji} + c_{ij} = 0 \quad \text{in} \quad \Omega_1 \tag{2.2.8}$$

$$S_i = \sigma_{ij}n_j \text{ on } \Gamma_1, \tag{2.2.9}$$

where $n = \{n_i\}$ is the outward unit normal vector to Γ_1,

$$\sigma_{ij} = t_{ij} - b_{ij} - c_{ij} \quad \text{in} \quad \Omega_1 \tag{2.2.10}$$

and $\sigma = \{\sigma_{ij}\}$ is the Cauchy stress tensor. In the framework of a classical continuum theory we have $b_{ij} = c_{ij}$. This is not the case, e.g. in polar continua and in electromagnetic continua.

Until now we have seen how the application of the postulates $[P_1]$ and $[P_2]$ permit us to derive the basic equations of the mechanics of continua. For the study of the dynamic behavior of a given continuous structure we have to take into account the kinematical and the statical constraints imposed on the structure in the application of the principle of virtual power. In the case of higher order gradient theories we would obtain higher order tensors [152] and accordingly we may extend the classical notion of force and call the elements of U^* generalized forces. Analogously the elements of U are called generalized velocities. From (2.2.8) and Green Gauss Theorem, we deduce that

$$\begin{aligned}\int_{\Omega_1} \rho\frac{d\hat{v}_i}{dt}z dx + \int_{\Omega_1} \sigma_{ij}D_{ij}(z)dx \\ = \int_{\Omega_1} f_i z dx + \int_{\Gamma_1} S_i(z)ds,\end{aligned} \tag{2.2.11}$$

for all $z \in U$.

2.3   PRINCIPLE OF VIRTUAL WORK

For static problem, $\hat{v}_i \equiv 0$ and the principle of virtual power takes a form known as principle of virtual work. Then U is the space of generalized displacements u and U^* is the space of corresponding generalized forces. From (2.2.8) we obtain

$$\int_{\Omega_1} \sigma_{ij}D_{ij}(z)dx = \int_{\Omega_1} f_i z dx + \int_{\Gamma_1} S_i z ds. \tag{2.3.1}$$

2.4   CONVEX SUPERPOTENTIALS

Let us now consider a mechanical system Σ on which the triplet $\{U, \langle \cdot, \cdot \rangle, U^*\}$, is defined and suppose that only certain subsets of U and

U^* are admissible for the mechanical system. Then a multivalued mapping $A : U \to U^*$ such that

$$f \in A(v) \, \forall \, v \in X \subset U \qquad (2.4.1)$$

introduces a law or a constraint on Σ. In equilibrium problems v will be replaced by the generalized displacement u.

Let us consider further the equilibrium of a system Σ subjected to certain forces f_i, $i = 1, \ldots, n$, and laws or constraints which are defined on X_j by the operators A_j, $j = 1, 2, \ldots, m$. From the principle of virtual work we find that at the position of equilibrium

$$\sum_{i=1}^{n} f_i + \sum_{j=1}^{m} \overline{f}_j = 0 \qquad (2.4.2)$$

$$\overline{f}_j \in A_j(u) \, \forall \, u \in X_j \quad j = 1, 2, \ldots, m.$$

Accordingly, at the position of equilibrium $u \in \cap_j X_j$ and

$$-\sum_{i=1}^{n} f_i \in \sum_{j=1}^{m} A_j(u). \qquad (2.4.3)$$

Of special interest is the case of subdifferential laws or constraints. Then $A = -\partial\Phi$ where Φ is a convex, l.s.c. and proper functional on U. The functional Φ is called convex superpotential, after Moreau [279]. Then (2.4.1) takes the form

$$-f \in \partial\Phi(u). \qquad (2.4.4)$$

By definition, for $u \in U$ and for $f \in U^*$, (2.4.4) is equivalent to the variational inequality.

$$\Phi(h) - \Phi(u) \geq -\langle f, h - u \rangle, \, \forall \, h \in U.$$

Suppose further that Φ is the indicator Ψ_K of a convex closed subset K of U. Then

$$-f \in \partial\Psi_K(u) \qquad (2.4.5)$$

and the constraint corresponding to (2.4.5) is called an "ideal unilateral constraint". According to (2.4.5) $-f$ is an element of the outward normal cone to K at u. The term "unilateral" results if one considers for $u \in K$ the variational inequality

$$\langle f, u^* - u \rangle \geq 0, \, \forall \, u^* \in K, \qquad (2.4.6)$$

which results by definition from (2.4.5). Indeed, if $u^* - u$ is an admissible variation of u (in the sense that it satisfies (2.4.6)), then the same does not hold for the variation $u - u^*$. Only if K is a linear subspace of U, (2.4.6) holds as an equality and then $f \in K^\perp$. In order to illustrate (2.4.6) let us assume that a material point with mass m is subjected to a force $f \in \mathbb{R}^3$ and is constrained to belong in a convex closed subset K of $\mathbb{R}^3$. If $K = \{x : x \in \mathbb{R}^3, F(x) \leq 0\}$, where F is a continuously differentiable function on $\mathbb{R}^3$ referred to a Cartesian coordinate system $\{0; X_1; X_2; X_3\}$ and the contact of the material point with the boundary of K is frictionless, then the reaction force R is given by the relation

$$R = -\lambda \, \nabla \, F(x), \quad \lambda > 0, \tag{2.4.7}$$

where λ is an unknown proportionality factor. If the material point is in $int\{K\}$, then the reaction force is zero, i.e., $\lambda = 0$; otherwise $\lambda \geq 0$. This type of constraint is described by the relation

$$-R \in \partial \Psi_K(x). \tag{2.4.8}$$

The material point is in equilibrium, if and only if

$$f + R = 0. \tag{2.4.9}$$

Thus

$$f \in \partial \Psi_K(x), \tag{2.4.10}$$

and conversely. For $x \in K$, (2.4.10) is equivalent to the variational inequality

$$f_i(h_i - x_i) \leq 0, \ \forall \, h \, \in \, K, \tag{2.4.11}$$

which is the expression of the principle of virtual work.

Let us now consider a system Σ acted upon by forces f_i where $i = 1, \ldots, n$, and reactions $f_j, j = 1, 2, \ldots, m$, which are derived (see (2.4.4)) from the superpotentials Φ_j defined on the space U of generalized displacements. Then, the condition of equilibrium (2.4.3) reads

$$\sum_{i=1}^{n} f_i \in \sum_{j=1}^{m} \partial \Phi_j(u). \tag{2.4.12}$$

A solution $u \in D(\partial \Phi_1) \cap \ldots \cap D(\partial \Phi_m)$ of (2.4.12) satisfies (see Proposition 1.3.8)

$$\sum_{i=1}^{n} f_i \in \partial \Phi_0(u), \quad \Phi_0(u) = \sum_{i=1}^{m} \Phi_j(u), \tag{2.4.13}$$

the converse being generally not true. A combination of Proposition 1.3.9 and Proposition 1.3.10 supplies the following sufficient condition for the equivalence of (2.4.12) and (2.4.13). If (i) the gradients of l $(0 \leq$

$l \leq m$) of the superpotentials Φ_j, exist for every $u \in U$ and if (ii) a $u_0 \in U$ exists such that from the remaining $m - l$ functionals, $m - l - 1$ are finite on U and continuous at u_0 and (iii) the $(m - l)$-th functional is finite at u_0, then every solution of (2.4.12) is a solution of (2.4.13), and conversely. Then (2.4.13) is equivalent to the problem

$$\Pi(u) = \min\{\Pi(u^*) : u^* \in U\}, \qquad (2.4.14)$$

where

$$\Pi(u^*) = \Phi_0(u^*) - \sum_{i=1}^{n} \langle f_i, u^* \rangle \qquad (2.4.15)$$

is the potential energy of the system considered. Note that if U is the space of generalized velocities and u in (2.4.12) is replaced by v, then (2.4.12) and (2.4.14) describe the motion of the mechanical system Σ where the corresponding inertial forces have been neglected.

2.5   NONCONVEX SUPERPOTENTIALS

With respect to a mechanical system Σ characterized by the triplet $\{U, \langle ., . \rangle, U^*\}$, a mechanical law or constraint is considered between the generalized forces f and the generalized displacements u of the form

$$-f \in \partial\Phi(u), \qquad (2.5.1)$$

where Φ is a locally Lipschitz functional defined on U. We shall call Φ a nonconvex superpotential. This mechanical law is by definition equivalent to the inequality

$$\Phi^0(u; h - u) \geq \langle -f, h - u \rangle, \ \forall \ h \in U \qquad (2.5.2)$$

for $u \in U$.

Suppose now that forces f_i, $i = 1, \ldots, n$, act on the system Σ which is subjected to the nonconvex superpotential laws or constraints $-\bar{f}_j \in \partial\Phi_j(u)$, $j = 1, \ldots, m$. Then the condition of equilibrium (2.4.2) implies that

$$\sum_{i=1}^{n} f_i \in \sum_{j=1}^{m} \partial\Phi_j(u). \qquad (2.5.3)$$

Obviously, if

$$0 \in \partial\Pi(u), \quad \Pi(u) = -\sum_{i=1}^{n} \langle f_i, u \rangle + \sum_{j=1}^{m} \Phi_j(u), \qquad (2.5.4)$$

then (2.5.3) holds, but the converse is not always true; u is a substationarity point of Π, where Π is the potential energy of the system Σ. Accordingly, for the present mechanical system, u is an equilibrium configuration, if u is a substationarity point of Π. Equivalent to (2.5.4) is the inequality

$$\Pi^0(u; h) \geq 0 \ \forall \ h \ \in \ U. \tag{2.5.5}$$

In Chapter 1 we have given the generalized gradients of several types of functions. Such types of functions permit the formulation of several classes of mechanical problems in terms of nonconvex superpotentials and thus in terms of hemivariational inequalities. In particular, the superpotentials resulting by integrating discontinuous functions $\beta \in L^\infty_{\mathrm{loc}}(\mathbb{R})$ play an important role in the formulation of hemivariational inequalities for several types of mechanical problems. The superpotential law which will be given next further illustrates the possibilities offered to Mechanics by the introduction of the notion of generalized gradient. We would like to point out that the aforementioned nonconvex superpotential laws and constraints also hold in the framework of generalized velocities.

2.6   MONOTONE UNILATERAL BOUNDARY CONDITIONS

We may now introduce subdifferential boundary conditions for a deformable body. These boundary conditions include the boundary conditions of classical elasticity as special cases. We denote by Ω an open bounded subset of $\mathbb{R}^3$ which is occupied by a deformable body. The forthcoming definitions hold both for small and large deformation theory. The boundary of Ω is denoted by Γ. The points $x \in \Omega, x = \{x_i\}, i = 1, 2, 3$, are referred to a Cartesian coordinate system. We denote by $S = \{S_i\}$ the stress vector on Γ. It is $S_i = \sigma_{ij} n_j$, where $\sigma = \{\sigma_{ij}\}$ is an appropriately defined stress tensor depending on the frame chosen and $n = \{n_i\}$ is the outward unit normal vector on Γ. The vector S is decomposed into a normal component S_N and a tangential component S_T with respect to Γ, i.e.

$$S = S_N n + S_T, \tag{2.6.1}$$

$$S_N = \sigma_{ij} n_j n_i, \tag{2.6.2}$$

and

$$S_{T_i} = \sigma_{ij} n_j - (\sigma_{ij} n_i n_j) n_i. \tag{2.6.3}$$

Analogously to S_N and S_T, u_N and u_T denote the normal and the tangential components of the displacement vector u with respect to Γ, i.e. $u = u_N n + u_T$.

A maximal monotone operator $\beta_i : \mathbb{R} \rightarrow 2^{\mathbb{R}}$ is introduced and a boundary condition of the form

$$-S_i \in \beta_i(u_i) \tag{2.6.4}$$

is considered in the i-th direction. Here $U = U^* = \mathbb{R}$. Then (Proposition 1.3.15) a convex, l.s.c. and proper functional j_i on $\mathbb{R}$ may be determined up to an additive constant such that

$$\beta_i = \partial j_i. \tag{2.6.5}$$

Then (2.6.4) is written as

$$-S_i \in \partial j_i(u_i). \tag{2.6.6}$$

This relation is a subdifferential boundary condition and is understood pointwise, i.e., as a relation between $-S_i(x) \in \mathbb{R}$ and $u_i(x) \in \mathbb{R}$ at every point $x \in \Gamma$. The graph of β_i, referred to a Cartesian system $\{O; x; y\}$, is a complete nondecreasing curve in $\mathbb{R}^2$ which is generally multivalued; thus the graph may include segments parallel to both coordinate axes. The "Superpotential" j_i is a local superpotential and expresses the potential of the constraint [279]. The boundary condition (2.6.4) may be considered as the material law of a fictive spring of zero length at x in the ith-direction.

Analogously to (2.6.4), a boundary condition of the form

$$-S_N \in \beta_N(u_N) = \partial j_N(u_N) \tag{2.6.7}$$

may be defined. Assume now that $U = U^* = \mathbb{R}^3$ and that j is a convex, l.s.c., proper functional on $\mathbb{R}^3$. Then a boundary condition of the form

$$-S \in \partial j(u) \tag{2.6.8}$$

is defined pointwise on Γ, i.e., as a monotone relation between $S(x)$ and $u(x)$. Similarly to (2.6.8), a subdifferential law

$$-S_T \in \partial j_T(u_T) \tag{2.6.9}$$

may be considered.

In dynamic mechanical problems, similar boundary conditions may be defined between S and the partial time derivative of the displacement $\partial u / \partial t$, or the velocity v.

We give some examples to illustrate these boundary conditions.

i) The classical boundary conditions $u_i = 0$ can be put in the form (2.6.4) through the operator

$$\beta_i(u_i) = \left\{ \begin{array}{ll} \mathbb{R} & \text{if } u_i = 0 \\ \emptyset & \text{otherwise,} \end{array} \right. \tag{2.6.10}$$

or through the functional $j_i(u_i) = \Psi_{\{0\}}(u_i)$. The boundary conditions $S_i = C_i$ is written in the form (2.6.4) or (2.6.6) with $\beta_i(u_i) = -C_i$ (C_i given) or $j_i(u_i) = -C_i u_i$ (no summation) for every $u_i \in \mathbb{R}$.

ii) Let Q be a nonempty closed convex subset of $\mathbb{R}^3$ containing Γ and suppose that Γ is constrained to stay in the box Q. Then we express that any point x of Γ after deformation $x + u(x)$ still belongs to Q. Setting

$$Q(x) = -x + Q,$$

we have thus

$$u(x) \in Q(x), \tag{2.6.11}$$

for all x of Γ. If a frictionless contact occurs between a material point x of Γ and the boundary of the box Q then there is a reaction force S_R satisfying the normal reaction law

$$S_R(x) \in -N_{Q(x)}(u(x)). \tag{2.6.12}$$

The relations (2.6.11) and (2.6.12) can also be formulated through the subdifferential law

$$-S_R(x) \in \partial\Psi_{Q(x)}(u(x)).$$

iii) The Winkler contact boundary condition

$$-S_N = ku_N, \; k \text{ const} > 0 \tag{2.6.13}$$

may be expressed in the form (2.6.7) by setting

$$\beta_N(u_N) = ku_N, \quad j_N(u_N) = \frac{1}{2}ku_N^2.$$

This law, describes in a simplified manner the interaction between a deformable body and the soil and is used in practical civil engineering.

iv) The foregoing boundary condition does not describe the case in which the body loses contact with the support [325], [328]. To do so we should consider the following law:

$$\text{if } u_N < 0, \quad \text{then } S_N = 0; \tag{2.6.14a}$$

$$\text{if } u_N \geq 0, \quad \text{then} \quad S_N + k u_N = 0; \quad k \text{ const} > 0. \qquad (2.6.14b)$$

Relation (2.6.14a) corresponds to the case of lack of contact and (2.6.14b) to the case of contact. The regions of contact and noncontact are not known a priori; thus (2.6.14a)-(2.6.14b) lead to a free B.V.P. The respective operator β_N (resp. superpotential j_N) is given by

$$\beta_N(u_N) = \begin{cases} k u_N & \text{if } u_N \geq 0 \\ 0 & \text{if } u_N < 0 \end{cases} \qquad (2.6.14c)$$

and

$$j_N(u_N) = \begin{cases} \dfrac{1}{2} k u_N^2 & \text{if } u_N \geq 0 \\ 0 & \text{if } u_N < 0. \end{cases} \qquad (2.6.14d)$$

Note that $j_N(u_N)$ can be written also as $\frac{1}{2}k(u_N^+)^2$, where we recall that u_N^+ denotes the positive part of u_N, i.e., $u_N^+ = \sup\{0, u_N\}$. Relations (2.6.14) are called conditions of unilateral contact (Fig. 2.6.1) for a linear Winkler law, whereas (2.6.13) is the condition of bilateral contact. In Fig. 2.6.1 the graph AOE corresponds to (2.6.14a)-(2.6.14b).

We can consider generally the operators

$$\beta_N(u_N) = \begin{cases} \beta_1(u_N - h) & \text{if } u_N \geq h \\ 0 & \text{if } u_N < h. \end{cases} \qquad (2.6.15)$$

Here β_1 is assumed to be a maximal monotone operator on $\mathbb{R}$ such that $0 \in \beta_1(0)$. The relation (2.6.15) leads to unilateral contact boundary conditions, but with a nonlinear Winkler law and a support at a given distance $h = h(x)$ from the body under consideration. In Fig. 2.6.1 the diagram AOCFG corresponds to (2.6.15), where the segment OC has a length h. The relations are not sufficient to formulate a B.V.P., but they must be combined with a boundary condition concerning S_T or u_T or both, e.g. $S_T = C_T$, where $C_T = C_T(x)$ is given, or $u_T = 0$, or, more generally, (2.6.9). It is also possible for β_N to change from point to point, in which case $\beta_N = \beta_N(x, u_N(x))$. Note that the uncoupling of the contact conditions in the tangential and in the normal directions is a considerable simplification of the mechanical problem. The case of coupled contact conditions will be examined at the end of this Section.

v) If the support is rigid, then the boundary conditions of Signorini hold [121], [130]-[131], [388]. They read

$$\begin{aligned} &\text{if } u_N < 0, \quad \text{then } S_N = 0; \\ &\text{if } u_N = 0, \quad \text{then } S_N \leq 0, \end{aligned} \qquad (2.6.16)$$

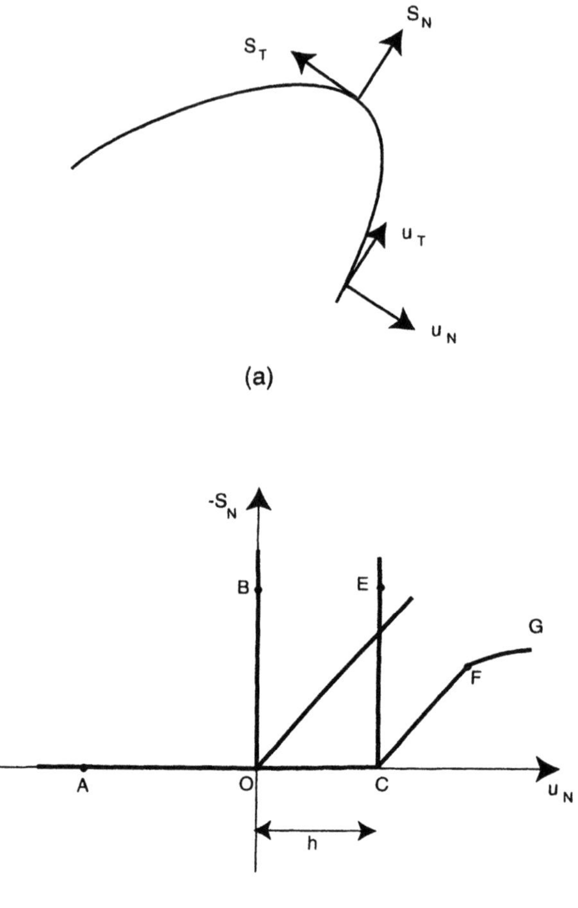

Figure 2.6.1.   Unilateral contact boundary conditions.

or equivalently

$$S_N \leq 0, \quad u_N \leq 0, \quad \text{and} \quad S_N u_N = 0. \tag{2.6.17}$$

This last form is called linear complementarity form (see Section 1.1.9). In Fig. 2.6.1 the graph AOB corresponds to the boundary conditions (2.6.17). The respective operator β_N is

$$\beta_N(u_N) = \begin{cases} 0 & \text{if } u_N < 0 \\ [0, +\infty) & \text{if } u_N = 0 \\ \emptyset & \text{if } u_N > 0 \end{cases} \tag{2.6.18}$$

and the corresponding superpotential

$$j_N(u_N) = \begin{cases} 0 & \text{if } u_N \leq 0 \\ \infty & \text{if } u_N > 0. \end{cases} \tag{2.6.19}$$

If the support is at a distance h from the boundary of the body, then u_N has to be replaced by $u_N - h$ (cf. in Fig. 2.6.1 the graph AOCD). To describe the contact with the possibility of debonding (or detachment) between two deformable bodies we consider an interface condition analogous to (2.6.16) on the assumption that the boundary displacements are sufficiently small. As the two bodies cannot penetrate one another, we assume that the sum of the displacements $u_N^{(1)}$ and $u_N^{(2)}$ of the two bodies and of the existing normal distance between them $h = h(x)$ must be greater than, or equal to the approach u^0 of the two bodies in the normal direction due to a rigid body displacement. We denote by $\bar{u}_N$ the quantity

$$\bar{u}_N := u_N^{(1)} + u_N^{(2)} + h - u^0,$$

and let R_N be the respective contact force. The contact conditions read:

$$\begin{aligned} \text{if } \bar{u}_N > 0, \quad &\text{then } R_N = 0; \\ \text{if } \bar{u}_N = 0, \quad &\text{then } R_N \geq 0. \end{aligned} \tag{2.6.20}$$

vi) The next example concerns the static version of Coulomb's friction boundary condition [120], [283]. We consider the following boundary conditions for $U = U^* = \mathbb{R}^3$ (if $\Omega \subset \mathbb{R}^3$):

$$\text{if } |S_T| < \mu|S_N|, \text{ then } u_{T_i} = 0, \quad i = 1, 2, 3 \tag{2.6.21a}$$

$$\text{if } |S_T| = \mu|S_N|, \text{ then there exists } \lambda \geq 0$$

$$\text{such that } u_{T_i} = -\lambda S_{T_i}, \quad i = 1, 2, 3. \tag{2.6.21b}$$

Here $\mu = \mu(x) > 0$ denotes the coefficient of friction and $|\cdot|$ the usual $\mathbb{R}^3$-norm. If Ω is a two-dimensional body, then Γ is a curve, and thus S_T, u_T may be referred to a local right-handed coordinate system (n, τ) on Γ where τ denotes the unit vector tangential to Γ. Then (2.6.21a)-(2.6.21b) can be put in the form

$$-S_T \in \beta_T(u_T), \tag{2.6.22}$$

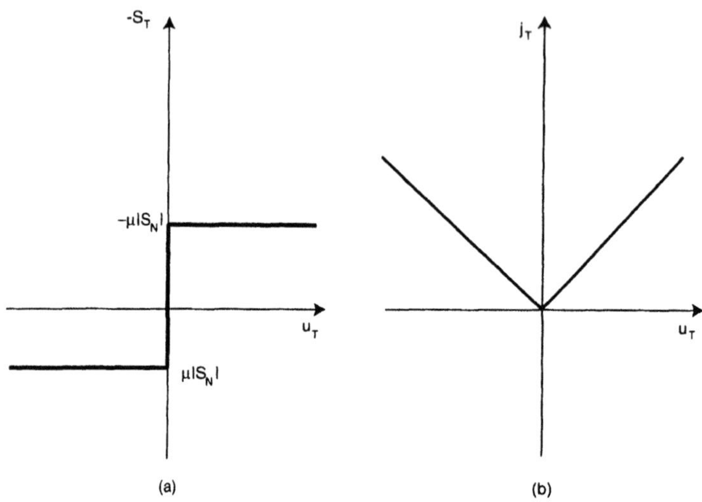

Figure 2.6.2.   The friction boundary condition.

where (Fig. 2.6.2)

$$\beta_T(u_T) = \begin{cases} \overline{B}(0, \mu \mid S_N \mid) & \text{if } u_T = 0 \\[2mm] \mu|S_N|d(u_T) & \text{if not,} \end{cases} \qquad (2.6.23)$$

where $\overline{B}(0, \mu \mid S_N \mid)$ denotes the closed ball of center 0 and radius $\mu \mid S_N \mid$ and $d(u_T)$ denotes the unit vector $d(u_T) = u_T / \mid u_T \mid$. Assume further that $S_N = C_N$, where C_N is given, and denote $\mu|C_N|$ by S_{T_0}. Then

$$\beta_T(u_T) = \partial(S_{T_0}|u_T|). \qquad (2.6.24)$$

If Ω is a three-dimensional body, then (2.6.21a) -(2.6.21b) can be written only in the form (2.6.9) with

$$j_T(u_T) = S_{T_0}|u_T|. \qquad (2.6.25)$$

Note that the subdifferential formulation gives rise, for $u_T, S_T \in \mathbb{R}^3$ to the variational inequality

$$j_T(h) - j_T(u_T) \geq -S_{T_i}(h_i - u_{T_i}), \ \forall \, h \, \in \, \mathbb{R}^3 \qquad (2.6.26)$$

In dynamic or quasistatic problems, a friction law of the form

$$-S_T \in \partial j_T(v_T) = \partial(S_{T_0}|v_T|) \qquad (2.6.27)$$

can be considered (Coulomb's law of friction). Here v_T denotes the tangential velocity which is equal to $\partial u_T/\partial t$ if the displacements are sufficiently small.

It is possible to combine the friction boundary condition with the unilateral contact boundary condition. Then we obtain the following relations:

$$\begin{aligned}
&\text{if } u_N < 0, \quad \text{then } S_N = 0, \quad S_{T_i} = 0, \quad i = 1,2,3 \\
&\text{if } u_N \geq 0, \quad \text{then } S_N + ku_N = 0,
\end{aligned} \qquad (2.6.28)$$

where k is a constant > 0 and (2.6.21a)-(2.6.21b) hold. In this case, it is not possible to write the boundary conditions in the subdifferential form. A generalization of (2.6.25) is obtained if the superpotential j_T takes the form

$$j_T(u_T) = |C_N|(\mu_a^2 u_{Ta}^2 + \mu_b^2 u_{Tb}^2)^{1/2}. \qquad (2.6.29)$$

Here a and b are two orthogonal directions termed orthotropy directions, which are defined at every point on the surface of the body, u_{Ta} and u_{Tb} are the components of the displacement u_T in purely static problems (cf. [121]) with respect to a local coordinate system (a, b), and μ_a and μ_b are the two corresponding friction coefficients. In quasistatic or dynamic genuine friction B.V.Ps, u_T must be replaced by the corresponding velocity component v_T. The resulting friction law is called orthotropic friction law (cf. in this context also [336]). Note that if we have two deformable bodies in contact, the interface friction condition can be described by the same laws given above with the only difference that the tangential displacement u_T must be replaced by the relative tangential displacement $[u_T]$, or relative velocity $[v_T]$.

2.7 MONOTONE INTERIOR UNILATERAL CONDITIONS

Here, we give some subdifferential boundary conditions arising in the theory of plates.

i) Let Ω be an open bounded subset $\mathbb{R}^2$ defined by the middle surface of a plate. We denote by Γ the boundary of Ω. The points of Ω are referred to a fixed Cartesian coordinate system $Ox_1x_2x_3$. The x_1- and x_2-axes coincide with the middle surface of the plate, and the x_3-axis with the direction of the normal to the middle surface. The positive direction of the x_3-axis is upwards. The displacements of the plate in its plane are denoted by u_1, u_2 and vertical to its plane by w. By M_n and K_n we denote respectively the bending moment and the total or Kirchhoff shearing force [155] on the boundary of the plate, and we introduce boundary conditions of the form

$$M_n \in \beta_1 \left(\frac{\partial w}{\partial n} \right) = \partial j_1 \left(\frac{\partial w}{\partial n} \right), \tag{2.7.1}$$

$$-K_n \in \beta_2(w) = \partial j_2(w). \tag{2.7.2}$$

ii) Another type of subdifferential relation can be formulated in the theory of plates. Assume that the load vector f at every point $x \in \Omega_0 \subset \Omega$ consists of a part $\bar{f}$, which is given, and of another part $\bar{\bar{f}}$ related to the displacement of that point by a relation of the form

$$- \bar{\bar{f}} \in \beta_3(w) = \partial j_3(w). \tag{2.7.3}$$

Here β_3 and j_3 have the same properties as β_i and j_i in (2.6.4), (2.6.5). As an application let us consider a plate which at points $x_0 \in \Omega_0 \subset \Omega$, $\bar{\Omega}_0 \cap \Gamma = \emptyset$, is at a distance $h = h(x)$ from a deformable support. It is assumed that the support causes a reaction force which is proportional to its deformation (Winkler support). We may then write the relation

$$- \bar{\bar{f}} \in \beta_3(w) \quad \text{in} \quad \Omega_0 \subset \Omega,$$

and

$$\bar{\bar{f}} = 0 \quad \text{in} \quad \Omega \backslash \Omega_0. \tag{2.7.4}$$

Here β_3 is a maximal monotone operator defined by

$$\beta_3(w) = \begin{cases} k(w - h), \ k \text{ const} > 0 & \text{if } w \geq h \\ 0 & \text{if } w < h. \end{cases} \tag{2.7.5}$$

iii) Note here that besides the boundary conditions (2.6.7) and (2.6.9), where the actions normally and tangentially to the boundary of a deformable body $\Omega \subset \mathbb{R}^3$ are considered separately, laws of the form

$$-S_N \in \partial j_N(u_N; S_T) \tag{2.7.6}$$

$$-S_T \in \partial j_T(u_T; S_N) \tag{2.7.7}$$

can be also considered. For instance, the unilateral contact with friction is a boundary condition of this type.

iv) The obstacle problem constitutes a fundamental unilateral problem from plane elasticity. If the body Ω is constrained to lie above another body represented by $\{(x_1, x_2, x_3) \in \mathbb{R}^3 : x_3 \leq \Psi(x_1, x_2)\}$ then we express that the admissible displacement u satisfies

$$u(x) \geq \Psi(x), \qquad (2.7.8)$$

for all $x \in \Omega$. As soon as frictionless contact occurs then a normal reaction force S_R appears and we express that

$$u(x) \in Q(x) \qquad (2.7.9)$$

and

$$-S_R(x) \in N_{Q(x)}(u(x)), \qquad (2.7.10)$$

where

$$Q(x) = \{z \in \mathbb{R} : z \geq \Psi(x)\}. \qquad (2.7.10)$$

The relations (2.7.9) and (2.7.10) are equivalent to the variational inequality:

$$u(x) \in Q(x) : S_R(x)^T(h - u(x)) \geq 0, \ \forall \, h \, \in \, Q(x).$$

2.8 NONMONOTONE UNILATERAL BOUNDARY CONDITIONS

In this Section we deal with boundary conditions expressed in the form

$$-S_N \in \partial j_N(u_N), \qquad (2.8.1a)$$

$$-S_T \in \partial j_T(u_T) \qquad (2.8.1b)$$

or

$$-S \in \partial j(u). \qquad (2.8.2)$$

Here ∂ is the generalized gradient and j_N, j_T and j are locally Lipschitz functionals defined on $\mathbb{R}$, $\mathbb{R}^3$ and $\mathbb{R}^3$ respectively. Relations (2.8.1), (2.8.2) may be considered both in the framework of a small or a large deformation theory. The nonconvex superpotential j_N in $\Omega \subset \mathbb{R}^3$ is formulated by integrating an appropriate function $\beta \in L^\infty_{loc}(\mathbb{R})$. The more general nonconvex superpotentials j and j_T for $\Omega \subset \mathbb{R}^3$ are formulated, by "extending" to $\mathbb{R}^3$ certain one-dimensional nonmonotone multivalued laws, e.g. by considering maximum type functions, etc. Let us first give some simple examples.

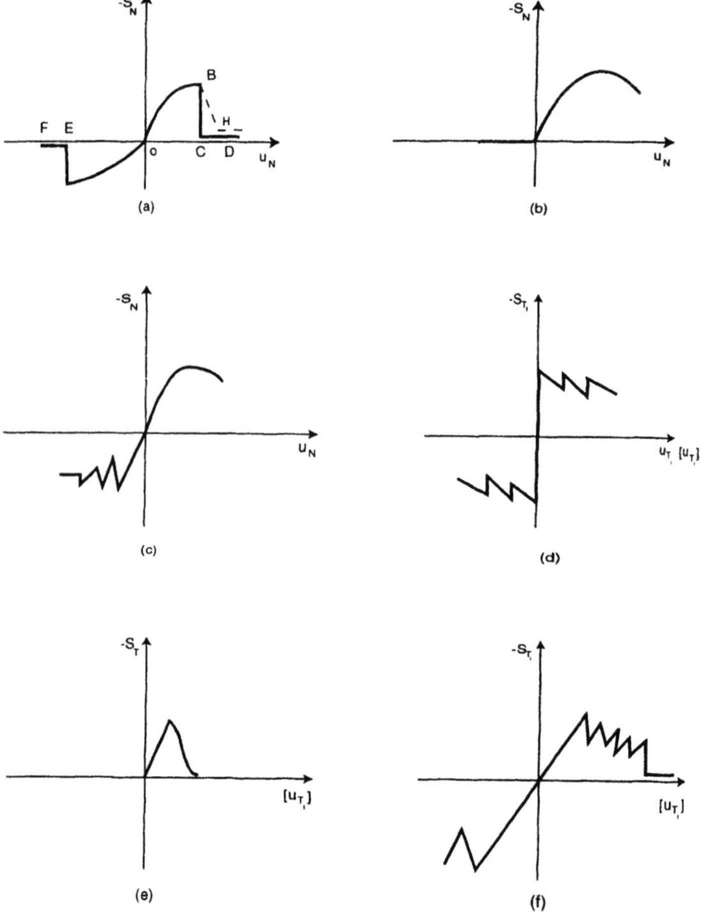

Figure 2.8.1.   Nonmonotone unilateral boundary conditions.

i) The diagram depicted in Fig. 2.8.1a concerns the behavior of the normal forces at any interface point in an adhesive contact problem. The adhesive material between the body and the support may sustain a small tensile force before rupture. Then debonding takes place, which may obey the brittle type diagram FEABCD or the semibrittle diagram FGAOBHD. Note that the vertical branches (i.e. the multivaluedness) are complete, i.e. for an appropriate loading the reaction and the normal boundary displacement u_N (resp. interface relative displacement $[u_N]$) can assume a value on the vertical branch.

ii) In Fig. 2.8.1b a normal contact law between a deformable body and a support of a granular material or concrete is depicted. If debonding takes place the stresses are zero.

iii) In Fig. 2.8.1c the same contact law this time without debonding is presented. Here the support is assumed to be a reinforced concrete support which obeys in tension Scanlon's zig-zag law [134]. An analogous multivalued zig-zag diagram in tension holds for a composite material. The fact that we may have an equilibrium state for which the stress and strain of an interface point may assume a value on the complete vertical branch of the boundary stress-strain law has been experimentally verified (cf. e.g. [34], [323] and the literature given there in).

iv) In Fig. 2.8.1d,e certain nonmonotone friction laws are depicted. The first comes from geomechanics and rock interface analysis, whereas the second arises between reinforcement and concrete in a concrete structure. Finally the law of Fig. 2.8.1f appears in the tangential direction of an adhesive interface and describes the partial cracking and crushing of the adhesive bonding material.

All these laws can be put in the form (2.8.1) where j_N and j_T are non-convex superpotentials resulting by integrating the diagrams describing the boundary conditions. In the aforementioned boundary conditions j_N and j_T are locally Lispchitz and thus the generalized gradients ∂j_N and ∂j_T are described by means of the directional differentials $j_N^0(\cdot;\cdot)$ and $j_T^0(\cdot;\cdot)$. Nonmonotone laws may hold either between a deformable body and a rigid support or between two deformable bodies. In the latter case u, u_N and u_T must be replaced by the corresponding relative displacements $[u], [u_N]$ and $[u_T]$, e.g. in the nonmonotone friction laws of Fig. 2.8.1d,e,f.

v) We also may have nonmonotone boundary conditions for the contact between a body Ω and a Winkler-type support which may sustain only limited values of efforts [305]. It means that the rupture of the support is assumed to occur in those boundary points of $\Gamma_0 \subset \Gamma$ in which the limit effort is accomplished. On neglecting plasticity effects and assuming that the tangential forces are known such contact can be formulated by the conditions

$$S_N = 0 \text{ if } u_N < 0, \tag{2.8.3a}$$

$$S_N = -k_0 u_N \text{ if } 0 \le u_N < \varepsilon, \tag{2.8.3b}$$

$$-k_0\varepsilon \le S_N \le 0 \text{ if } u_N = \varepsilon, \tag{2.8.3c}$$

$$S_N = 0 \text{ if } u_N > \varepsilon, \tag{2.8.3d}$$

$$S_T = C_T, \tag{2.8.3e}$$

where C_T is given on Γ_0. The condition (2.8.3a) holds in the noncontact region and (2.8.3b) describes the part of Γ_0 where the contact occurs and is idealized by the Winkler law. In formula (2.8.3c) we deal with the destruction of the support while (2.8.3d) holds in a region where the support has been destructed. The maximal value of reactions that can be maintained by the support is given by $k_0\varepsilon$ and it is accomplished whenever $u_N = \varepsilon$. Here k_0 is interpreted as the Winkler coefficient. The graph of this nonmonotone Winkler law (see Fig. 2.8.2a) can be put in the form (2.8.1a) with (see Fig. 2.8.2b)

$$
j_N(u_N) = \begin{cases} 0 & \text{if} \quad u_N < 0 \\[2mm] \frac{1}{2}k_0 u_N^2 & \text{if} \quad 0 \leq u_N < \varepsilon \\[2mm] \frac{1}{2}k_0\varepsilon^2 & \text{if} \quad u_N \geq \varepsilon. \end{cases}
$$

2.9   NONMONOTONE INTERIOR UNILATERAL CONDITIONS

Here we consider again the problems of Section 2.7 and interior boundary conditions of the form (2.7.3), i.e.

$$
-\overline{\overline{f}} \in \beta_3(w). \tag{2.9.1}
$$

In Section 2.7 we have discussed the case of monotone graph β_3. Here we assume that $b_3 : \mathbb{R} \to \mathbb{R}$ is a locally bounded measurable function, i.e. $b_3 \in L^\infty_{loc}(R)$ and we set

$$
\underline{b}_3(\zeta) = \lim_{\mu \to 0^+} \operatorname*{ess\,inf}_{|\zeta-z|<\mu} \{b(z)\} \tag{2.9.2}
$$

and

$$
\overline{b}_3(\zeta) = \lim_{\mu \to 0^+} \operatorname*{ess\,sup}_{|\zeta-z|<\mu} \{b(z)\} \tag{2.9.3}
$$

so as to define the multivalued graph

$$
\beta_3 = [\underline{b}_3, \overline{b}_3] \tag{2.9.4}
$$

filling in the graph of b_3. We have given in Section 1.2.3 mathematical conditions ensuring the existence of a locally Lipschitz function j_3 such that

$$
\beta_3 = \partial j_3.
$$

This approach constitutes a fundamental way to formulate unilateral boundary conditions.

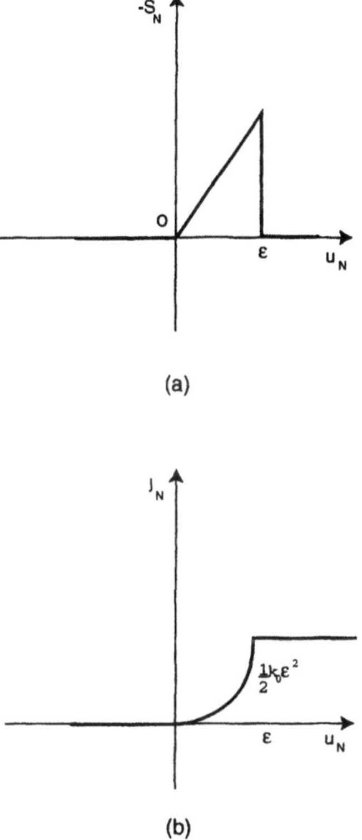

Figure 2.8.2.   Nonmonotone Winkler law.

2.10   MATHEMATICAL MODELLING OF UNILATERAL CONDITIONS AND UNILATERAL B.V.PS IN FUNCTIONAL SPACES

The aim of this Section is to provide the notions and main results in functional analysis that are necessary for the mathematical formulations of the unilateral conditions and B.V.Ps discussed in this Chapter and later in this book.

It is indeed now necessary for the formulation and the study of B.V.Ps in Unilateral Mechanics to give a precise mathematical sense to the data and the unknowns of the problem. For instance, one needs to introduce a "difference operator" with a large domain of definition in order to formulate the partial differential equations related to the conservation laws and geometric equations used to model the considered problem.

In this sense, Sobolev spaces are introduced. In order to deal with the boundary conditions associated with our problem, we need to speak about the values which certain elements (e.g. $u, u_N, u_T, \frac{\partial u}{\partial n}$) take on the boundary Γ of Ω. Fore these reasons a regularity concept on the body Ω is introduced and trace theorems as well as Green-Gauss formulae are discussed. Dynamic problems need to precise the sense of the operators associated to the terms $\frac{du}{dt}$ and $\frac{d^2u}{dt^2}$. Finally, until now we have considered subdifferential boundary conditions of a pointwise nature and in order to use it together with variational principles in Mechanics, we need to consider appropriate extensions of these boundary conditions to Sobolev spaces. This important material is now discussed. The references [2], [24], [25], [27], [31], [67], [99], [128], [210], [220], [251], [313], [235] have been used to prepare this Section.

2.10.1   REGULAR SETS AND SOBOLEV SPACES

Let us first discuss the mathematical representation of the body Ω. The body Ω is usually mathematically defined via a nonempty bounded and open subset of $\mathbb{R}^n (n = 1, 2, 3)$. The boundary $\partial\Omega$ of Ω is usually denoted by Γ. A concept of regularity is now introduced. Let $\Omega \subset \mathbb{R}^n$ $(n \geq 1)$ be a bounded open set. One says that Ω is of class $C^{k,\mu}$ $(0 < \mu \leq 1, k \in \mathbb{N})$ and one notes $\Omega \in C^{k,\mu}$, if there exist two real numbers $b > 0$ and $h > 0, m$ systems of coordinates $(x'_\alpha, x_{\alpha_n}) :=$ $(x_{\alpha_1}, \cdots, x_{\alpha_{n-1}}, x_{\alpha_n})$, $\alpha = 1, \cdot, m$ and m real functions φ_α such that, with $\Delta_\alpha = \{x'_\alpha \in \mathbb{R}^{n-1} :\mid x_{\alpha_i} \mid < b,\ i = 1, \cdots, n-1\}$, $\varphi_\alpha \in C^{k,\mu}(\overline{\Delta}_\alpha)$ and with $\Gamma_\alpha = \{(x'_\alpha,\ x_{\alpha_n}) \in \mathbb{R}^n : x'_\alpha \in \Delta_\alpha$ and $\varphi_\alpha(x'_\alpha) = x_{\alpha_n}\}$, $V_\alpha^+ = \{(x'_\alpha, x_{\alpha_n}) \in \mathbb{R}^n : x'_\alpha \in \Delta_\alpha$ and $\varphi_\alpha(x'_\alpha) < x_{\alpha_n} < \varphi_\alpha(x'_\alpha) + h\}$ and $V_\alpha^- = \{(x'_\alpha, x_{\alpha_n}) \in \mathbb{R}^n : x'_\alpha \in \Delta_\alpha$ and $\varphi_\alpha(x'_\alpha) - h < x_{\alpha_n} < \varphi_\alpha(x'_\alpha)\}$ we have $\Gamma = \cup_{\alpha=1}^m \Gamma_\alpha$ and, for every $\alpha = 1, \cdots, m$, $V_\alpha^+ \subset \Omega$ and $V_\alpha^- \subset \mathbb{R}^n \setminus \overline{\Omega}$. In other words, the boundary $\partial\Omega$ is a $n-1$ $C^{k,\mu}$ manifold and Ω is located on one side of Γ.

The formulation of problems in Mechanics needs a "differentiation operator" with a large domain of definition and which coincides with the usual derivative on the class of "smooth" functions. This tool is given by the differentiation in the sense of distributions [99], [424] and in order to formulate our problems in a suitable framework, one introduces the Sobolev space $W^{k,p}(\Omega)$ of order $(k, p) \in \mathbb{N} \times [1, +\infty]$ as the linear space of functions in $L^p(\Omega)$ whose distributional partial derivative $D^\alpha u, (\alpha \in \mathbb{N}^n)$ i.e. $D^\alpha u \in \mathcal{D}^*(\Omega)$ and

$$\langle D^\alpha u, \varphi \rangle = (-1)^{|\alpha|} \langle u, D^\alpha \varphi \rangle, \ \forall\, \varphi \, \in\, \mathcal{D}(\Omega),$$

of all orders $\mid \alpha \mid := \alpha_1 + \alpha_2 + ... + \alpha_n$ such that $0 \leq \mid \alpha \mid \leq k$ are in $L^p(\Omega)$, that is

$$W^{k,p}(\Omega) = \{v \in \mathcal{D}^*(\Omega) : D^\alpha v \in L^p(\Omega), 0 \leq \mid \alpha \mid \leq k\}.$$

The spaces $W^{k,p}(\Omega)$, $(k \in \mathbb{N}, 1 \leq p \leq +\infty)$ are Banach spaces with the norms

$$\parallel u \parallel_{k,p} = [\int_\Omega \sum_{|\alpha| \leq k} \mid D^\alpha u(x) \mid^p dx]^{\frac{1}{p}}, 1 \leq p < \infty,$$

$$\parallel u \parallel_{k,\infty} = \max_{1 \leq |\alpha| \leq k} \mid D^\alpha u(x) \mid_{0,\infty},$$

$$\mid w \mid_{0,\infty} = \operatorname*{ess\,sup}_{x \in \Omega} \mid w(x) \mid.$$

We usually also use the notation $\mid . \mid_{0,p}$ $(=\parallel . \parallel_{0,p})$ to denote the norm in $L^p(\Omega)$ $(= W^{0,p}(\Omega))$.

If $1 < p < +\infty$ then the spaces $W^{k,p}(\Omega)$ are reflexive Banach spaces and if $p = 2$ they are Hilbert spaces with the scalar product

$$(u,v)_{k,2} = \sum_{|\alpha| \leq k} (D^\alpha u, D^\alpha v)_{0,2},$$

and where $(.,.)_{0,2}$ denotes the scalar product in $L^2(\Omega)$. If $1 \leq p < +\infty$ then the spaces $W^{k,p}(\Omega)$ are separable. We also denote

$$W_0^{k,p}(\Omega) = \overline{C_0^\infty(\Omega)}^{W^{k,p}(\Omega)},$$

i.e. the closure of $C_0^\infty(\Omega)$ in $W^{k,p}(\Omega)$. For $1 \leq p < +\infty$, the dual of $W_0^{k,p}(\Omega)$ is denoted by $W^{-k,q}(\Omega), \frac{1}{p} + \frac{1}{q} = 1$, $W^{-1,2}(\Omega) = H^{-1}(\Omega)$. More general fractional Sobolev spaces $W^{s,p}(\Omega)$ where $s \geq 0$ can also be defined. Denoting by $[s]$ the integer part of s, one defines

$$W^{s,p}(\Omega) = \{u \in W^{[s],p}(\Omega) : \frac{\mid D^\alpha u(x) - D^\alpha u(y) \mid}{\mid x - y \mid^{s-[s]+n/p}} \in L^p(\Omega \times \Omega), \mid \alpha \mid = [s]\}.$$

The norm on $W^{s,p}(\Omega)$ is given by

$$\parallel u \parallel_{s,p} = \left(\parallel u \parallel_{[s],p}^p + \sum_{|\alpha|=[s]} \int_\Omega \int_\Omega \frac{\mid D^\alpha u(x) - D^\alpha u(y) \mid^p}{\mid x - y \mid^{n+p(s-[s])}} dx dy \right)^{\frac{1}{p}}.$$

The duality product between the space $W^{k,p}(\Omega)$ and its dual is denoted by $\langle .,. \rangle_{k,p}$. Important properties of Sobolev spaces are now listed.

Proposition 2.10.1   Let $k \in \mathbb{N}, 1 \leq p \leq +\infty$ and Ω a subset of $\mathbb{R}^n (n \geq 1)$.

A) Suppose that Ω is open in $\mathbb{R}^n$. If $k \geq 0$ and $1 \leq p < +\infty$ then

$$C^k(\Omega) \cap W^{k,p}(\Omega) \text{ is dense in } W^{k,p}(\Omega).$$

B) Suppose that Ω is open, bounded and of class $C^{0,1}$. If $k, p \in \mathbb{N}$ then

$$\mathcal{D}(\overline{\Omega}) \text{ is dense in } W^{k,p}(\Omega).$$

C) Suppose that Ω is open, bounded and of class $C^{0,1}, k \geq 1$ and $1 \leq p < +\infty$. Then the following embedding hold:

(C.1) if $n > kp$ then
$$W^{k,p}(\Omega) \hookrightarrow L^q(\Omega),$$

continuously for $1 \leq q \leq np/(n-kp)$ and compactly for $1 \leq q < np/(n-kp)$.

(C.2) If $n = kp$ then
$$W^{k,p}(\Omega) \hookrightarrow L^q(\Omega),$$

compactly for every $1 \leq q < +\infty$.

(C.3) If $n < kp$ then
$$W^{k,p}(\Omega) \hookrightarrow C^{0,\mu}(\overline{\Omega}),$$

compactly for
$$
\begin{array}{llll}
\mu & = & k - \frac{n}{p} & \text{if} \quad k - \frac{n}{p} < 1, \\
\mu & < & 1 & \text{if} \quad k - \frac{n}{p} = 1, \\
\mu & = & 1 & \text{if} \quad k - \frac{n}{p} > 1.
\end{array}
$$

Remark 2.10.2   i) In the case of vector-valued functions $v : \Omega \to \mathbb{R}^N$, we use the notation

$$W^{k,p}(\Omega; \mathbb{R}^N) = \Pi_{i=1}^N W^{k,p}(\Omega)$$

and

$$\| v \|_{k,p} = \left(\sum_{i=1}^N \| v_i \|_{k,p}^2 \right)^{\frac{1}{2}}.$$

The scalar product in $W^{1,2}(\Omega; \mathbb{R}^N)$ and the duality product between $W^{k,p}(\Omega; \mathbb{R}^N)$ and its dual are also denoted by $(.,.)_{k,2}$ and $\langle .,. \rangle_{k,p}$ respectively. If the formulation of a given problem involves at the same time scalar-valued functions and vector-valued functions then we will denote the duality products by using precise notations of the form $\langle .,.\rangle_{X^*,X}$ in which the considered space X is specified. The same remark holds for the norm and the scalar product if X is a Hilbert space. ii) Setting

$$u^+ = max\{u,0\}, u^- = max\{-u,0\}, \mid u \mid = u^+ + u^-,$$

one can prove for instance that if $u \in W^{1,2}(\Omega)$ then $u^+, u^-, \mid u \mid \in W^{1,2}(\Omega)$ and

$$\nabla u^+ = \Xi_{\{u>0\}} \nabla u, \nabla u^- = \Xi_{\{u<0\}} \nabla u, \nabla \mid u \mid = sign(u)\nabla u,$$

where for a subset $A \subset \mathbb{R}$,

$$\Xi_A(x) = \begin{cases} 1 & \text{if} \quad x \in A \\ 0 & \text{if} \quad x \notin A, \end{cases}$$

and for a real $a \in \mathbb{R}$,

$$sign(a) = \begin{cases} 1 & \text{if} \quad a > 0 \\ -1 & \text{if} \quad a < 0 \\ 0 & \text{if} \quad a = 0. \end{cases}$$

iii) The spaces $W^{k,2}(\Omega)$ are also denoted by $H^k(\Omega)$ and one has for $m \geq k, m, k \in \mathbb{N}$ the inclusions

$$\mathcal{D}(\Omega) \subset H_0^m(\Omega) \subset H_0^k(\Omega) \subset L^2(\Omega) \subset H^{-k}(\Omega) \subset H^{-m}(\Omega) \subset \mathcal{D}^*(\Omega),$$

each space being dense in the following spaces.

iv) A more general statement of (C.3) in Proposition 2.10.1 is:

(C.3)' If $n < (k-m)p$, $m \in \mathbb{N}$, $m \leq k$ then

$$W^{k,p}(\Omega) \hookrightarrow C^{m,\mu}(\overline{\Omega}),$$

compactly for

$$\begin{aligned} \mu &= k-m-\tfrac{n}{p} & \text{if} \quad k-m-\tfrac{n}{p} < 1, \\ \mu &< 1 & \text{if} \quad k-m-\tfrac{n}{p} = 1, \\ \mu &= 1 & \text{if} \quad k-m-\tfrac{n}{p} > 1. \end{aligned}$$

2.10.2   TRACE THEOREMS

In order to deal with the boundary conditions associated with our problems, we need to speak about the values which certain elements take on the boundary Γ of Ω, as well as about the normal derivative of these values. Let Ω be an open bounded subset of $\mathbb{R}^n$. For $\Omega \in C^{[s],1}$, $s \geq 0$ and $1 \leq p < +\infty$, we set

$$W^{s,p}(\Gamma) = \{u \in L^p(\Gamma) : u(x'_\alpha, \varphi_\alpha(x'_\alpha)) \in W^{s,p}(\Delta_\alpha), \alpha = 1, \cdots, m\},$$

with the notations in Section 2.10.1. The space $W^{s,p}(\Gamma)$ is a reflexive Banach space with the norm

$$\| u \|_{s,p;\Gamma} = [\sum_{\alpha=1}^{m} \| u(x'_\alpha, \varphi_\alpha(x'_\alpha)) \|^p_{W^{s,p}(\Delta_\alpha)}]^{\frac{1}{p}}.$$

The vector-valued functions space $W^{s,p}(\Gamma; \mathbb{R}^N)$ is defined by the formula

$$W^{s,p}(\Gamma; \mathbb{R}^N) = [W^{s,p}(\Gamma)]^N$$

and Remark 2.10.2 i) holds here too. If $\Omega \in C^{0,1}$ then we may define the unit outward normal vector in local coordinates by

$$n(x'_\alpha, \varphi_\alpha(x'_\alpha)) = (-\nabla' \varphi_\alpha(x'), 1)[1 + | \nabla' \varphi_\alpha(x') |^2]^{-\frac{1}{2}},$$

with $\nabla' \varphi(x') = (\varphi_{x_1}(x'), \cdots, \varphi_{x_{n-1}}(x'))$. Here $n \in L^\infty(\Gamma; \mathbb{R}^n)$ while if $\Omega \in C^{k,\mu}(k \geq 1, 0 \leq \mu \leq 1)$ then $n \in C^{k-1,\mu}(\Gamma; \mathbb{R}^n)$. One will speak about the value of $u \in W^{k,p}(\Omega)$ on Γ by means of the concept of trace $\gamma(u)$ which coincides with the classical restriction as soon as u is smooth enough.

Proposition 2.10.3 (Trace Theorem I). Let Ω be a bounded open set such that $\Omega \in C^{0,1}, k \geq 1$ an integer and $p \in [1, +\infty)$. Then

(A) if $kp < n$ and $1 \leq q \leq (n-1)p/(n-kp)$ then there exists a unique linear and continuous operator $\gamma : W^{k,p}(\Omega) \to L^q(\Gamma)$ such that $\gamma(W^{k,p}(\Omega)) = W^{k-\frac{1}{p},p}(\Gamma)$ and

$$\gamma(u) = u_{|\Gamma}, \forall u \in \mathcal{D}(\overline{\Omega}).$$

If in addition $p > 1$ then γ is compact provided that $q < (n-1)p/(n-kp)$.

(B) If $kp \geq n$ then (A) holds for arbitrary $q \geq 1$.

(C) If $kp > n$ then $\gamma(u)$ is the classical restriction.

(D) $\gamma(W^{k,p}(\Omega))$ is dense in $L^p(\Gamma)$.

The operator γ in Proposition 2.10.3 is called the trace operator. Let $\Omega \in C^{0,1}$ be given and $u \in C^k(\overline{\Omega})$, $k \in \mathbb{N}, k \geq 1$. For $0 \leq j \leq k$, one defines the *jth*-derivative of u with respect to the normal n by the formula

$$\frac{\partial^j u}{\partial n^j} = \sum_{|\alpha|=j} \frac{j!}{\alpha!}(D^\alpha u_{|\Gamma})n^\alpha, \text{ a.e. on } \Gamma,$$

where $\alpha! := \alpha_1!\alpha_2!...\alpha_n!$ and $n^\alpha := n_1^{\alpha_1}n_2^{\alpha_2}...n_n^{\alpha_n}$. The following result holds.

Proposition 2.10.4 (Trace Theorem II). Let Ω be a bounded open set of class $C^{m,1}, m \in \mathbb{N}$, and assume $k - \frac{1}{p}$ is not an integer, $k \leq m + 1$, $0 < k - \frac{1}{p} - m < 1$. Then there exists a linear and continuous operator $\overline{\gamma} : W^{k,p}(\Omega) \to [L^p(\Gamma)]^m$; $u \to \overline{\gamma}(u) = (\gamma(u), \gamma_1(u), \cdots, \gamma_m(u))$ such that

$$\begin{aligned}
\gamma(W^{k,p}(\Omega)) &= W^{k-\frac{1}{p},p}(\Gamma); \\
\gamma_j(W^{k,p}(\Omega)) &= W^{k-j-\frac{1}{p},p}(\Gamma), j = 1, \cdots, m; \\
\gamma(u) &= u_{|\Gamma}, \forall u \in \mathcal{D}(\overline{\Omega})
\end{aligned}$$

and

$$\gamma_j(u) = \frac{\partial^j u}{\partial n^j}, \forall u \in \mathcal{D}(\overline{\Omega}).$$

Moreover

$$Ker \ \overline{\gamma} = \{u \in W^{k,p}(\Omega) : \gamma(u) = 0 \text{ and } \gamma_j(u) = 0, j = 1, \cdots, m\}$$

$$= W_0^{k,p}(\Omega).$$

2.10.3   TRACE RESULTS FOR ELASTICITY PROBLEMS

Further specific trace results are given in this Section. They play a key role in the study of unilateral problems in elasticity. Let Ω be a nonempty, bounded and open subset of $\mathbb{R}^n(n = 2, 3)$. For every $v \in L^2(\Gamma)$ we may consider the decomposition

$$v = v_T + v_N n, \ v_N = v.n, \ v_T = v - v_N n,$$

with $v_N \in L^2(\Gamma)$ and $v_T \in L^2_T(\Gamma; \mathbb{R}^n)$, where

$$L^2_T(\Gamma) = \{v \in L^2(\Gamma; \mathbb{R}^n) : v_N = 0\}.$$

One has

$$(u, v)_{0,2} = (u_N, v_N)_{0,2;N} + (u_T, v_T)_{0,2;T},$$

where

$$(u, v)_{0,2} = \int_\Gamma u.v ds,$$

$$(u_N, v_N)_{0,2;N} = \int_\Gamma u_N v_N ds,$$

and

$$(u_T, v_T)_{0,2;T} = \int_\Gamma u_T.v_T ds.$$

Let us now suppose that $\Omega \in C^{1,1}$. Then for $v \in H^{\frac{1}{2}}(\Gamma; \mathbb{R}^n)$ one has

$$v_N = v.n \in H^{\frac{1}{2}}(\Gamma)$$

and

$$v_T = v - v_N n \in H^{\frac{1}{2}}_T(\Gamma; \mathbb{R}^n),$$

with

$$H^{\frac{1}{2}}_T(\Gamma; \mathbb{R}^n) = \{v \in H^{\frac{1}{2}}(\Gamma; \mathbb{R}^n) : v_N = 0\}.$$

The space $H^{\frac{1}{2}}_T(\Gamma; \mathbb{R}^n)$ is closed in $H^{\frac{1}{2}}(\Gamma; \mathbb{R}^n)$ and

$$H^{\frac{1}{2}}(\Gamma; \mathbb{R}^n) = H^{\frac{1}{2}}_T(\Gamma; \mathbb{R}^n) \oplus \{v \in H^{\frac{1}{2}}(\Gamma) : v = v_N n, v_N \in H^{\frac{1}{2}}(\Gamma)\}.$$

Note also that the spaces $H^{\frac{1}{2}}(\Gamma; \mathbb{R}^n)$ and $H^{\frac{1}{2}}_T(\Gamma; \mathbb{R}^n)$ are densely and continuously embedded in $L^2(\Gamma; \mathbb{R}^n)$. Normal and tangential trace operators are now defined.

Proposition 2.10.5 Let Ω be an open bounded and connected subset of class $C^{1,1}$ of $\mathbb{R}^n$. There exist uniquely determined linear continuous maps $\gamma_N : H^1(\Omega; \mathbb{R}^n) \to L^2(\Gamma)$ and $\gamma_T : H^1(\Omega; \mathbb{R}^n) \to L_T(\Gamma; \mathbb{R}^n)$ such that

$$\gamma(v) = \gamma_N(v)n + \gamma_T(v), \ \forall \, v \, \in \, H^1(\Omega; \mathbb{R}^3),$$

$$\gamma_N(H^1(\Omega; \mathbb{R}^n)) = H^{\frac{1}{2}}(\Gamma),$$

$$\gamma_T(H^1(\Omega; \mathbb{R}^n)) = H^{\frac{1}{2}}_T(\Gamma; \mathbb{R}^n)$$

and

$$\gamma_N(v) \;=\; v_N, \; \forall \, v \in C^\infty\,(\overline{\Omega})$$
$$\gamma_T(v)_i \;=\; v_{T_i}, \; \forall \, v \in C^\infty\,(\overline{\Omega}).$$

Moreover

$$\gamma_N(H^1(\Omega)) \;=\; H^{\frac{1}{2}}(\Gamma)$$

and

$$\gamma_T(H^1(\Omega; \mathbb{R}^n)) \;=\; H_T^{\frac{1}{2}}(\Gamma; \mathbb{R}^n).$$

If $h \in H_T^{-\frac{1}{2}}(\Gamma; \mathbb{R}^n)$ then one may write

$$h = h_T + h_N n$$

with

$$h_T \in H_T^{-\frac{1}{2}}(\Gamma; \mathbb{R}^n), \; h_N \in H^{\frac{1}{2}}(\Gamma),$$

and thus

$$\langle h, \gamma v \rangle_{H^{-\frac{1}{2}}(\Gamma;\mathbb{R}^n), H^{\frac{1}{2}}(\Gamma;\mathbb{R}^n)} = \langle h_T, \gamma_T v \rangle_{H_T^{-\frac{1}{2}}(\Gamma;\mathbb{R}^n), H_T^{\frac{1}{2}}(\Gamma;\mathbb{R}^n)}$$

$$+ \langle h_N, \gamma_N v \rangle_{H^{-\frac{1}{2}}(\Gamma), H^{\frac{1}{2}}(\Gamma)}.$$

Boundary conditions are usually specified on a part of the boundary instead of the whole boundary. For this reason it is useful to denote, for a nonempty open subset D of Γ, by γ_D the map which associates to $v \in H^{\frac{1}{2}}(\Omega; \mathbb{R}^n)$ the restriction of $\gamma(v) \in H^{\frac{1}{2}}(\Gamma; \mathbb{R}^n)$ to D, i.e.

$$\gamma_D : H^1(\Omega; \mathbb{R}^n) \to H^{\frac{1}{2}}(D; \mathbb{R}^n); u \to \gamma(u)_{|D}.$$

Complementarily to γ_D it is now natural to consider the trace operator on $\Sigma = \Gamma \backslash \overline{D}$, i.e.

$$\gamma_\Sigma : H^1(\Omega; \mathbb{R}^n) \to H^{\frac{1}{2}}(\Sigma; \mathbb{R}^n), u \to \gamma(u)_{|\Sigma}$$

defined in a manner analogous to γ_D.

Remark 2.10.6 Let Ω be an open bounded connected and of class $C^{0,1}$ subset of $\mathbb{R}^n (n \geq 2)$. Suppose also that

$$\Gamma = \overline{D}_1 \cup \overline{D}_2,$$

where D_1 and D_2 are two disjoint open subsets of Γ, i.e.

$$D_1 \cap D_2 = \emptyset.$$

Set

$$V_0 = \{u \in H^p(\Omega; \mathbb{R}^n) : \gamma_{D_1}(u) = 0 \text{ a.e. on } D_1\}$$

and suppose that $H_{n-1}(D_1) > 0$. Then

$$\left(\int_\Omega \sum_{i=1}^n \mid D^i u \mid^2 dx\right)^{\frac{1}{2}}$$

is an equivalent norm on V_0.

Let us now set

$$V_D = \{u \in H^1(\Omega; \mathbb{R}^n) : \gamma_D(u) = 0 \text{ a.e. on } D\}$$

and

$$H_{00}^{\frac{1}{2}}(\Sigma; \mathbb{R}^n) = \gamma_\Sigma(V_D).$$

The space $H_{00}^{\frac{1}{2}}(\Sigma; \mathbb{R}^n)$ is provided with the topology determined by the norm

$$\| z \|_{H_{00}^{\frac{1}{2}}} = \inf\{\| v \|_{1,2}: v \in V_D \text{ and } z = \gamma_\Sigma(v)\}.$$

The use of the space $H_{00}^{\frac{1}{2}}(\Sigma)$ is necessary because the operation of restriction to $\Sigma \subset \Gamma$ of the elements of $H^{-\frac{1}{2}}(\Gamma; \mathbb{R}^n)$ is not valued on $H^{-\frac{1}{2}}(\Sigma; \mathbb{R}^n)$ but in the larger space $(H_{00}^{\frac{1}{2}}(\Sigma; \mathbb{R}^n))^*$. The following result constitutes a suitable version of Proposition 2.10.5 for mixed problems.

Proposition 2.10.7 Let $\Omega \subset \mathbb{R}^n$ be an open bounded and connected set of class $C^{1,1}$. Let D be a nonempty open subset of Γ. Then there exist uniquely determined linear continuous maps $\gamma_{\Sigma,N} : V_D \to L^2(\Sigma)$ and $\gamma_{\Sigma,T} : V_D \to L_T(\Sigma; \mathbb{R}^n)$ such that

$$\begin{aligned}
\gamma_\Sigma(v) &= \gamma_{\Sigma,N}(v)n + \gamma_{\Sigma,T}(v), \\
\gamma_{\Sigma,N}(V_D) &= H_{00}^{\frac{1}{2}}(\Sigma),
\end{aligned}$$

and

$$\gamma_{\Sigma,T}(V_D) = H_{00T}^{\frac{1}{2}} := \{v \in H_{00}^{\frac{1}{2}}(\Sigma; \mathbb{R}^n) : v_N = 0\}.$$

2.10.4   GREEN-GAUSS FORMULAE

In this Section, we consider some general Green-Gauss formulae which play a key role in the analysis of unilateral problems in elasticity.

Let us first discuss the Green's formula for the elliptic operator

$$L. \equiv - \sum_{i,j=1}^{n} \frac{\partial}{\partial x_i}\left(a_{ij}\frac{\partial}{\partial x_j}.\right) + \sum_{i=1}^{n} b_i \frac{\partial}{\partial x_i}. + \sum_{i=1}^{n} \frac{\partial}{\partial x_i}(c_i.) + d.,$$

where

$$a_{ij} \in L^{\infty}(\Omega); b_j \in L^{\infty}(\Omega); c_i \in L^{\sigma}(\Omega); d \in L^{\sigma}(\Omega),$$

with

$$\sigma = \begin{cases} n & \text{if} \quad n > 2 \\ 2 + \varepsilon & \text{if} \quad n = 2 \ (\varepsilon > 0) \\ 2 & \text{if} \quad n = 1. \end{cases}$$

One sets

$$H_L^1(\Omega) = \{u \in H^1(\Omega) : Lu \in L^2(\Omega)\}.$$

To the operator L, we associate the bilinear form

$$a(u,v) = \int_{\Omega}\left(\sum_{i,j=1}^{n} a_{ij}\frac{\partial u}{\partial x_j}\frac{\partial v}{\partial x_j} + \sum_{i=1}^{n} b_i\frac{\partial u}{\partial x_i}v - \sum_{i=1}^{n} c_i u\frac{\partial v}{\partial x_i} + duv\right)dx.$$

The following result is known as the Green's formula.

Proposition 2.10.8   Let $\Omega \subset \mathbb{R}^3$ be a bounded connected open set with boundary Γ of class $C^{0,1}$. Then there exists a uniquely determined linear continuous mapping $\gamma_a : H^1(\Omega) \to H^{-\frac{1}{2}}(\Gamma)$ such that

$$a(u,v) = \int_{\Omega} (Lu)v dx + \langle \gamma_a(u), \gamma(v) \rangle_{\frac{1}{2}}, \ \forall \, u \in H_L^1(\Omega), \ v \in H^1(\Omega)$$

Note that if u, a_{ij} and c_i are sufficiently regular then $\gamma_a(u)$ reduces to the conormal derivative of u, i.e.

$$\gamma_a(u) = \frac{\partial u}{\partial \nu_a}$$

$$= \sum_{i,j=1}^{n} a_{ij}\mid_{\Gamma} \frac{\partial u}{\partial x_j}\mid_{\Gamma} \cos(n,x_i) - \sum_{i=1}^{n} c_i \mid_{\Gamma} u \mid_{\Gamma} \cos(n,x_i).$$

More generally, the following Green's formula is suitable for the study of mixed problems.

Proposition 2.10.9 Let $\Omega \subset \mathbb{R}^3$ be a bounded connected open set with boundary of class $C^{0,1}$. Let Γ_0 be a connected open set of Γ with $H_{n-1}(\Gamma_0) > 0$ and $\Gamma_1 = \Gamma \backslash \overline{\Gamma}_0$. Then

$$a(u, v) = \int_\Omega (Lu)v dx + \langle \gamma_{a|_{\Gamma_1}}(u), \gamma_{|_{\Gamma_1}}(v) \rangle_{00;\Gamma_1},$$

for all $u \in H_L^1(\Omega)$ and $v \in V_{\Gamma_0} = \{v \in H^1(\Omega) : \gamma_{|_{\Gamma_0}}(u) = 0 \text{ a.e. on } \Gamma_0\}$. Here $\langle \, , \, \rangle_{00;\Gamma_1}$ denotes the duality product between $H_{00}^{\frac{1}{2}}(\Gamma_1)$ and its dual.

The following Green-Gauss formulae are related to the spaces of stress tensors and their boundary values. We set

$$\mathcal{S} = \{\tau = [\tau_{ij}] \in [L^2(\Omega)]^9 : \tau_{ij} = \tau_{ji}, i, j = 1, \cdots n\},$$

$$\mathcal{T} = \{\tau \in \mathcal{S} : \tau_{ij,j} \in L^2(\Omega), i = 1, \cdots n\}.$$

Proposition 2.10.10 Let $\Omega \subset \mathbb{R}^n$ be an open, bounded and connected set of class $C^{0,1}$. There exists a uniquely determined linear continuous mapping $\pi : \mathcal{T} \to H^{\frac{1}{2}}(\Gamma; \mathbb{R}^n)$ such that

$$\int_\Omega \tau_{ij} v_{i,j} dx + \int_\Omega \tau_{ij,j} v_i dx = \langle \pi(\tau), \gamma(v) \rangle_{H^{-\frac{1}{2}}(\Gamma;\mathbb{R}^n), H^{\frac{1}{2}}(\Gamma;\mathbb{R}^n)},$$

$$\forall \, \tau \in \mathcal{T}, \, v \in H^1(\Omega; \mathbb{R}^n),$$

and

$$(\pi(\tau))_i = \tau_{ij|_\Gamma} n_j \text{ if } \tau \in C^1(\overline{\Omega}).$$

In order to take into account separately the effects due to the forces acting in the normal directions and the effects due to the forces acting in the tangential directions, we record the following decomposition result.

Proposition 2.10.11 Let $\Omega \subset \mathbb{R}^n$ be an open bounded and connected set of class $C^{1,1}$. There exist uniquely determined maps $\pi_N : \mathcal{T} \to H^{\frac{1}{2}}(\Gamma)$ and $\pi_T : \mathcal{T} \to H_T^{-\frac{1}{2}}(\Gamma; \mathbb{R}^n)$ such that

$$\langle \pi(\tau), \gamma(v) \rangle_{H^{-\frac{1}{2}}(\Gamma;\mathbb{R}^n), H^{\frac{1}{2}}(\Gamma;\mathbb{R}^n)} = \langle \pi_N(\tau), \gamma_N(v) \rangle_{H^{-\frac{1}{2}}(\Gamma), H^{\frac{1}{2}}(\Gamma)}$$

$$+ \langle \pi_T(\tau), \gamma_T(v) \rangle_{H^{-\frac{1}{2}}(\Gamma;\mathbb{R}^n), H^{\frac{1}{2}}(\Gamma;\mathbb{R}^n)}$$

and

$$\begin{aligned}
\pi_N(\tau) &= \tau_{ij}n_jn_i, \ \forall \ \tau \ \in \ C^1(\overline{\Omega}), \\
\pi_T(\tau)_i &= \tau_{ij}n_j - \pi_N(\tau)n_i, \ \forall \ \tau \ \in \ C^1(\overline{\Omega}).
\end{aligned}$$

Finally, for mixed problems the following result is of particular interest.

Proposition 2.10.12 Let $\Omega \subset \mathbb{R}^n$ be an open bounded and connected set of class $C^{0,1}$. Then there exists a uniquely determined linear continuous operator $\pi_\Sigma : \mathcal{T} \to (H^{\frac{1}{2}}_{00}(\Sigma))^*$ such that

$$\int_\Omega \tau_{ij}v_{i,j}dx + \int_\Omega \tau_{ij,j}v_idx = \langle \pi_\Sigma(\tau), \gamma_\Sigma(v)\rangle_{00;\Sigma}, \ \forall \ \tau \ \in \ \mathcal{T}, \ v \ \in \ V_D,$$

and

$$(\pi_\Sigma(\tau))_i = \tau_{ij|_\Sigma}n_j \ \text{if} \ \tau \in C^1(\Omega).$$

Moreover, if $\Omega \in C^{1,1}$ then π_Σ can be decomposed into operators $\pi_{\Sigma,N}$ an $\pi_{\Sigma,T}$ such that

$$\langle \pi_{\Sigma,N}(\tau), \gamma_\Sigma(v)\rangle_{H^{-\frac{1}{2}}(\Sigma;\mathbb{R}^n),H^{\frac{1}{2}}(\Sigma;\mathbb{R}^n)} = \langle \pi_{\Sigma,N}(\tau), \gamma_{\Sigma,N}(v)\rangle_{H^{-\frac{1}{2}}(\Sigma),H^{\frac{1}{2}}(\Sigma)}$$

$$+\langle \pi_{\Sigma,T}(\tau), \gamma_{\Sigma,T}(v)\rangle_{(H^{-\frac{1}{2}}_{00T}(\Sigma;\mathbb{R}^n))^*,H^{\frac{1}{2}}_{00T}(\Sigma;\mathbb{R}^n)}, \forall \ \tau \ \in \ \mathcal{T}, \ v \ \in \ V_D.$$

Moreover

$$\begin{aligned}
\pi_{\Sigma,N}(\tau) &= \tau_{ij|_\Sigma}n_jn_i, \ \forall \ \tau \ \in \ C^1(\overline{\Omega}) \\[2ex]
(\pi_{\Sigma,T}(\tau))_i &= \tau_{ij|_\Sigma}n_j - \pi_{\Sigma,N}(\tau)n_i, \ \forall \ \tau \ \in \ C^1(\overline{\Omega}).
\end{aligned}$$

2.10.5   BANACH-VALUED DISTRIBUTION AND $W^{1,P}$-SPACES

Let X be a Banach space, (a,b) an open interval in $\mathbb{R}$. We set $dt = d\mathcal{L}_1$ where $\mathcal{L}_1$ denotes here the Lebesgue measure on (a,b).

Definition 2.10.13 One says that

$$f \in L^p(a,b;X) \ (1 \le p < +\infty)$$

if $f : (a, b) \to X$ is measurable for dt and

$$| f |_{L^p(a,b;X)} := \left(\int_a^b \| f \|_X^p \, dt \right)^{\frac{1}{p}} < +\infty.$$

Definition 2.10.14   One says that

$$f \in L^\infty(a, b; X)$$

if $f : (a, b) \to X$ is bounded almost everywhere on (a, b). We set

$$| f |_{L^\infty(a,b;X)} = \inf\{M : \| f(t) \|_X \le M \text{ a.e. on } (a, b)\}.$$

Provided that no confusion can occur then the notation $\| \cdot \|_\infty$ will also be used to denote shortly the norm in $L^\infty(a, b; X)$. In the result below, we list some fundamental properties related to these spaces.

Proposition 2.10.15

1) $L^p(a, b; X)$ $(1 \le p \le +\infty)$ is a Banach space,

2) $L^p(a, b; X)$ $(1 \le p < \infty)$ is separable if and only if X is separable,

3) if $u \in L^p(a, b; X)$ $(1 \le p \le \infty)$ and $f \in X^*$ then

$$\langle f, \int_a^b u(t)dt \rangle = \int_a^b \langle f, u(t) \rangle dt.$$

One says that u is a X-valued distribution if $u \in \mathcal{L}(\mathcal{D}(a, b), X)$. Concepts of "time-derivatives" are now introduced in the distributional sense.

Let $\mathcal{V}$ be a real Banach space and $\mathcal{H}$ a real Hilbert space such that

$$\mathcal{V} \hookrightarrow \mathcal{H} \hookrightarrow \mathcal{V}^*$$

with dense and continuous embeddings.

Definition 2.10.16 Let $u \in L^p(a, b; \mathcal{V})$. One defines the weak derivative of u with respect to t by the formula

$$\frac{du}{dt}(\varphi) = - \int_a^b u(t)\varphi'(t)dt, \varphi \in \mathcal{D}(a, b).$$

We have $\frac{du}{dt} \in \mathcal{L}(\mathcal{D}(a,b);\mathcal{V})$ and we say that $\frac{du}{dt} \in L^2(a,b;\mathcal{H})$(resp. $\frac{du}{dt} \in L^2(a,b;\mathcal{V}^*)$) if there exists $v \in L^2(a,b;\mathcal{H})$(resp. $v \in L^2(a,b;\mathcal{V}^*)$) such that

$$\int_a^b (v(t),w)_{\mathcal{H}}\varphi(t)dt =$$

$$- \int_a^b (u(t),w)_{\mathcal{H}}\varphi'(t)dt, \ \forall \ w \ \in \ \mathcal{V}, \ \varphi \ \in \ \mathcal{D}(a,b),$$

respectively

$$\int_a^b \langle v(t),w\rangle_{\mathcal{V}^*,\mathcal{V}}\varphi(t)dt =$$

$$- \int_a^b (u(t),w)_{\mathcal{H}}\varphi'(t)dt, \ \forall \ w \ \in \ \mathcal{V}, \ \varphi \ \in \ \mathcal{D}(a,b) \ ).$$

Suppose now in addition that $\mathcal{V}$ and $\mathcal{H}$ are separable Hilbert spaces and set $(1 \leq p \leq +\infty)$:

$$W^{1,p}(a,b;\mathcal{V},\mathcal{V}^*) = \{u : u \in L^p(a,b;\mathcal{V}), \frac{du}{dt} \in L^p(a,b;\mathcal{V}^*)\}.$$

The case $p = 2$ is of particular interest for our purposes and one has the following properties.

Proposition 2.10.17   Let $-\infty < a < b < +\infty$ be given.

1) $W^{1,2}(a,b;\mathcal{V},\mathcal{V}^*) \hookrightarrow C^0([a,b];\mathcal{H})$ continuously,

2) If $u,v \in W^{1,2}(a,b;\mathcal{V},\mathcal{V}^*)$ then

$$\int_a^b \langle\frac{du}{dt},v\rangle_{\mathcal{V}^*,\mathcal{V}}dt + \int_a^b \langle\frac{dv}{dt},u\rangle_{\mathcal{V}^*,\mathcal{V}} = (u(b),v(b))_{\mathcal{H}} - (u(a),v(a))_{\mathcal{H}}.$$

3) If $u,v \in W^{1,2}(a,b;\mathcal{V},\mathcal{V}^*)$ and $z \in \mathcal{V}$ then

$$\langle\frac{du}{dt},z\rangle_{\mathcal{V}^*,\mathcal{V}} = \frac{d}{dt}\langle u(.),z\rangle \text{ in } \mathcal{D}'(a,b).$$

4) If $\mathcal{V} \hookrightarrow \mathcal{H}$ compactly then $W^{1,2}(a,b;\mathcal{V},\mathcal{V}^*) \hookrightarrow L^2(a,b;\mathcal{H})$ compactly.

Let $\mathcal{H}$ be a real Hilbert space. We set $(1 \leq p \leq +\infty)$

$$W^{1,p}(a,b;\mathcal{H}) = \{u : u \in L^p(a,b;\mathcal{H}), \frac{du}{dt} \in L^p(a,b;\mathcal{H})\}.$$

Let $-\infty < a < b < +\infty$ be given. We recall here that $W^{1,1}(a,b;\mathcal{H})$ can be identified with the space of $\mathcal{H}$-valued absolutely continuous functions on $[a,b]$ and that if $u \in W^{1,1}(a,b;\mathcal{H})$ then u is almost everywhere differentiable on (a,b) and

$$v(t) = v(a) + \int_a^t \frac{dv}{ds}(s)ds, \ a \le t \le b.$$

Note also that the space $W^{1,\infty}(a,b;\mathcal{H})$ can be identified with the space of $\mathcal{H}$-valued Lipschitz continuous functions on $[a,b]$.

Similarly the second derivative $\frac{d^2u}{dt^2}$ is defined in the distributional sense, i.e.

$$\frac{d^2u}{dt^2}(\varphi) = \int_a^b u(t)\varphi'' dt, \ \forall \, \varphi \in \mathcal{D}(a,b),$$

and we say that $\frac{d^2u}{dt^2} \in L^2(a,b;\mathcal{H})$ (resp. $\frac{du}{dt} \in L^2(a,b;\mathcal{V}^*)$) if there exists $v \in L^2(a,b;\mathcal{H})$ (resp. $v \in L^2(a,b;\mathcal{V}^*)$) such that

$$\int_a^b (v(t),w)_{\mathcal{H}}\varphi(t)dt = \int_a^b (u(t),w)_{\mathcal{H}}\varphi(t)'' dt, \ \forall \, \varphi \in \mathcal{D}(a,b),$$

(respectively

$$\int_a^b \langle v(t),w \rangle_{\mathcal{V}^*,\mathcal{V}}\varphi(t)dt = \int_a^b (u(t),w)_{\mathcal{H}}\varphi''(t)dt, \ \forall \, \varphi \in \mathcal{D}(a,b) \,).$$

One sets

$$\tilde{W}^{1,2}(a,b;\mathcal{V},\mathcal{V}^*) = \{u : u \in L^2(a,b;\mathcal{V}), \frac{du}{dt} \in W^{1,2}(a,b;\mathcal{V},\mathcal{V}^*)\}$$

and we recall that

$$\tilde{W}^{1,2}(a,b;\mathcal{V},\mathcal{V}^*) \hookrightarrow C^0([0,T];\mathcal{V}),$$

continuously. Note that we often use also the notations

$$u' = \frac{du}{dt},$$

$$u'' = \frac{d^2u}{dt^2}.$$

2.10.6   FORMULATIONS OF UNILATERAL CONDITIONS AND SUBSTATIONARITY PROBLEMS

Let us here discuss the use of the previous material on a simple but illustrative pilot problem. Further problems will be discussed later in Section 2.11 and Chapters 6, 7, 8, 9 and 10.

Let us consider an elastic body Ω subjected to a body force f and unilateral conditions of the form

$$S_N \in -\partial j_N(u_N), \tag{2.10.1}$$

$$S_T = C_T, \tag{2.10.2}$$

where j_N is a locally Lipschitz function and C_T is given. The displacement field $u : \Omega \to \mathbb{R}^3$ of the body is described by the partial differential equations:

$$\sigma_{ij,j} + f_i = 0 \text{ on } \Omega, \tag{2.10.3}$$

$$S_i = \sigma_{ij}n_j \text{ on } \partial\Omega, \tag{2.10.4}$$

$$\sigma_{ij} = C_{ijkl}\varepsilon_{kl} \text{ on } \Omega. \tag{2.10.5}$$

The notations of Section 2.1 have here been used again. The symmetric tensor $[C_{ijkl}]$ is given.

Let us in a first step assume that all the data are sufficiently smooth to justify differentiations, integrations and the use of Green-Gauss formula. Multiplying (2.10.3) by $(v - u)$ and integrating over Ω, we obtain

$$\int_\Omega \sigma_{ij,j}(v_i - u_i)dx + \int_\Omega f_i(v_i - u_i)dx = 0.$$

From Green-Gauss formula, we obtain

$$\int_\Omega \sigma_{ij}\varepsilon_{ij}(v - u)dx = \int_\Omega f_i(v_i - u_i)dx + \int_\Gamma \sigma_{ij}n_j(v_i - u_i)ds.$$

Using now the decomposition

$$\int_\Gamma \sigma_{ij}n_j(v_i - u_i)ds = \int_\Omega \sigma_{ij}n_jn_i(v_N - u_N)ds$$

$$+ \int_\Gamma (\sigma_{ij}n_j - (\sigma_{ij}n_jn_i)n_i)(v_{T_i} - u_{T_i})ds,$$

we obtain

$$\int_\Omega \sigma_{ij}\varepsilon_{ij}(v - u)dx = \int_\Omega f_i(v_i - u_i)dx + \int_\Gamma S_N(v_N - u_N)ds$$

$$+ \int_\Gamma S_T.(v_T - u_T)ds. \tag{2.10.6}$$

From (2.10.1) we deduce that

$$j_N^0(u_N; v - u_N) + S_N(v - u_N) \geq 0 \text{ on } \Gamma$$

and thus

$$\int_\Gamma j_N^0(u_N; v - u_N)ds + \int_\Gamma S_N(v - u_N)ds \geq 0.$$

From (2.10.6) we obtain the inequality problem:

$$\int_\Omega \sigma_{ij}\varepsilon_{ij}(v - u)dx + \int_\Gamma j_N^0(u_N; v - u_N)ds$$

$$\geq \int_\Gamma S_T.(v_T - u_T)ds + \int_\Omega f.(v - u)dx. \qquad (2.10.7)$$

The relation (2.10.7) constitutes a "formal" inequality expression of the principle of virtual work of our problem. Let us now in a second step precise the sense of each term appearing in (2.10.7). We suppose that Ω is a nonempty open bounded and connected subset of $\mathbb{R}^3$ with a boundary of class $C^{1,1}$ and we assume that $u, v \in H^1(\Omega; \mathbb{R}^3)$. If we suppose that $C_{ijkl} \in L^\infty(\Omega)$, $\forall\, i, j, k, l \in \{1, 2, 3\}$ then the term

$$\sigma_{ij}\varepsilon_{ij}(v - u) = C_{ijkl}\varepsilon_{kl}(u)\varepsilon_{ij}(v - u)$$

belongs to $L^1(\Omega)$ and the expression

$$\int_\Omega \sigma_{ij}(u)\varepsilon_{ij}(v - u)dx$$

is well-defined. If we suppose that $f \in L^2(\Omega; \mathbb{R}^3)$ then

$$f.(v - u) \in L^1(\Omega)$$

and the contribution

$$\int_\Omega f.(v - u)dx$$

is well-defined. It is now necessary to give a precise sense to u_T, v_T, u_N and v_N. This is done by using the normal and tangential trace operators

$$\gamma_N : H^1(\Omega; \mathbb{R}^3) \to H^{\frac{1}{2}}(\Gamma) \subset L^2(\Gamma)$$

and

$$\gamma_T : H^1(\Omega; \mathbb{R}^3) \to H_T^{\frac{1}{2}}(\Gamma; \mathbb{R}^3) \subset L^2(\Gamma; \mathbb{R}^3).$$

Then if we suppose that $S_T \equiv C_T \in L^2(\Gamma; \mathbb{R}^3)$, we have

$$S_T.(\gamma_T(v) - \gamma_T(u)) \in L^1(\Gamma)$$

and the expression

$$\int_T S_T.(\gamma_T(v) - \gamma_T(u))ds$$

is well-defined. Note that if $S_T \in H^{-\frac{1}{2}}(\Gamma; \mathbb{R}^3)$ we may then consider the expression

$$\langle S_T, \gamma_T(v-u)\rangle_{H^{-\frac{1}{2}}(\Gamma;\mathbb{R}^3), H^{\frac{1}{2}}(\Gamma;\mathbb{R}^3)}$$

to model the work exerted by S_T.

The functional $j_N : \mathbb{R} \to \mathbb{R}$ is locally Lipschitz. It is now necessary to impose on j_N conditions ensuring that

$$j_N^0(z(.); h(.)) \in L^1(\Gamma),$$

for all $z, h \in L^2(\Gamma)$. The results of Section 1.2.3 (1.2.15) and (1.2.16) can be used in this sense. For example if j is Lipschitz continuous the result is true and we may consider the expression

$$\int_\Gamma j_N^0(\gamma_N u(x); \gamma_N v(x) - \gamma_N u(x)) ds.$$

So, a precise mathematical formulation of (2.10.7) gives the inequality:

$$\int_\Omega \sigma_{ij}\varepsilon_{ij}(v-u)dx \;+\; \int_\Gamma j_N^0(\gamma_N u(x); \gamma_N v(x) - \gamma_N u(x))ds$$

$$\geq \int_\Gamma S_T.\gamma_T(v-u)ds + \int_\Gamma f.(v-u)dx.$$

Let us now introduce the energy functionals

$$\Phi(u) = \frac{1}{2}\int_\Omega C_{ijkl} \mid \varepsilon_{ij}(u) \mid^2 dx, u \in H^1(\Omega; \mathbb{R}^3)$$

and

$$\Pi_f(u) = \langle \overline{f}, u\rangle_{1,2}, u \in H^1(\Omega; \mathbb{R}^3),$$

where $\overline{f} \in (H^1(\Omega; \mathbb{R}^3))^*$ is defined by the formula

$$\langle \overline{f}, v\rangle = \int_\Omega f(x).v(x)dx, \ \forall \, v \in H^1(\Omega; \mathbb{R}^3).$$

We see that $\Phi, \Pi_f \in C^1(H^1(\Omega; \mathbb{R}^3); \mathbb{R})$,

$$\langle \Phi'(u), v-u\rangle_{1,2} = \int_\Omega \sigma_{ij}(u)\varepsilon_{ij}(v-u)dx, \ \forall \, u, v \in H^1(\Omega; \mathbb{R}^3)$$

and

$$\langle \Pi_f'(u), v-u\rangle_{1,2} = \int_\Omega f(x).(v(x) - u(x))dx, \ \forall \, u, v \in H^1(\Omega; \mathbb{R}^3).$$

Let us now set

$$\Pi_T(u) = \langle \overline{S}_T, u \rangle_{1,2}, u \in H^1(\Omega; \mathbb{R}^3),$$

where $\overline{S}_T \in (H^1(\Omega; \mathbb{R}^3))^*$ is defined by

$$\langle \overline{S}_T, v \rangle_{1,2} = \int_\Gamma S_T.\gamma_T(v - u)ds, \ \forall \ v \ \in \ H^1(\Omega; \ \mathbb{R}^3).$$

We have

$$\langle \Pi_T'(u), v - u \rangle_{1,2} = \int_\Gamma S_T.\gamma_T(v - u)ds, \ \forall \ u, \ v \ \in \ H^1(\Omega; \ \mathbb{R}^3).$$

Finally, we set

$$\Pi_N(z) = \int_\Gamma j_N(z)ds, \ z \in L^2(\Gamma)$$

and

$$\hat{\Pi}_N = \Pi_N(\gamma_N(u)), u \in H^1(\Omega; \mathbb{R}^3).$$

From Theorem 1.2.13 and Theorem 1.2.20, we obtain

$$\hat{\Pi}_N^0(u; v-u) \leq \int_\Gamma j_N^0(\gamma_N u(x); \gamma_N v(x) - \gamma_N u(x))dx, \ \forall \ u, \ v \ \in \ H^1(\Omega; \ \mathbb{R}^3).$$

The total energy of the system is given by

$$E = \Phi + \Pi_T + \hat{\Pi}_N + \Pi_f$$

and from Proposition 2.10.1(ii), we deduce that if u satisfies the stationarity principle

$$0 \in \partial E(u), u \in H^1(\Omega; \mathbb{R}^3) \tag{2.10.8}$$

then

$$\langle \Phi'(u), v - u \rangle_{1,2} \ + \ \hat{\Pi}_N^0(u; v - u)$$

$$\geq \ \langle \Pi_T'(u), v - u \rangle_{1,2} + \langle \Pi_f'(u), v - u \rangle_{1,2}, \tag{2.10.9}$$

for all $v \in H^1(\Omega; \mathbb{R}^3)$. From (2.10.9) it results that u satisfies the inequality problem (2.10.7). So, a solution of the substationarity problem is related to the inequality problem in the sense that any solution of (2.10.8) is a solution of (2.10.7) too. The critical point concepts of Section 1.2.4 and Section 1.3.5 have been defined in this sense.

Suppose that instead of (2.10.1)-(2.10.2) we need to extend a subdifferential contact boundary condition of the form

$$-S(x) \in \partial j(u(x))$$

where j is a proper, convex and l.s.c. function. Then we introduce the functional

$$\Psi(u) = \begin{cases} \int_\Gamma j(u(x))d\Gamma & \text{if } j(u) \in L^1(\Gamma) \\ +\infty & \text{otherwise} \end{cases} ,$$

which is convex, l.s.c. and proper on $[L^2(\Gamma)]^3$ (see Proposition 1.3.19) and the total energy of the system reads here

$$E = \Phi + \hat{\Psi} + \Pi_f$$

with

$$\hat{\Psi}(u) = \Psi(\gamma(u)), \ \forall u \in H^1(\Omega; \mathbb{R}^3),$$

and where γ denotes the trace operator from $H^1(\Omega; \mathbb{R}^3)$ onto $L^2(\Gamma; \mathbb{R}^3)$.

2.10.7   MODELS OF NONCONVEX ENERGY INTEGRAL FUNCTIONALS

In Section 1.2.4, we have pointed out that solutions of hemivariational inequalities can be obtained by asking to contain 0 the Clarke's subdifferential of some locally Lipschitz energy functional of the form (1.2.17) thanks to the relation (1.2.18) of Theorem 1.2.20 and Chain Rules. Their roles in the formulation of B.V.Ps have already been discussed in Section 2.10.6. In this Section, we discuss the most important forms of energy functionals we will encounter later in this book.

Let us firstly recall that if $F : X \to Y$ is continuously differentiable with X and Y Banach spaces and let $J : Y \to \mathbb{R}$ be a locally Lipschitz function, then one has (see Theorem 1.2.13 and formula (1.2.8))

$$(J \circ F)^0 \le g^0(F(u); F'(u)v), \ \forall u, v \in X. \tag{2.10.10}$$

If in addition F is linear then

$$(J \circ F)^0 \le g^0(F(u); F(v)), \ \forall u, v \in X. \tag{2.10.11}$$

Let $(T, \mathcal{T}, \mu)$ be a positive complete measure space such that $\mu(T) < +\infty$. Let $j : T \times \mathbb{R}^m \to \mathbb{R}$ be a function such that $j(x, \cdot) : \mathbb{R}^m \to \mathbb{R}$ is locally Lipschitz whenever $x \in T$, $j(\cdot, y) : T \to \mathbb{R}$ is measurable whenever $y \in \mathbb{R}^m$, $j(., e) \in L^1(T; \mathbb{R}^m)$ for some $e \in L^p(T; \mathbb{R}^m)$ $(1 \le p < +\infty)$ and satisfies either (1.2.15) or (1.2.16).

Suppose that Ω is a nonempty open bounded and connected subset of class $C^{0,1}$ in $\mathbb{R}^n$.

Example 2.10.18 Let $T = \Omega$ with the corresponding Lebesgue measure of dimension n. Let $1 \leq p < +\infty$ and set

$$J_\Omega : L^p(\Omega; \mathbb{R}^m) \to \mathbb{R}; \quad v \to J_\Omega(v) := \int_\Omega j(x, v(x))dx. \qquad (2.10.12)$$

Let us now denote by i the linear and continuous embedding

$$i : W^{1,p}(\Omega, \mathbb{R}^m) \hookrightarrow L^p(\Omega; \mathbb{R}^m)$$

and set

$$\hat{J}_\Omega := J|_{W^{1,p}(\Omega,\mathbb{R}^m)} = J_\Omega \circ i.$$

From the rule (2.10.11) we obtain

$$J_\Omega^0(i(u); i(v)) \geq \hat{J}_\Omega^0(u; v), \ \forall \, u, \, v \, \in \, W^{1,p}(\Omega, \, \mathbb{R}^m).$$

Using now Theorem 1.2.20 (and identifying $i(u)(x)$ with $u(x)$ on Ω for $u \in W^{1,p}(\Omega; \mathbb{R}^m)$), we obtain the relation

$$\int_\Omega j_y^0(x, u(x); v(x))dx \geq \hat{J}_\Omega^0(u; v), \ \forall \, u, \, v \, \in \, W^{1,p}(\Omega; \, \mathbb{R}^m). \quad (2.10.13)$$

The relation (2.10.13) implies that if $-\zeta \in \partial \hat{J}_\Omega(u)$ then $\zeta \in [W^{1,p}(\Omega; \mathbb{R}^m)]^*$ and

$$\langle \zeta, v \rangle_{1,p} + \int_\Omega j_y^0(x, u(x); v(x))dx \geq 0, \ \forall \, v \, \in \, W^{1,p}(\Omega; \, \mathbb{R}^m).$$

Note that Corollary 1.2.14 implies that $\zeta \in L^q(\Omega; \mathbb{R}^m)$ $(q^{-1} + p^{-1} = 1)$ since the embedding from $W^{1,p}(\Omega; \mathbb{R}^m)$ into $L^p(\Omega; \mathbb{R}^m)$ is dense. Thus for $\zeta \in -\partial \hat{J}_\Omega(u)$, we have

$$\int_\Omega \zeta(x)v(x)dx + \int_\Omega j_y^0(x, u(x); v(x))dx \geq 0, \ \forall \, v \, \in \, W^{1,p}(\Omega; \, \mathbb{R}^m).$$

Example 2.10.19 Let $T = \Omega_0$ for a measurable part of Ω such that $\overline{\Omega_0} \subset \Omega$ and $\mathcal{L}_n(\Omega_0) > 0$. Let $+\infty > p \geq 1$ and consider the functional

$$J_{\Omega_0} : L^p(\Omega_0; \mathbb{R}^m) \to \mathbb{R}; \quad v \to \int_{\Omega_0} j(x, v(x))dx. \qquad (2.10.14)$$

We denote by i_0 the linear and continuous mapping which associate to $u \in W^{1,p}(\Omega; \mathbb{R}^m)$ the restriction of $i(u)$ (see Example 2.10.18) to Ω_0. We set

$$\hat{J}_{\Omega_0}(\cdot) = \int_{\Omega_0} j(x, i_0(\cdot))dx = J_{\Omega_0} \circ i_0.$$

Using (2.10.11) we obtain

$$J^0_{\Omega_0}(i_0(u), i_0(v)) \geq \hat{J}^0_{\Omega_0}(u, v), \ \forall\, u,\, v\, \in\, W^{1,p}(\Omega;\, \mathbb{R}^m).$$

Then using Theorem 1.2.20 (and identifying $i_0(u)(x)$ with $u(x)$ on Ω_0 for $u \in W^{1,p}(\Omega; \mathbb{R}^m)$) we obtain the relation

$$\int_{\Omega_0} j^0_y(x, u(x); v(x))dx \geq \hat{J}^0_{\Omega_0}(u, v), \ \forall\, u,\, v\, \in\, W^{1,p}(\Omega;\, \mathbb{R}^m). \quad (2.10.15)$$

Example 2.10.20   Let $T = \Gamma := \partial\Omega$ with the Hausdorff measure of dimension $n - 1$. Let $1 \leq p < +\infty$ and set

$$J_\Gamma : L^p(\Gamma; \mathbb{R}^m) \to \mathbb{R}, \quad v \to \int_\Gamma j(x, v(x))ds. \quad (2.10.16)$$

Let us recall that $\gamma : W^{1,p}(\Omega; \mathbb{R}^m) \to L^p(\Gamma, \mathbb{R}^m)$ denotes the usual trace operator. We set for $u \in W^{1,p}(\Omega; \mathbb{R}^m)$

$$\hat{J}_\Gamma(\cdot) = \int_\Gamma j(x, \gamma(\cdot))ds = J \circ \gamma.$$

As in the previous example, we obtain for all $u, v \in W^{1,p}(\Omega; \mathbb{R}^m)$ the relation

$$\int_\Gamma j^0_y(x, \gamma u(x); \gamma v(x))ds \geq \hat{J}^0_\Gamma(u; v). \quad (2.10.17)$$

Example 2.10.21   Let $T = \Gamma$, $1 \leq p < +\infty$ and suppose that Γ_0 is a nonempty open subset of Γ. We consider the functional

$$J_{\Gamma_0} : L^p(\Gamma_0; \mathbb{R}^m) \to \mathbb{R}; \quad v \to \int_{\Gamma_0} j(x, v(x))ds. \quad (2.10.18)$$

Let us denote by γ_{Γ_0} the continuous and linear mapping which associate to each $u \in W^{1,p}(\Omega; \mathbb{R}^m)$ the restriction of $\gamma(u)$ (see Example 2.10.21) to Γ_0. We set

$$\hat{J}_{\Gamma_0} = J_0 \circ \gamma_{\Gamma_0}$$

and we obtain as usually the relation

$$\int_{\Gamma_0} j^0_y(x, \gamma_{\Gamma_0} u(x); \gamma_{\Gamma_0} v(x))ds \geq \hat{J}_{\Gamma_0}(u; v),$$

$$\forall\, u,\, v\, \in\, W^{1,p}(\Omega;\, \mathbb{R}^m). \tag{2.10.19}$$

Let $\{\Omega_i\}_{i=1,\cdots,m}$ be a family of nonempty open bounded subsets of $\mathbb{R}^n$. Let $j_i : \Omega_i \times \mathbb{R} \to \mathbb{R}$ $(i = 1, \cdots, m)$ be a family of m functions satisfying either (1.2.15) or (1.2.16) as specified in the outset of this Section.

Example 2.10.22 Let Ω'_α be a measurable set such that $\overline{\Omega'_\alpha} \subset \Omega_\alpha \cap \Omega_{\alpha+1}$ and $\mathcal{L}_n(\Omega'_\alpha) > 0$ $(\alpha = 1, \cdots, m-1)$. We consider the functional

$$J : L^p(\Omega_1) \times L^p(\Omega_2) \times \cdots \times \mathcal{L}^p(\Omega_N) \to \mathbb{R};$$

$$v = (v_1, v_2, \cdots, v_N) \to J(v) = \sum_{\alpha=1}^{m-1} J_\alpha(v) \tag{2.10.20}$$

where

$$J_\alpha(v) = \int_{\Omega'_\alpha} j_\alpha(x, v(x))dx.$$

Let i_α be the continuous embedding from $W^{1,p}(\Omega_\alpha)$ into $L^p(\Omega_\alpha)$ and let $i_{\alpha,\beta}$ $(\beta = \alpha, \alpha+1)$ be the linear and continuous mapping which associate to $u \in W^{1,p}(\Omega_\beta)$ the restriction of $i_\beta(u)$ to Ω'_α. We set

$$\hat{J}(v) = \sum_{\alpha=1}^{m-1} \hat{J}_\alpha(v),$$

where

$$\hat{J}_\alpha(v) = J_\alpha \circ L_\alpha$$

with

$$L_\alpha : \times_{j=1}^m W^{1,p}(\Omega_j) \to L^p(\Omega'_\alpha); v \to i_{\alpha,\alpha}(v_\alpha) - i_{\alpha,\alpha+1}(v_{\alpha+1}).$$

The map L_α is linear and continuous and omitting to write the "$i_{\alpha,\beta}$", (which is justified by natural identifications) we obtain with the notation

$$[v]^\alpha = v_\alpha - v_{\alpha+1} \tag{2.10.21}$$

that for all $u, v \in \times_{j=1}^m W^{1,p}(\Omega_j)$

$$\sum_{\alpha=1}^{m-1} \int_{\Omega'_\alpha} j^0_{\alpha,y}(x, [u(x)]^\alpha; [v(x)]^\alpha)dx \geq \sum_{\alpha=1}^{m-1} \hat{J}^0_\alpha(u; v).$$

Moreover, using Proposition 1.2.10(ii), we get

$$\sum_{\alpha=1}^{m-1} \int_{\Omega'_\alpha} j^0_{\alpha,y}(x, [u(x)]^\alpha; [v(x)]^\alpha)dx \geq \hat{J}^0(u; v). \tag{2.10.22}$$

Let $j_N : \Gamma \times \mathbb{R} \to \mathbb{R}$ and $j_T : \Gamma \times \mathbb{R}^3 \to \mathbb{R}$ be functions as specified at the beginning of this Section. Suppose also that Ω satisfy the conditions mentioned above with $n = 3$. Here we suppose also that Ω is of class $C^{1,1}$.

Example 2.10.23 Let $T = \Gamma := \partial\Omega$ and $1 \le p < +\infty$. We set

$$J_N : L^p(\Gamma) \to \mathbb{R}, \quad v \to \int_\Gamma j_N(x, v(x))ds \qquad (2.10.23)$$

and

$$J_T : L^p(\Gamma; \mathbb{R}^3) \to \mathbb{R}, \quad v \to \int_\Gamma j_T(x, v(x))ds. \qquad (2.10.24)$$

Let us recall that we denote by $\gamma_N : W^{1,p}(\Omega; \mathbb{R}^3) \to L^p(\Gamma)$ and $\gamma_T : W^{1,p}(\Omega; \mathbb{R}^3) \to L^p(\Gamma; \mathbb{R}^3)$ the usual normal and tangential trace mappings. We set

$$\hat{J}_N = J_N \circ \gamma_N$$

and

$$\hat{J}_T = J_T \circ \gamma_T .$$

We obtain for all $u, v \in W^{1,p}(\Omega; \mathbb{R}^3)$ the inequalities

$$\int_\Gamma j_{N,y}^0(x, \gamma_N u(x); \gamma_T v(x))ds \ge \hat{J}_N^0(u; v) \qquad (2.10.25)$$

and

$$\int_\Gamma j_{T,y}^0(x, \gamma_T u(x); \gamma_T v(x))ds \ge \hat{J}_T^0(u; v). \qquad (2.10.26)$$

2.10.8   MODELS OF CONVEX ENERGY INTEGRAL FUNCTIONALS

In this Section, we list as in Section 2.10.7 the most important forms of energy functionals encountered in the applications. Here the functionals considered are convex and related to monotone unilateral conditions. Let us first recall that if $F : X \to Y$ is linear and continuous with X and Y real Banach spaces, $\Phi : Y \to \mathbb{R} \cup \{+\infty\}$ is a proper convex and l.s.c. function, and a point $y_0 = F(x_0)$, $x_0 \in X$ exists at which Φ is finite and continuous then one has (Proposition 1.3.11) that if $W \in \partial\Phi \circ F(u) \subset X^*$ then there exists $w \in \partial\Phi(F(u)) \subset Y^*$ such that

$$W = F^*(w).$$

Let $\varphi : \mathbb{R}^m \to \mathbb{R} \cup \{+\infty\}$ $(m \in \mathbb{N}, m \geq 1)$ be a proper convex and l.s.c. function. Suppose that Ω is a nonempty open bounded and connected subset of $\mathbb{R}^n$ $(n \in \mathbb{N}, n \geq 1)$. Moreover we suppose that Ω is of class $C^{0,1}$ and we denote by Γ the boundary of Ω. Let $+\infty > p \geq 1$ be given.

Example 2.10.24   One sets

$$\Phi_\Omega : L^p(\Omega; \mathbb{R}^m) \to \mathbb{R} \cup \{+\infty\};$$

$$v \to \Phi_\Omega(v) := \begin{cases} \int_\Omega \varphi(v(x))dx & \text{if} \qquad \varphi(v(.)) \in L^1(\Omega) \\ \\ +\infty & \text{if not} \end{cases}$$

and let

$$\hat{\Phi}_\Omega : W^{1,p}(\Omega; \mathbb{R}^m) \to \mathbb{R} \cup \{+\infty\}; v \to \hat{\Phi}_\Omega(v) := \Phi(i(v)),$$

where i has been defined in Example 2.10.18. The function Φ_Ω is proper convex and l.s.c. thanks to Proposition 1.3.19 and so is $\hat{\Phi}_\Omega$ as the composite of a proper convex l.s.c. and a linear and continuous mapping. Suppose that there exists some point $y_0 = i(x_0)$, $x_0 \in W^{1,p}(\Omega; \mathbb{R}^m)$ at which Φ_Ω is finite and continuous. Then if $W \in \partial \hat{\Phi}_\Omega(u)$ for some $u \in W^{1,p}(\Omega; \mathbb{R}^m)$ then there exists $w \in L^q(\Omega; \mathbb{R}^m)(\frac{1}{p} + \frac{1}{q} = 1)$ such that

$$W = i^*(w)$$

and (from Theorem 1.3.21) there exists $A \subset \Omega$ such that $\mathcal{L}_n(A) = 0$ and

$$\varphi(y) - \varphi(u(x)) \geq w(x)^T(y - u(x)), \ \forall \, y \in \mathbb{R}^m, \ \forall \, x \in \Omega \backslash A.$$

Example 2.10.25   Let Ω_0 be a measurable part of Ω such that $\overline{\Omega}_0 \subset \Omega$ and $\mathcal{L}_n(\Omega_0) > 0$. One sets

$$\Phi_{\Omega_0} : L^p(\Omega_0; \mathbb{R}^m) \to \mathbb{R} \cup \{+\infty\};$$

$$v \to \Phi_{\Omega_0}(v) := \begin{cases} \int_{\Omega_0} \varphi(v(x))dx & \text{if} \qquad \varphi(v(.)) \in L^1(\Omega_0) \\ \\ +\infty & \text{if not} \end{cases}$$

and

$$\hat{\Phi}_{\Omega_0} : W^{1,p}(\Omega; \mathbb{R}^m) \to \mathbb{R} \cup \{+\infty\}; v \to \hat{\Phi}_{\Omega_0}(v) := \Phi(i_0(v)),$$

where i_0 has been defined in Example 2.10.19. Suppose that there exists some point $y_0 = i_0(x_0)$, $x_0 \in W^{1,p}(\Omega; \mathbb{R}^m)$ at which Φ_{Ω_0} is finite and continuous. The functional $\hat{\Phi}_{\Omega_0}$ is proper, convex and l.s.c. and if $w \in \partial\hat{\Phi}_{\Omega_0}(u)$ for some $u \in W^{1,p}(\Omega; \mathbb{R}^m)$ then there exists $w \in L^q(\Omega_0; \mathbb{R}^m)(\frac{1}{p} + \frac{1}{q} = 1)$ such that

$$W = i_0^*(w)$$

and there exists $A_0 \subset \Omega_0$ such that $\mathcal{L}_n(A_0) = 0$ and

$$\varphi(y) - \varphi(u(x)) \geq w(x)^T(y - u(x)), \ \forall\, y \in \mathbb{R}^m, \ \forall\, x \in \Omega_0 \backslash A_0.$$

Example 2.10.26   We consider the functional

$$\Phi_\Gamma : L^p(\Gamma; \mathbb{R}^m) \to \mathbb{R} \cup \{+\infty\};$$

$$v \to \Phi_\Gamma(v) := \begin{cases} \int_\Gamma \varphi(v(x))ds & \text{if} & \varphi(v(.)) \in L^1(\Gamma) \\ +\infty & \text{if not} \end{cases}$$

and

$$\hat{\Phi}_\Gamma : W^{1,p}(\Omega; \mathbb{R}^m) \to \mathbb{R} \cup \{+\infty\}; v \to \hat{\Phi}_\Gamma(v) := \Phi(\gamma(v)).$$

Suppose that there exists some point $y_0 = \gamma(x_0)$, $x_0 \in W^{1,p}(\Omega; \mathbb{R}^m)$ at which Φ_Γ is finite and continuous. The functional $\hat{\Phi}_\Gamma$ is proper convex and l.s.c. and if $W \in \partial\hat{\Phi}_\Gamma(u)$ for some $u \in W^{1,p}(\Omega; \mathbb{R}^m)$ then there exists $w \in L^q(\Gamma; \mathbb{R}^m)$ such that

$$W = \gamma^* w$$

and there exists $A \subset \Gamma$ such that $H_{n-1}(A) = 0$ and

$$\varphi(y) - \varphi(u(x)) \geq w(x)^T(y - u(x)), \ \forall\, y \in \mathbb{R}^m, \ \forall\, x \in \Gamma \backslash A.$$

Example 2.10.27   Let Γ_0 be a nonempty open subset of Γ. We set

$$\Phi_{\Gamma_0} : L^p(\Gamma_0; \mathbb{R}^m) \to \mathbb{R} \cup \{+\infty\};$$

$$v \to \Phi_{\Gamma_0}(v) := \begin{cases} \int_{\Gamma_0} \varphi(v(x))ds & \text{if} & \varphi(v(.)) \in L^1(\Gamma_0) \\ +\infty & \text{if not} \end{cases}$$

and

$$\hat{\Phi}_{\Gamma_0} : W^{1,p}(\Omega) \to \mathbb{R}\cup\{+\infty\}; v \to \hat{\Phi}_{\Gamma_0}(v) := \Phi_{\Gamma_0}(\gamma_{\Gamma_0}(v)),$$

where γ_{Γ_0} has been defined in Example 2.10.21. Suppose that there exists some point $y_0 = \gamma_{\Gamma_0}(x_0)$, $x_0 \in W^{1,p}(\Omega; \mathbb{R}^m)$ at which Φ_{Γ_0} is finite and continuous. The functional $\hat{\Phi}_{\Gamma_0}$ is proper, convex and l.s.c. and if $W \in \partial\hat{\Phi}_{\Gamma_0}(u)$ for some $u \in W^{1,p}(\Omega; \mathbb{R}^m)$ then there exists $w \in L^q(\Gamma_0; \mathbb{R}^m)$ such that

$$W = \gamma_{\Gamma_0}^* w$$

and there exists $A_0 \subset \Gamma_0$ such that $H_{n-1}(A_0) = 0$ and

$$\varphi(y) - \varphi(u(x)) \geq w(x)^T(y - u(x)), \ \forall \ y \ \in \ \mathbb{R}^m, \ \forall \ x \ \in \ \Gamma_0 \backslash A_0.$$

Example 2.10.28   Here we suppose additionally that Ω is of class $C^{1,1}$. Let $\varphi_N : \mathbb{R} \to \mathbb{R}\cup\{+\infty\}$ and $\varphi_T : \mathbb{R}^3 \to \mathbb{R}\cup\{+\infty\}$ be two proper convex and l.s.c. functions. We set

$$\Phi_N : L^p(\Gamma) \to \mathbb{R}\cup\{+\infty\};$$

$$v \to \begin{cases} \int_\Gamma \varphi_N(v(x))ds & \text{if} & \varphi_N(v(.)) \in L^1(\Gamma) \\ +\infty & \text{if not} \end{cases}$$

and

$$\Phi_T : L^p(\Gamma; \mathbb{R}^3) \to \mathbb{R}\cup\{+\infty\};$$

$$v \to \begin{cases} \int_\Gamma \varphi_T(v(x))ds & \text{if} & \varphi_T(v(.)) \in L^1(\Gamma) \\ +\infty & \text{if not.} \end{cases}$$

Let us now consider the functionals

$$\hat{\Phi}_N : W^{1,p}(\Omega; \mathbb{R}^3) \to \mathbb{R}\cup\{+\infty\}; v \to \hat{\Phi}_N(v) = \Phi_N(\gamma_N(v))$$

and

$$\hat{\Phi}_T : W^{1,p}(\Omega; \mathbb{R}^3) \to \mathbb{R}\cup\{+\infty\}; v \to \hat{\Phi}_T(v) = \Phi_T(\gamma_T(v)).$$

We suppose that there exist points $y_0 = \gamma_N(x_0)$ and $z_0 = \gamma_T(s_0)$, $x_0, s_0 \in W^{1,p}(\Omega; \mathbb{R}^m)$ at which respectively Φ_N and Φ_T are finite and continuous. If $W_N \in \partial\hat{\Phi}_N(u)$ and $W_T \in \partial\hat{\Phi}_T(u)$ for some $u \in W^{1,p}(\Omega;\mathbb{R}^3)$ then there exists $w_N \in L^q(\Gamma)$ and $w_T \in L^q(\Gamma;\mathbb{R}^3)(\frac{1}{p}+\frac{1}{q}=1)$ such that

$$W_N = \gamma_N^* w_N, W_T = \gamma_T^* w_T,$$

and there exist $A_N, A_T \subset \Gamma$ such that $H_{n-1}(A_N) = 0$, $H_{n-1}(A_T) = 0$,

$$\varphi_N(y) - \varphi_N(\gamma_N u(x)) \geq w_N(y - \gamma_N u(x)), \ \forall \, y \, \in \, \mathbb{R}, \ \forall \, x \, \in \, \Gamma\backslash A_N,$$

and

$$\varphi_T(z) - \varphi_T(\gamma_T u(x)) \geq w_T^T(z - \gamma_T u(x)), \ \forall \, z \, \in \, \mathbb{R}^3, \ \forall \, x \, \in \, \Gamma\backslash A_T.$$

2.11 VARIATIONAL AND HEMIVARIATIONAL INEQUALITIES IN UNILATERAL MECHANICS

In the present Section we present the method for the formulation of variational inequalities, hemivariational inequalities and related variational expressions.

Subsequently dynamic problems and eigenvalue problems for hemivariational inequalities are studied, as well as certain variational formulations for the loading and unloading problem.

2.11.1 CONTACT PROBLEMS IN LINEAR ELASTICITY

A variational formulation is a statement that a solution of an operator equation subjected to certain boundary and/or initial conditions makes an expression involving variations of the quantities of the problems equal to zero or nonnegative. Thus we may distinguish between the bilateral or equality problems and the unilateral or inequality problems. Moreover in Mechanics we call the variational formulations, variational "principles". Further we shall derive certain variational principles for a deformable body. A first pilot problem has been detailed in Section 2.10.6. Let $\Omega \subset \mathbb{R}^3$ be an open bounded connected subset occupied by a deformable body in its undeformed state. On the assumption of small deformations we can write the relation.

$$\int_\Omega \sigma_{ij}(u)\varepsilon_{ij}(v-u)dx = \int_\Omega f_i(v_i - u_i)dx$$

$$+ \int_{\Gamma} \sigma_{ij} n_j (v_i - u_i) ds, \ \forall v \in V, \tag{2.11.1}$$

for $u \in V$. Here V denotes the function space of the displacements which will be defined further. Relation (2.11.1) is obtained from the operator equations of the problem by applying the Green-Gauss theorem, and is the expression of the principle of virtual work for the body when considered free, i.e., with no constraints on its boundary Γ. Note that for the derivation of (2.11.1) we have multiplied the equilibrium equation

$$\sigma_{ij,j} + f_i = 0 \tag{2.11.2}$$

where the f_i is the volume force vector, by $v_i - u_i$ and then we have integrated over Ω. On the assumption of "appropriately smooth" functions, we have applied the Green-Gauss theorem by taking into account the strain-displacement relation

$$\varepsilon_{ij} = \frac{1}{2}(u_{i,j} + u_{j,i}). \tag{2.11.3}$$

Let us assume further that the body is linear elastic, i.e. that

$$\sigma_{ij} = C_{ijhk}\varepsilon_{hk} \tag{2.11.4}$$

where $C = \{C_{ijhk}\}$, $i, j, h, k = 1, 2, 3$, is the elasticity tensor which satisfies the well-known symmetry and ellipticity properties

$$C_{ijhk} = C_{jihk} = C_{khij}, \tag{2.11.5}$$

$$C_{ijhk}\varepsilon_{ij}\varepsilon_{hk} \geq c\varepsilon_{ij}\varepsilon_{ij}, \ \forall \varepsilon = \{\varepsilon_{ij}\} \in \mathbb{R}^9, \ \varepsilon_{ij} = \varepsilon_{ji} \quad (c > 0). \tag{2.11.6}$$

We denote the bilinear form of linear elasticity by $a(\cdot, \cdot)$, i.e.

$$a(u, v) = \int_{\Omega} C_{ijhk}\varepsilon_{ij}(u)\varepsilon_{hk}(v)dx. \tag{2.11.7}$$

Note also that instead of (2.11.1) we can write the relation

$$\int_{\Omega} \sigma_{ij}\varepsilon_{ij}(v - u)dx = \int_{\Omega} f_i(v_i - u_i)dx + \int_{\Gamma} S_N(v_N - u_N)ds$$

$$+ \int_{\Gamma} S_{T_i}(v_{T_i} - u_{T_i})ds, \ \forall v \in V. \tag{2.11.8}$$

Here we have splitted the last term in (2.11.1) into the work of the normal and the work of the tangential tractions to the boundary. From (2.11.1) or (2.11.8) we shall derive certain variational formulations.

Let us assume first that on Γ the classical boundary conditions $S_N = 0$ and $u_{T_i} = 0$, $i = 1, 2, 3$, hold. Then (2.11.8) with (2.11.7) leads to the following variational equality: Find $u \in V_0 = \{v \in V : v_{T_i} = 0 \text{ on } \Gamma\}$ such that

$$a(u, v) = \int_\Omega f_i v_i dx, \ \forall \, v \in V_0. \tag{2.11.9}$$

Let us assume now that on Γ the general monotone multivalued boundary condition (2.6.8) holds. Introducing into (2.11.1) the inequality

$$j(v) - j(u) \geq S_i(v_i - u_i), \ \forall \, v = \{v_i\} \in \mathbb{R}^3, \tag{2.11.10}$$

holding by definition at every point of Γ due to (2.6.8), we obtain together with (2.11.7) the following variational inequality: Find $u \in V$ with $j(u) \in L^1(\Gamma)$, such that

$$a(u, v - u) \ + \ \int_\Gamma (j(v) - j(u))ds \geq \int_\Omega f_i(v_i - u_i)dx, \tag{2.11.11}$$

$$\forall \, v \in V \text{ with } j(v) \in L^1(\Gamma).$$

Let us assume further that on Γ the nonmonotone, possibly multivalued boundary condition (2.8.2) holds where j is a suitable locally Lipschitz functional (of the type considered in (1.2.15) or (1.2.16). Combining (2.11.1) with the inequality

$$j^0(u; v - u) \geq S_i(v_i - u_i), \ \forall \, v = \{v_i\} \in \mathbb{R}^3, \tag{2.11.12}$$

which defines on Γ the condition (2.8.2) we obtain with (2.11.7) the following hemivariational inequality: Find $u \in V$ such as to satisfy the inequality

$$a(u, v - u) + \int_\Gamma j^0(u; v - u)ds \geq \int_\Omega f_i(v_i - u_i)dx, \ \forall \, v \in V. \tag{2.11.13}$$

If instead of (2.8.2), (2.8.1a)-(2.8.1b) hold on Γ then (2.11.8) gives rise to the following hemivariational inequality: Find $u \in V$ such that

$$a(u, v - u) + \int_\Gamma j_N^0(u_N; v_N - u_N)ds + \int_\Gamma j_T^0(u_T; v_T - u_T)ds$$

$$\geq \int_{\Omega} f_i(v_i - u_i)dx, \ \forall \, v \, \in \, V. \qquad (2.11.14)$$

In order to investigate in which sense the solution of the hemivariational inequalities satisfies the operator equations of the problem, e.g. the equation of equilibrium, and the boundary conditions of the problem we need a more rigorous formulation of the above variational expressions. Let us place ourselves in the mathematical framework of Sobolev spaces. The notations and concepts introduced in Section 2.10 are now used. For

$$u \in V = H^1(\Omega; \mathbb{R}^3), \quad f_i \in L^2(\Omega), \ C_{ijkl} \in L^\infty(\Omega) \qquad (2.11.15)$$

we have instead of (2.11.1) the relation

$$\int_{\Omega} \sigma_{ij}\varepsilon_{ij}(v - u)dx = \int_{\Omega} f_i(v_i - u_i)dx$$

$$+\langle \sigma_{ij}n_j, v_i - u_i \rangle_{H^{-\frac{1}{2}}(\Gamma), H^{\frac{1}{2}}(\Gamma)}, \ \forall \, v \, \in \, H^1(\Omega; \mathbb{R}^3), \qquad (2.11.16)$$

where

$$\sigma_{ij} \in L^2(\Omega), \quad \varepsilon_{ij} \in L^2(\Omega),$$

as a consequence of the choice of V in (2.11.15) and where we suppose that

$$\sigma_{ij}n_j \in H^{-1/2}(\Gamma; \mathbb{R}^3).$$

Note here that with our assumptions, the bilinear form is nonnegative and continuous. The following result characterizes the rigid-body displacements.

Proposition 2.11.1   Let Ω be an open bounded connected subset of class $C^{0,1}$ in $\mathbb{R}^3$. Then the following two conditions are equivalent:

(i) $a(u, u) = 0$

(ii) $u(x) = a + b \wedge x; \ a, b \in \mathbb{R}^3, \ \forall \, x \, \in \, \Omega.$

Remark 2.11.2   i) Note here that if $\Omega \subset \mathbb{R}^2$ then one has a similar result which asserts that $a(u, u) = 0$ if and only if $u_1(x) = a_1 - bx_2$ and $u_2(x) = a_2 + bx_1; \ a_1, a_2, b \in \mathbb{R}, \ \forall x \in \Omega.$ ii) The set

$$R = \{u \in H^1(\Omega; \mathbb{R}^3) : a(u, u) = 0\}$$

constitutes the space of rigid-body displacements. Suppose that V_0 is a closed linear subspace of $H^1(\Omega; \mathbb{R}^3)$ such that $V_0 \cap R = \{0\}$ then there exists $c > 0$ such that

$$a(u, u) \geq c \parallel u \parallel_{1,2}^2, \ \forall\, u \ \in \ V_0.$$

This last inequality is called the Korn's first inequality. One has also the Korn's second inequality: There exist $\mu > 0, c > 0$ such that

$$a(u, u) + \mu \mid u \mid_{0,2}^2 \geq c \parallel u \parallel_{1,2}^2, \ \forall\, u \ \in \ H^1(\Omega;\ \mathbb{R}^3).$$

Having in mind our deformable body, that means that if it is fixed on a nonempty measurable part of its boundary Γ_U $(H_2(\Gamma_U) > 0)$ then the Korn's first inequality is satisfied on

$$V_0 = \{u \in H^1(\Omega; \mathbb{R}^3) : \gamma(u) = 0 \text{ a.e. on } \Gamma_U\}.$$

If $\Gamma_U = \emptyset$ then the Korn's second inequality holds on $H^1(\Omega; \mathbb{R}^3)$. These inequalities can be used straightforwardly to check if the bilinear form is coercive or semicoercive.

The last term in (2.11.16) is the work produced by the traction $S_i = \sigma_{ij}n_j$ for the displacement variation $v_i - u_i$. If $\sigma_{ij}n_j \in L^2(\Gamma)$ then this term reduces to the corresponding integral in (2.11.1). Assuming that $S \in H_T^{-\frac{1}{2}}(\Gamma; \mathbb{R}^3)$ and Ω is of class $C^{1,1}$ then instead of (2.11.8) we have the relation

$$\int_\Omega \sigma_{ij}\varepsilon_{ij}(v - u)dx = \int_\Omega f_i(v_i - u_i)dx + \langle S_N, \gamma_N(v - u)\rangle_{H^{-\frac{1}{2}}(\Gamma), H^{\frac{1}{2}}(\Gamma)}$$

$$+ \langle S_T, \gamma_T(v - u)\rangle_{H_T^{-\frac{1}{2}}(\Gamma;\mathbb{R}^3), H_T^{\frac{1}{2}}(\Gamma;\mathbb{R}^3)}, \ \forall\, v \ \in \ H^1(\Omega;\ \mathbb{R}^3). \qquad (2.11.17)$$

In the case of (2.11.9) we have $V_0 = \{v \in H^1(\Omega; \mathbb{R}^3) : v_T = 0 \text{ a.e. on } \Gamma\}$.

Let us now consider the boundary condition (2.6.8). Then using (2.11.10) we are led to the following variational inequality: Find $u \in H^1(\Omega; \mathbb{R}^3)$ such that

$$a(u, v - u) + \hat{\Phi}_\Gamma(v) - \hat{\Phi}_\Gamma(u) \geq \int_\Omega f_i(v_i - u_i)dx,$$

$$\forall v \in H^1(\Omega;\ \mathbb{R}^3), \qquad (2.11.18)$$

where $\hat{\Phi}_\Gamma$ is defined as in Example 2.10.26 (with the data $n = m = 3, p = 2, \varphi = j$).

Let us now discuss in which sense the solution of (2.11.18) satisfies the equations and boundary conditions of the problem. Let us first suppose that $\sigma_{ij,j} \in L^2(\Omega)$. Setting in (2.11.18) $v_i - u_i = \pm\varphi_i$, $i = 1, 2, 3$, where φ is an infinitely differentiable function from the space $C_0^\infty(\Omega)$ we obtain from (2.11.18) by means of (2.11.3), ((2.11.4) and the Green-Gauss theorem (Proposition 2.10.10) the relation

$$\int_\Omega (\sigma_{ij,j} + f_i)\varphi_i dx = 0, \ \forall \ \varphi_i \in C_0^\infty(\Omega) \tag{2.11.19}$$

which implies that (2.11.2) holds in the sense of the distributions, i.e. as an equality in the space of distributions $\mathcal{D}'(\Omega)$. Then the assumption $f_i \in L^2(\Omega)$ implies that (2.11.2) holds in $L^2(\Omega)$. Accordingly we can multiply it by $v - u \in H^1(\Omega; \mathbb{R}^3)$ and apply the Green-Gauss theorem (Proposition 2.10.10) to obtain the inequality

$$\hat{\Phi}_\Gamma(v) - \hat{\Phi}_\Gamma(u) \geq -\langle \pi(\sigma), \gamma(v - u) \rangle_{H^{-\frac{1}{2}}(\Gamma;\mathbb{R}^3), H^{\frac{1}{2}}(\Gamma;\mathbb{R}^3)},$$

$$\forall \ v \in H^1(\Omega; \mathbb{R}^3). \tag{2.11.20}$$

This inequality implies that the solution of the variational inequality (2.11.18) satisfies the boundary condition in the weak form.

Let us further assume that the general nonmonotone possibly multi-valued boundary condition (2.8.2) holds. Using (2.11.12) we are led to the following hemivariational inequality: Find $u \in H^1(\Omega; \mathbb{R}^3)$ such as to satisfy

$$a(u, v - u) + \int_\Gamma j^0(\gamma u; \gamma v - \gamma u) ds$$

$$\geq \int_\Omega f_i(v_i - u_i) dx, \ \forall \ v \in H^1(\Omega; \mathbb{R}^3). \tag{2.11.21}$$

Conversely we can show as before that a solution of (2.11.12), if any exists, satisfies the equations of equilibrium (2.11.2) in the sense of distributions and thus in the sense of $L^2(\Omega)$, because $f_i \in L^2(\Omega)$. Indeed setting in (2.11.21) $v_i - u_i = \pm\varphi_i \in C_0^\infty(\Omega)$ implies (2.11.19), etc. Then we proceed as before by applying the Green-Gauss theorem, and we obtain that the solution of (2.11.21) satisfies the inequality

$$\int_\Gamma j^0(\gamma u; \gamma v - \gamma u) ds \geq -\langle \pi(\sigma), \gamma(v - u) \rangle_{H^{-\frac{1}{2}}(\Gamma;\mathbb{R}^3), H^{\frac{1}{2}}(\Gamma;\mathbb{R}^3)},$$

$$\forall \ v \in H^1(\Omega; \mathbb{R}^3) \tag{2.11.22}$$

which is a weak formulation of the boundary condition (2.11.2).

A B.V.P. is called unilateral if it leads to variational, or hemivariational, or variational-hemivariational inequality formulations. We call the unilateral problems "inequality problems" too. Note at this point that the term "unilateral boundary conditions" has been initially used and is until now in use, in order to characterize boundary conditions involving inequalities. As Fourier has noticed [238] the inequality form of the principle of virtual or complementary virtual work is due to the fact that the variations of certain variables involved into the problem are "irreversible". For instance, if (2.4.6) held for $u, u^* \in V$, where V is a vector space then the substitution $u^* - u = \pm w$ would lead to a variational equality. But since $u, u^* \in K$, where K is a closed convex set, we cannot set $u^* - u = \pm w$, i.e., the variation $u^* - u$ is irreversible. Irreversible variations are called "unilateral" variations. Unilateral are the variations also in (2.11.18), unless $\nabla \hat{\Phi}_\Gamma$ exists everywhere. Indeed in this case (2.11.18) is equivalent to the variational equality

$$a(u,w) + \langle \nabla \hat{\Phi}_\Gamma(u), w \rangle_{1,2} = \int_\Omega f_i w_i dx, \ \forall \, w \, \in H^1(\Omega; \mathbb{R}^3), \quad (2.11.23)$$

as it results easily by setting in (2.11.18) $v = u \pm \lambda w, \lambda \to 0^+$. The converse easily results by setting in (2.11.23) $w = v - u$ and by applying the inequality ($\hat{\Phi}$ is convex)

$$\hat{\Phi}_\Gamma(v) - \hat{\Phi}_\Gamma(u) \geq \langle \nabla \hat{\Phi}_\Gamma(u), v - u \rangle_{1,2}, \ \forall \, v \, \in H^1(\Omega; \mathbb{R}^3). \quad (2.11.24)$$

Analogously we may argue in the case of hemivariational inequalities.

Example 2.11.3   The Signorini-Fichera problem . Let us consider a linear elastic body which in its undeformed state occupies a bounded connected and open subset Ω of $\mathbb{R}^3$. The set Ω is referred to a fixed Cartesian coordinate system $\{0; x_1; x_2; x_3\}$ and we assume that $\Omega \in C^{1,1}$. We assume that on Γ the Signorini-Fichera boundary conditions (2.6.17) together with the tangential boundary conditions $S_T = C_T \in L^2(\Gamma; \mathbb{R}^3)$ hold. We suppose also that the body is subjected to a volumic force field $f \in L^2(\Omega; \mathbb{R}^3)$. The B.V.P. gives rise to the variational inequality for the displacement field $u : \Omega \to \mathbb{R}^3$;

$$u \in K = \{u \in H^1(\Omega; \mathbb{R}^3) : \gamma_N(u) \leq 0 \text{ a.e. on } \Gamma\}$$

and

$$a(u, v - u) \geq \int_\Omega f.(v - u)dx$$

$$+ \int_\Gamma C_T(\gamma_T(v) - \gamma_T(u))ds, \ \forall \, v \, \in \, K. \qquad (2.11.25)$$

If different kinds of unilateral effects occur on different parts of the boundary, let us say $\Gamma_1, \Gamma_2, \cdots, \Gamma_s$ satisfying

$$\Gamma = \bigcup_{i=1}^{s} \overline{\Gamma}_i$$

and

$$\Gamma_i \cap \Gamma_j = \emptyset, \ \forall \, i, j \, \in \, \{1, \cdots, s\}, \ i \neq j,$$

then it is convenient to split the expression $\int_\Gamma S.(v-u)ds$ in (2.11.1) as the sum of all the partial expressions $\int_{\Gamma_i} S.(v-u)ds$. If for some $j \in \{1, \cdots, s\}$ the boundary conditions are expressed via the normal and tangential components of S and u then it is also convenient to split the expression $\int_{\Gamma_j} S.(v - u)ds$ as follows

$$\int_{\Gamma_j} S_N(v_N - u_N)ds + \int_{\Gamma_j} S_T.(v_T - u_T)ds.$$

Suppose for instance that $s = 3$. Assume also that

$$S \in -\partial\varphi_1(u) \text{on} \Gamma_1$$

$$S \in -\partial j_2(u) \text{on} \Gamma_2$$

$$S_N \in -\partial j_N(u_N) \text{ on } \Gamma_3,$$

and

$$S_T \in \partial\varphi_T(u_T) \text{ on } \Gamma_3,$$

where φ_1 and φ_T are proper convex l.s.c. functionals while j_2 and j_N are suitable locally Lipschitz functionals (of the types considered in (1.2.15) or (1.2.16). Then we obtain

$$\int_{\Gamma_1} \varphi_1(\gamma_{\Gamma_1}v)ds \; - \; \int_{\Gamma_1} \varphi_1(\gamma_{\Gamma_1}u)ds$$

$$+ \; \int_{\Gamma_2} j_2^0(\gamma_{\Gamma_2}u; \gamma_{\Gamma_2}v - \gamma_{\Gamma_2}u)ds$$

$$+ \; \int_{\Gamma_3} j_N^0(\gamma_{\Gamma_3,N}u; \gamma_{\Gamma_3,N}v - \gamma_{\Gamma_3,N}u)ds$$

$$+ \; \int_{\Gamma_3} \varphi_T(\gamma_{\Gamma_3,T}v)ds - \int_{\Gamma_3} \varphi_T(\gamma_{\Gamma_3,T}u)ds$$

$$\geq \; \int_{\Gamma} S.(\gamma v - \gamma u)ds,$$

$\forall \, v \, \in \, H^1(\Omega; \, \mathbb{R}^3)$, with $\varphi_1(\gamma_{\Gamma_1}v) \in L^1(\Gamma_1)$ and $\varphi_T(\gamma_{\Gamma_3,T}v) \in L^1(\Gamma_3)$.

From (2.11.1), we obtain the variational-hemivariational inequality problem: Find $u \, \in \, H^1(\Omega; \mathbb{R}^3)$ such that $\varphi_1(\gamma_{\Gamma_1}u) \, \in \, L^1(\Gamma_1)$, $\varphi_T(\gamma_{\Gamma_3,T}u) \in L^1(\Gamma_3)$ and

$$a(u, v - u) \; + \; \int_{\Gamma_1} \varphi_1(\gamma_{\Gamma_1}v)ds - \int_{\Gamma_1} \varphi_1(\gamma_{\Gamma_1}u)ds$$

$$+ \; \int_{\Gamma_2} j_2^0(\gamma_{\Gamma_2}u; \gamma_{\Gamma_2}v - \gamma_{\Gamma_2}u)ds$$

$$+ \; \int_{\Gamma_3} j_N^0(\gamma_{\Gamma_3,N}u; \gamma_{\Gamma_3,N}v - \gamma_{\Gamma_3,N}u)ds$$

$$+ \; \int_{\Gamma_3} \varphi_T(\gamma_{\Gamma_3,T}v)ds - \int_{\Gamma_3} \varphi_T(\gamma_{\Gamma_3,T}u)ds$$

$$\geq \; \int_{\Omega} f.(\gamma v - \gamma u)dx,$$

$$\forall \, v \, \in \, H^1(\Omega; \, \mathbb{R}^3), \text{with } \varphi_1(\gamma_{\Gamma_1}v) \in L^1(\Gamma_1)$$

$$\text{and } \varphi_T(\gamma_{\Gamma_3,T}v) \in L^1(\Gamma_3). \tag{2.11.26}$$

Example 2.11.4 Consider again the linear elastic body of Example 2.11.3. The boundary Γ of the body Ω is assumed to consist of the three open disjoint parts Γ_1, Γ_2 and Γ_3, i.e. $\Gamma = \overline{\Gamma_1} \cup \overline{\Gamma_2} \cup \overline{\Gamma_3}$. We assume that on the part Γ_1 of the boundary of the body the constraint

$$u(x) \in Q(x), \ \forall \, x \, \in \, \Gamma_1,$$

is satisfied (see (2.6.11), where $Q(x)$ is defined as follows

$$Q(x) := \{ y \in \mathbb{R}^3 : f(x, y) \le 0 \}.$$

Here $f : \Gamma_1 \times \mathbb{R}^3 \to \mathbb{R}$ is continuous and convex in the second variable. Moreover, we assume that $f(x, 0) \le 0, \forall x \in \Gamma_1$. Thus $Q(x)$ is a non-empty, closed and convex subset for all $x \in \Gamma_1$. Now, the formulation of these unilateral constraints has to encompass the associated forces of reactions $r(x)$. We assume a normal reaction law of the form (i.e. without frictional effects; see (2.6.12))

$$-r(x) \in N_{Q(x)}(u(x)), \ \forall \, x \, \in \, \Gamma_1.$$

Recall that the previous relation is equivalent to

$$u(x) \in Q(x) : r(x)^T (v - u(x)) \ge 0, \ \forall \, v \, \in \, Q(x), \ \forall \, x \, \in \, \Gamma_1. \quad (2.11.27)$$

The part Γ_2 of the boundary is assumed to be fixed, that is

$$u(x) = 0, \ \forall \, x \, \in \, \Gamma_2. \quad (2.11.28)$$

On Γ_3, we consider conditions of adhesive contact, i.e. a unilateral contact relation between the body and a Winkler-type support which may sustain only limited values of traction, and any compression. These conditions take the following form (see also (2.8.3))

$$\begin{aligned}
&\text{if} \quad -\varepsilon < u_N, \ \text{then} \ S_N + k u_N = 0, \\
&\text{if} \quad u_N = -\varepsilon, \ \text{then} \ 0 \le S_N \le k\varepsilon \\
&\text{if} \quad u_N < -\varepsilon, \ \text{then} \ S_N = 0
\end{aligned}$$

where ε is a positive real number, k is a positive continuous function. Moreover, in the tangential direction the forces are given, i.e.

$$S_T = C_T.$$

We can express the relations holding on Γ_3 in the normal direction by means of the following set-valued relation:

$$-S_N \in \partial_y j(x, u_N(x)), \ \forall \, x \, \in \, \Gamma_3,$$

where

$$j(x, y) := \begin{cases} \frac{1}{2} k(x) y^2 & \text{if} -\varepsilon \le y \\[2mm] \frac{1}{2} k(x) \varepsilon^2 & \text{if} \ y \le -\varepsilon. \end{cases}$$

From the orthogonality between $S_N n$ and u_T and between S_T and $u_N n$, and the definition of the generalized gradient in the previous relation we obtain the hemivariational inequality

$$S_N v_N + S_T^T v_T + j_y^0(x, u_N(x); v_N)$$

$$-C_T^T v_T \geq 0, \forall\, v_N \in \mathbb{R},\ v_T \in \mathbb{R}^3,\ x \in \Gamma_3. \qquad (2.11.29)$$

The B.V.P. leads to the following variational-hemivariational inequality problem: Find u satisfying (2.11.27) and (2.11.28) such that

$$\int_\Omega C_{ijkl}\varepsilon_{ij}(u)\varepsilon_{kl}(v-u)dx + \int_{\Gamma_3} j_y^0(x, u_N; v_N - u_N)ds$$

$$\geq \int_\Omega f^T(v-u)dx + \int_{\Gamma_3} C_T^T(v_T - u_T)ds,$$

for all v satisfying (2.11.27) and (2.11.28).

Let us now define the functional framework. We set

$$V = \{u \in H^1(\Omega; \mathbb{R}^3) : \gamma_{\Gamma_2}u(x) = 0,\ \text{a.e. } x \in \Gamma_2\}$$

and

$$C = \{u \in V : \gamma_{\Gamma_1}u(x) \in Q(x),\ \text{a.e. } x \in \Gamma_1\}.$$

Let us now assume the following regularity properties for the data:

$$C_{ijkl} \in L^\infty(\Omega), f \in L^2(\Omega; \mathbb{R}^3), C_T \in L^2(\Gamma_3; \mathbb{R}^3), \Omega \in C^{1,1}.$$

We obtain the mathematical model: Find $u \in C$ such that

$$a(u, v-u) + \int_{\Gamma_3} j_y^0(x, \gamma_{\Gamma_3,N}u; \gamma_{\Gamma_3,N}v - \gamma_{\Gamma_3,N}u)ds$$

$$\geq \int_\Omega f^T(v-u)dx + \int_{\Gamma_3} C_T^T(\gamma_{\Gamma_3,T}v - \gamma_{\Gamma_3,T}u)ds,$$

$$\forall\, v \in V. \qquad (2.11.30)$$

2.11.2   DYNAMIC FRICTION PROBLEMS

In the case of dynamic B.V.Ps, the term f_i has to be replaced by $f_i - \rho u_i''$, where ρ is the density of Ω. Then we consider the d'Alembert's principle for $u : (0, T) \to H^1(\Omega; \mathbb{R}^3)$:

$$\int_\Omega \rho u''.(v - u')dx \;\; + \;\; a(u, v - u') = \int_\Omega f_i(v_i - u'_i)dx$$

$$+ \;\; \langle S_N, \gamma_N v - \gamma_N u' \rangle_{H^{-\frac{1}{2}}(\Gamma), H^{\frac{1}{2}}(\Gamma)}$$

$$+ \;\; \langle S_T, \gamma_T v - \gamma_T v' \rangle_{H_T^{-\frac{1}{2}}(\Gamma; \mathbb{R}^3), H_T^{\frac{1}{2}}(\Gamma; \mathbb{R}^3)} \qquad (2.11.31)$$

$$\forall\, v \;\in\; H^1(\Omega;\; \mathbb{R}^3), \text{ a.e. } t \in (0, T).$$

In order to write (2.11.31) we have assumed that

$$u'(t) \in H^1(\Omega; \mathbb{R}^3),\; u''(t) \in L^2(\Omega; \mathbb{R}^3),$$
$$S_T(t) \in H^{-\frac{1}{2}}(\Gamma; \mathbb{R}^3),\; S_N(t) \in H^{-\frac{1}{2}}(\Gamma),\; f(t) \in L^2(\Omega; \mathbb{R}^3),$$

a.e. $t \in (0, T)$. We suppose also that the displacement and velocity field at time $t = 0$ are given, i.e.

$$u(0) = u_0, u'(0) = u_1.$$

The variational principle (2.11.31) can be used together with unilateral boundary conditions in order to describe dynamic friction problems.

For example, let us again consider the linear elastic body of Example 2.11.3. One supposes here that on Γ friction boundary conditions (2.6.21a)-(2.6.21b) hold, that is

$$| S_T | \;\; < \;\; \mu\, | S_N | \Rightarrow u'_T = 0 \text{ on } \Gamma \times (0, T)$$
$$| S_T | \;\; = \;\; \mu\, | S_N | \Rightarrow \exists \lambda \geq 0 : u'_T = -\lambda S_T \text{ on } \Gamma \times (0, T)$$

and

$$S_N = C_N \text{ on } \Gamma \times (0, T).$$

Suppose that $C_N(t) \in L^2(\Omega; \mathbb{R}^3)$, a.e. $t \in (0, T)$. The dynamic B.V.P. gives rise to the variational inequality:

$$\int_\Omega \rho u''.(v - u')dx + a(u, v - u') + \Phi(\gamma_T v) - \Phi(\gamma_T u')$$

$$\geq \int_\Omega f.(v - u')dx + \int_\Gamma C_N(\gamma_N v - \gamma_N u')ds, \qquad (2.11.32)$$

$$\forall\, v \;\in\; H^1(\Omega;\; \mathbb{R}^3), \text{ a.e. } t \in (0, T)$$

and where

$$\Phi(v) = \int_\Gamma \mu\, | C_N | \, | v | \, ds.$$

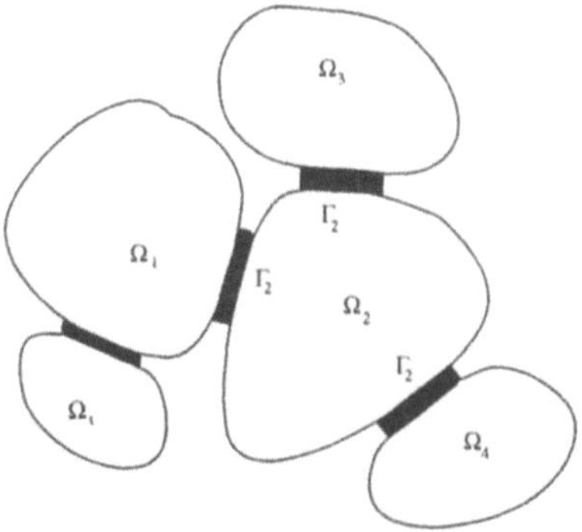

Figure 2.11.1.   Adhesive contact problem.

Suppose now that on Γ friction boundary conditions can be described via the superpotential relation

$$S_T \in -\partial j_T(u_T)$$

for some suitable locally Lipschitz function (of the type considered in (1.2.15) or (1.2.16)). Then instead of (2.11.32), the dynamic B.V.P. gives rise to the hemivariational inequality:

$$\int_\Omega \rho u''(v - u')dx + a(u, v - u') + \int_\Gamma j_T^0(\gamma_T u; \gamma_T v - \gamma_T u_T')ds$$

$$\geq \int_\Omega f.(v - u')dx + \int_\Gamma C_N(\gamma_N v - \gamma_N u')ds,$$

$$\forall\, v \in H^1(\Omega;\, \mathbb{R}^3), \text{ a.e. } t \in (0, T). \tag{2.11.33}$$

This last inequality problem constitutes the inequality form of the d'Alembert's principle.

2.11.3   ADHESIVE CONTACT PROBLEMS

In this Section we shall derive certain hemivariational inequalities and variational-hemivariational inequalities with respect to the problem of adhesive contact of linear elastic and nonlinear elastic bodies.

Let $\Omega^{(m)}$, $m = 1, 2, \ldots, l$, be a set of deformable bodies, possibly with different elasticity properties, with the boundaries Γ_m, $m = 1, 2, \ldots, l$, assumed to be appropriately regular. Let $x = \{x_i\}$, $i = 1, 2, 3$, be a point of $\mathbb{R}^3$ and let $\sigma^{(m)} = \sigma_{ij}^{(m)}$ and $\varepsilon^{(m)} = \varepsilon_{ij}^{(m)}$, $i, j = 1, 2, 3$, be the stress and strain tensors of the m-body. We denote by $f^{(m)} = \{f_i^{(m)}\}$ and $u^{(m)} = \{u_i^{(m)}\}$ the volume force and the displacement vector in each body. If $n^{(m)} = \{n_i^{(m)}\}$ is the outward unit normal vector to $\Gamma^{(m)}$, the boundary force on $\Gamma^{(m)}$ is $S^{(m)} = \{S_i^{(m)}\} = \{\sigma_{ij}^{(m)} n_j^{(m)}\}$ (summation convention). Let $S_N^{(m)}$ and $S_T^{(m)}$ be the normal and tangential components of it respectively. The corresponding displacement components are $u_N^{(m)}$ and $u_T^{(m)}$. The boundary $\Gamma^{(m)}$ is divided into three non-overlapping parts $\Gamma_U^{(m)}, \Gamma_F^{(m)}$ and $\Gamma_S^{(m)}$. On $\Gamma_U^{(m)}$ the displacements are given; let us take for simplicity that

$$u_i^{(m)} = 0 \quad \text{on} \quad \Gamma_U^{(m)}. \tag{2.11.34}$$

On $\Gamma_F^{(m)}$ the forces are prescribed, i.e.,

$$S_i^{(m)} = F_i^{(m)} \quad \text{on} \quad \Gamma_F^{(m)} \tag{2.11.35}$$

and on $\Gamma_S^{(m)}$-which corresponds to the interface of structure m with other substructures - nonmonotone interface conditions hold describing slip and delamination effects. We write in the general case $\Omega^{(m)} \subset \mathbb{R}^3$ the interface conditions in the form

$$-S_N^{(m)} \in \partial j_{N(m)}(S^{(m)}; [u_N^{(m)}]) \tag{2.11.36}$$

$$-S_T^{(m)} \in \partial j_{T(m)}(S^{(m)}; [u_T^{(m)}]) \tag{2.11.37}$$

in the normal and in the tangential direction to the interface. The super-potentials $j_{N(m)}$ and $j_{T(m)}$ are assumed to be suitable locally Lipschitz functions (of the types considered in (1.2.15) or (1.2.16)) of the interlayer gap $[u_N^{(m)}]$ and slip $[u_T^{(m)}]$ respectively and of the interface traction $S^{(m)}$, which is also a function of $u^{(m)}$. Let us remark that $[u_N^{(m)}] = u_N^{(m)} + u_N^{(j)}$ (resp. $[u_T^{(m)}] = u_T^{(m)} + u_T^{(j)}$) on the part of $\Gamma_S^{(m)}$ in contact with $\Gamma^{(j)}$. Here, however, we assume that (2.11.36) and (2.11.37) are uncoupled, i.e. that $S^{(m)}$ is considered as having a given value, or that $j_{N(m)}$ and $j_{T(m)}$ do not depend on $S^{(m)}$. Then (2.11.36) and (2.11.37) are equivalent to the inequalities

$$j_{N(m)}^0([u_N^{(m)}]; v - [u_N^{(m)}]) \geq -S_N^{(m)}(v - [u_N^{(m)}]), \; \forall\, v \in \mathbb{R}, \tag{2.11.38}$$

$$j_{T(m)}^0([u_T^{(m)}]; v - [u_T^{(m)}]) \geq \sum_{i=1}^{3} -S_{T_i}^{(m)}(v_i - [u_{T_i}^{(m)}]), \forall v \in \mathbb{R}^3. \tag{2.11.39}$$

In the framework of small deformations and linear elastic behavior for $\Omega^{(m)}$, $m = 1, 2, \ldots, l$, we can write the relations

$$\sigma_{ij,j}^{(m)} + f_i^{(m)} = 0, \qquad (2.11.40)$$

$$\varepsilon_{ij}^{(m)} = \frac{1}{2}(u_{i,j}^{(m)} + u_{j,i}^{(m)}) = \varepsilon_{ij}(u^{(m)}), \qquad (2.11.41)$$

$$\sigma_{ij}^{(m)} = C_{ijhk}^{(m)}\varepsilon_{hk}^{(m)}. \qquad (2.11.42)$$

Hooke's tensor $C^{(m)} = \{C_{ijhk}^{(m)}\}$ satisfies the usual symmetry and ellipticity conditions (see (2.11.5) and (2.11.6)). We write the principle of virtual work for every body $\Omega^{(m)}$ in the form

$$\int_{\Omega^{(m)}} \sigma_{ij}^{(m)}\varepsilon_{ij}^{(m)}(v^{(m)} - u^{(m)})dx = \int_{\Omega^{(m)}} f_i^{(m)}(v_i^{(m)} - u_i^{(m)})dx$$

$$+ \int_{\Gamma_F{}^{(m)}} F_i^{(m)}(v_i^{(m)} - u_i^{(m)})ds + \int_{\Gamma_S{}^{(m)}} \left[S_N^{(m)}(v_N^{(m)} - u_N^{(m)}) \right.$$

$$\left. + S_{T_i}^{(m)}(v_{T_i}^{(m)} - u_{T_i}^{(m)}) \right]ds, \forall\, v \in U_{\text{ad}}^{(m)}, \qquad (2.11.43)$$

where $U_{\text{ad}}^{(m)}$ is the kinematically admissible set of $\Omega^{(m)}$, i.e.

$$U_{\text{ad}}^{(m)} = \{v^{(m)} \in U(\Omega^{(m)}; \mathbb{R}^3) : v_i^{(m)} = 0 \text{ on } \Gamma_U^{(m)}\}. \qquad (2.11.44)$$

Here $U(\Omega^{(m)}; \mathbb{R}^3)$ denotes a space of vector-valued functions defined on $\Omega^{(m)}$. Adding with respect to m all the expressions (2.11.43) and taking into account the interconnection of the bodies, we obtain the relation

$$\sum_{m=1}^{l} \int_{\Omega^{(m)}} \sigma_{ij}^{(m)}\varepsilon_{ij}^{(m)}(v^{(m)} - u^{(m)})dx = \sum_{m=1}^{l} \left[\int_{\Omega^{(m)}} f_i^{(m)}(v_i^{(m)} - u_i^{(m)})dx + \right.$$

$$\left. \int_{\Gamma_F^{(m)}} F_i^{(m)}(v_i^{(m)} - u_i^{(m)})ds \right] + \sum_{q=1}^{k} \left[\int_{\Gamma^{(q)}} S_N^{(q)}([v_N^{(q)}] - [u_N^{(q)}])ds \right.$$

$$\left. + \int_{\Gamma^{(q)}} S_{T_i}^{(q)}([v_{T_i}^{(q)}] - [u_{T_i}^{(q)}])ds \right], \forall\, v \in U_{\text{ad}}, \qquad (2.11.45)$$

where $U_{ad} = \times_{m=1}^{l} U_{ad}^{(m)}$. In (2.11.45) the integrals along the joints Γ_q, $q = 1,\ldots,k$, have been introduced. The new enumeration of the $\Gamma_S^{(m)}$-boundaries has the advantage that finally the energy of each joint appears (see Fig. 2.11.1). One should take into account that the variation of the energy of each constraint of the form (2.11.36) and (2.11.37) must appear only once in the last terms of (2.11.45). Further we introduce the elastic energy of the m-structure

$$a_m(u^{(m)}, v^{(m)}) = \int_{\Omega^{(m)}} C_{ijhk}^{(m)} \varepsilon_{ij}(u^{(m)}) \varepsilon_{hk}(v^{(m)}) dx \qquad (2.11.46)$$

and by taking into account (2.11.38), (2.11.39) and (2.11.46), we get from (2.11.45) the following hemivariational inequality: Find $u \in U_{ad}$ such as to satisfy

$$\sum_{m=1}^{l} a_m(u^{(m)}, v^{(m)} - u^{(m)}) + \sum_{q=1}^{k} \left[\int_{\Gamma_q} \left[j_{N(q)}^0 ([u_N^{(q)}]; [v_N^{(q)}] - [u_N^{(q)}]) \right. \right.$$

$$\left. + j_{T(q)}^0 ([u_T^{(q)}]; [v_T^{(q)}] - [u_T^{(q)}]) \right] ds \Bigg] \geq \sum_{m=1}^{l} \left[\int_{\Omega^{(m)}} f_i^{(m)} (v_i^{(m)} - u_i^{(m)}) dx \right.$$

$$\left. + \int_{\Gamma_F^{(m)}} F_i^{(m)} (v_i^{(m)} - u_i^{(m)}) ds \right], \forall \, v \in U_{ad}. \qquad (2.11.47)$$

This hemivariational inequality is the expression of the principle of virtual work in its inequality form for the structure under consideration.

To check in which sense a solution of (2.11.47) fullfils (2.11.40), the boundary conditions on $\Gamma_F^{(m)}$ $m = 1,\ldots,l$ and the interface relations (2.11.36), (2.11.37) we must make the functional setting of the problem more precise. So we assume that

$$f_i^{(m)} \in L^2(\Omega^{(m)}), F_i^{(m)} \in L^2(\Gamma_F^{(m)}), C_{ijhk}^{(m)} \in L^\infty(\Omega^{(m)}).$$

The space $U_{ad}^{(m)}$ is specified as

$$U_{ad}^{(m)} = \{ v^{(m)} \in H^1(\Omega^{(m)}; \mathbb{R}^3) : \gamma_{\Gamma_U^{(m)}} v^{(m)} = 0 \text{ a.e. on } \Gamma_U^{(m)} \}.$$

The inequality (2.11.47) is now formulated as

$$\sum_{m=1}^{l} a_m(u^{(m)}, v^{(m)} - u^{(m)})$$

$$+ \sum_{q=1}^{k} \left[\int_{\Gamma_q} \left[j_{N(q)}^0([\gamma_{\Gamma_q,N} u^{(q)}]; [\gamma_{\Gamma_q,N} v^{(q)}] - [\gamma_{\Gamma_q,N} u^{(q)}]) \right. \right.$$

$$\left. \left. + j_{T(q)}^0([\gamma_{\Gamma_q,T} u^{(q)}]; [\gamma_{\Gamma_q,T} v^{(q)}] - [\gamma_{\Gamma_q,T} u^{(q)}]) \right] ds \right]$$

$$\geq \sum_{m=1}^{l} \left[\int_{\Omega^{(m)}} f^{(m)} \cdot (v^{(m)} - u^{(m)}) dx \right.$$

$$\left. + \int_{\Gamma_F^{(m)}} F^{(m)} \cdot (\gamma_{\Gamma_F^{(m)}} v^{(m)} - \gamma_{\Gamma_F^{(m)}} u^{(m)}) ds \right], \forall\, v \in U_{\text{ad}}. \qquad (2.11.48)$$

We set in (2.11.48) $v_i^{(m)} - u_i^{(m)} = \pm \phi_i^{(m)}$ where $\phi_i^{(m)}$ belongs to the space $C_0^\infty(\Omega^{(m)})$ of infinitely differentiable functions with compact support in $\Omega^{(m)}$. Then from (2.11.48) by setting $v_i^{(m)} - u_i^{(m)} = \pm \psi_i^{(m)}$ for $m = n$ and $v_i^{(m)} - u_i^{(m)} = 0$ for $m \neq n$ we obtain

$$a_n(u^{(n)}, \phi^{(n)}) = \int_{\Omega^{(n)}} f_i^{(n)} \phi_i^{(n)} dx$$

since $\phi_i^{(n)} = 0$ on $\Gamma^{(n)}$. The previous relation means that (2.11.40) holds on $\Omega^{(n)}$ in the sense of distributions over $\Omega^{(n)}$. This procedure is repeated for $n = 1, 2, \ldots, l$. Suppose that $\sigma_{ij,j}^{(m)} \in L^2(\Omega^{(m)})$ and applying the Green-Gauss theorem (Proposition 2.10.12) to each body to obtain the equality

$$a_m(u^{(m)}, v^{(m)} - u^{(m)}) = \int_{\Omega^{(m)}} f_i^{(m)}(v_i^{(m)} - u_i^{(m)}) dx$$

$$+ \langle \pi_{\Sigma^{(m)},N}(\sigma^{(m)}), \gamma_{\Sigma^{(m)},N}(v^{(m)} - u^{(m)}) \rangle_{\Sigma^{(m)}}$$

$$+ \langle \pi_{\Sigma^{(m)},T}(\sigma^{(m)}), \gamma_{\Sigma^{(m)},T}(v^{(m)} - u^{(m)}) \rangle_{0,\Sigma^{(m)}}, \forall\, v^{(m)} \in U_{ad}^{(m)}, \quad (2.11.49)$$

with $\Sigma^{(m)} = \overline{\Gamma^{(m)} \backslash \Gamma_F^{(m)} \cup \Gamma_S^{(m)}}$. The notations

$$\langle \cdot, \cdot \rangle_{\Sigma^{(m)}} := \langle \cdot, \cdot \rangle_{H^{-\frac{1}{2}}(\Sigma^{(m)}), H^{\frac{1}{2}}(\Sigma^{(m)})}$$

and

$$\langle \cdot, \cdot \rangle_{0,\Sigma^{(m)}} := \langle \cdot, \cdot \rangle_{H_{00T}^{-\frac{1}{2}}(\Sigma^{(m)};\mathbb{R}^3), H_{00T}^{\frac{1}{2}}(\Sigma^{(m)};\mathbb{R}^3)}$$

have been here used. From (2.11.49) and (2.11.48) we obtain the inequality

$$\sum_{q=1}^{k}\left[\int_{\Gamma_q}\left[j_{N(q)}^0([\gamma_{\Gamma_q,N}u^{(q)}];[\gamma_{\Gamma_q,N}v^{(q)}]-[\gamma_{\Gamma_q,N}u^{(q)}])\right.\right.$$

$$\left.\left.+j_{T(q)}^0([\gamma_{\Gamma_q,T}u^{(q)}];[\gamma_{\Gamma_q,T}v^{(q)}]-[\gamma_{\Gamma_q,T}u^{(q)}])\right]ds\right]$$

$$-\sum_{m=1}^{l}\int_{\Gamma_F^{(m)}}F^{(m)}.(\gamma_{\Gamma_F^{(m)}}v^{(m)}-\gamma_{\Gamma_F^{(m)}}u^{(m)})ds$$

$$+\sum_{m=1}^{l}\langle\pi_{\Sigma^{(m)},N}(\sigma^{(m)}),\gamma_{\Sigma^{(m)},N}(v^{(m)}-u^{(m)})\rangle_{\Sigma^{(m)}}$$

$$+\sum_{m=1}^{l}\langle\pi_{\Sigma^{(m)},T}(\sigma^{(m)}),\gamma_{\Sigma^{(m)},T}(v^{(m)}-u^{(m)})\rangle_{0,\Sigma^{(m)}}$$

$$\geq 0,\ \forall\, v\,\in\,U_{ad}.\qquad(2.11.50)$$

For any $\varphi^{(m)}\in\mathcal{D}(\Gamma^{(m)};\mathbb{R}^3)$ there exists $\overline{\varphi}^{(m)}\in C^0(\overline{\Omega}^{(m)};\mathbb{R}^3)\cap U_{ad}^{(m)}$ such that $\overline{\varphi}^{(m)}=\varphi^{(m)}$ on $\Gamma^{(m)}$. Setting $v=u+\overline{\varphi}$ (i.e. $v^{(m)}=u^{(m)}+\overline{\varphi}^{(m)}$ ($m=1,\cdots,l$)) in (2.11.50) we obtain

$$\sum_{q=1}^{k}\left[\int_{\Gamma_q}\left[j_{N(q)}^0([\gamma_{\Gamma_q,N}u^{(q)}];[\varphi_N^{(q)}])+j_{T(q)}^0([\gamma_{\Gamma_q,T}u^{(q)}];[\varphi_T^{(q)}])\right]ds\right]$$

$$-\sum_{m=1}^{l}\int_{\Gamma_F^{(m)}}F^{(m)}.\varphi^{(m)}ds+\sum_{m=1}^{l}\langle\pi_{\Sigma^{(m)},N}(\sigma^{(m)}),\varphi_N^{(m)}\rangle_{\Sigma^{(m)}}$$

$$+\sum_{m=1}^{l}\langle\pi_{\Sigma^{(m)},T}(\sigma^{(m)}),\varphi_T^{(m)}\rangle_{0,\Sigma^{(m)}}\geq 0.$$

Suppose now formally that $\sigma_{ij,j}^{(m)}\in C^1(\overline{\Omega^{(m)}})$. The last relation being true for any $\varphi\in\mathcal{D}(\Gamma^{(m)};\mathbb{R}^3)$, we deduce that ($i=1,2,3;\ m=1,\cdots,l;\ q=1,\cdots,k$)

$$\int_{\Gamma_F^{(m)}}S_i^{(m)}\theta ds=\int_{\Gamma_F^{(m)}}F_i^{(m)}\theta ds,\ \forall\,\theta\,\in\,\mathcal{D}(\Gamma_F^{(m)}),$$

$$\int_{\Gamma_q} j^0_{N(q)}([\gamma_{N|\Gamma_q} u^{(q)}]; [\varphi_N^{(q)}])ds \geq -\int_{\Gamma_q} S_N^{(q)}[\varphi_N^{(q)}]ds,$$

$$\forall \varphi \in \mathcal{D}(\Gamma_q), \ \varphi = \varphi_N n^{(q)}$$

and

$$\int_{\Gamma_q} j^0_{T(q)}([u_T^{(q)}]; [\varphi_T^{(q)}])ds \geq -\sum_{i=1}^{3} \int_{\Gamma_q} S_{T_i}^{(q)}[\varphi_{T_i}^{(q)}]ds,$$

$$\forall \varphi \in \mathcal{D}(\Gamma_q), \ \varphi_N = 0.$$

These relations constitute a "weak" formulation of the boundary conditions.

Suppose now that the substructures $\Omega^{(m)}$ obey a general law of the form

$$\sigma^{(m)} = \frac{\partial w_{(m)}}{\partial F}(\varepsilon^{(m)})$$

where $w_{(m)} \in C^1(U_{ad}^{(m)}; \mathbb{R})$. Then the variational expression of the problem is the same as in (2.11.47) but now the term $\sum_{m=1}^{l} a_m(u^{(m)}, v^{(m)} - u^{(m)})$ has to be replaced by $\sum_{m=1}^{l} \int_{\Omega^{(m)}} (\frac{\partial w_{(m)}}{\partial F}(\varepsilon))_{ij} \varepsilon_{ij}(v^{(m)} - u^{(m)})dx$, where the integrand is supposed to be a $L^1(\Omega^{(m)})$-function.

2.11.4   LAMINATED VON KÁRMÁN PLATES

In this Section we will study the delamination effect for laminated plates undergoing large displacements (von Kármán plates). Delamination [211] is one of the main causes of strength-degradation. For a laminated plate the mechanical behavior of the interlayer binding material, together with the possibility of debonding is described by a nonmonotone, possibly multivalued law connecting the interlaminar bonding forces with the corresponding relative displacements. At the boundary of the plate monotone boundary conditions are assumed to hold, e.g. the Signorini-Fichera boundary condition or the plastic hinge boundary condition. The interlayer law (resp. the boundary law) is expressed through nonconvex (resp. convex) superpotentials leading to hemivariational (resp. variational) inequalities. Thus the whole problem gives rise, as we shall see further, to a hemivariational inequality concerning the bending of the laminae and to a variational inequality concerning the stretching of the plate.

Consider a laminated plate consisting of two laminae and the binding material between them (Fig 2.11.2b). In the undeformed state the

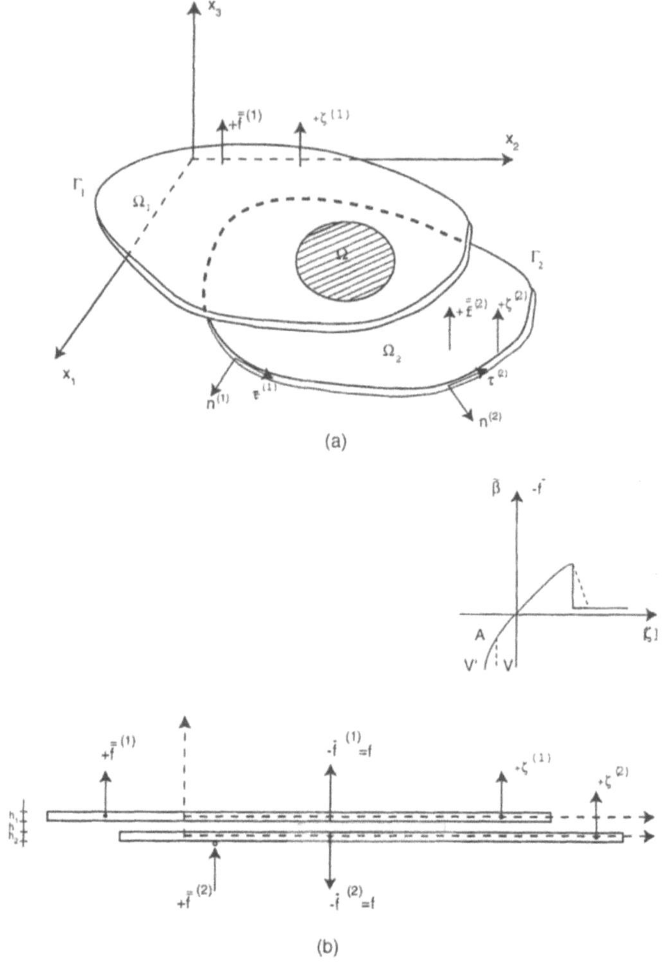

Figure 2.11.2.   Notation and geometry of the laminated plate.

middle surface of lamina j occupies a nonempty open, bounded and connected subset Ω_j of $\mathbb{R}^2$, referred to a fixed right-handed Cartesian coordinate system $\{O; x_1; x_2; x_3\}$. Let $\Gamma_j, j = 1, 2$, be the boundary of the j-th lamina. The set Ω_j is assumed to be of class $C^{1,1}$. Let also the interlaminar binding material occupy a subset Ω' such that $\Omega' \subset \Omega_1 \cap \Omega_2$ and $\bar{\Omega}' \cap \Gamma_1 = \emptyset, \bar{\Omega}' \cap \Gamma_2 = \emptyset$. The binding material and the two laminae together form one integral structural element. We denote by $\zeta^{(j)}(x)$ the vertical deflection of the point $x \in \Omega_j$ of the j-th lamina, and by $f^{(j)} = (0, 0, f_3^{(j)}(x))$ the distributed vertical load acting on the j-th lamina.

Further, let $u^{(j)} = \{u_1^{(j)}, u_2^{(j)}\}$ be the in-plane displacement of the j-th lamina. We assume that the j-th lamina has constant thickness h_j, while the interlaminar binding layer has constant thickness h. Moreover, we assume that each lamina obeys the von Kármán plate theory, (see e.g. [82]), i.e. it is a thin plate having large deflections. The following system of partial differential equations holds for von Kármán plates:

$$K_j \Delta\Delta\zeta^{(j)} - h_j(\sigma_{\alpha\beta}^{(j)}\zeta_{,\beta}^{(j)})_{,\alpha} = f^{(j)} \quad \text{in } \Omega_j, \tag{2.11.51}$$

$$\sigma_{\alpha\beta,\beta}^{(j)} = 0 \quad \text{in } \Omega_j, \tag{2.11.52}$$

and

$$\sigma_{\alpha\beta}^{(j)} = C_{\alpha\beta\gamma\delta}^{(j)}(\varepsilon_{\gamma\delta}^{(j)}(u^{(j)}) + \frac{1}{2}\zeta_{,\gamma}^{(j)}\zeta_{,\delta}^{(j)}) \quad \text{in } \Omega_j. \tag{2.11.53}$$

Here the subscripts $\alpha, \beta, \gamma, \delta = 1, 2$ correspond to the coordinate directions; the superscript or the subscript $j = 1, 2$ refers to the j-th lamina; $\{\sigma_{\alpha\beta}^{(j)}\}$, $\{\varepsilon_{\alpha\beta}^{(j)}\}$, and $\{C_{\alpha\beta\gamma\delta}^{(j)}\}$ denote the stress, strain and elasticity tensors in the plane of the plate. The components of $C^{(j)}$ are elements of $L^\infty(\Omega_j)$ and have the usual symmetry and ellipticity properties. Moreover, $K_j = E_j h_j^3/12(1 - \nu_j^2)$ is the bending rigidity of the j-th plate with E_j the modulus of elasticity of the j-th plate and $0 < \nu_j < \frac{1}{2}$ the Poisson ratio of the j-th plate. For the sake of simplicity, we consider here isotropic homogeneous plates of constant thickness. In laminated and layered plates, the interlaminar normal stress σ_{33} is one of the main causes for delamination effects (see e.g. [290]). Note that this is a simplification of the mechanical problem. In order to model the action of σ_{33}, $f^{(j)}$ is split into $\bar{f}^{(j)}$, which describes the interaction of the two plates, and $\bar{\bar{f}}^{(j)} \in L^2(\Omega_j)$, which represents the external loading applied on the j-th plate:

$$f^{(j)} = \bar{f}^{(j)} + \bar{\bar{f}}^{(j)} \quad \text{in } \Omega_j, \quad j = 1, 2. \tag{2.11.54}$$

If f denotes the stress in the interlaminar binding layer, the following holds (see Fig. 2.11.2)

$$f = \bar{f}^{(1)} = -\bar{f}^{(2)} \quad \text{in } \Omega'. \tag{2.11.55}$$

We introduce now a phenomenological law connecting f with the corresponding relative deflection of the plates (see Fig. 2.11.2) $[\zeta] = \zeta^{(1)} - \zeta^{(2)}$ (see also (2.10.21) and Example 2.10.22.

We assume that

$$-f \in \tilde{\beta}([\zeta]) \quad \text{in } \Omega' \tag{2.11.56}$$

where $\tilde{\beta}$ is the multivalued function defined by

$$\tilde{\beta}(z) = [\underline{\beta}(z), \overline{\beta}(z)]$$

for some $\beta \in L^\infty_{\text{loc}}(\mathbb{R})$ (see Section 2.9). We note here that cracking as well as crushing effects of either a brittle or semi-brittle nature can be accounted for by means of this law. The impenetrability restriction would imply a vertical branch AV (Fig. 2.11.2b) in the final compression state of the binding layer. However a slightly inclined branch AV$'$ is taken here into account in order to consider the compression of the laminae in the Ox_3-direction. The following relations complete in a natural way the definition of $\bar{f}^{(j)}$:

$$\bar{f}^{(j)} = 0 \quad \text{in } \Omega_j \backslash \Omega', \quad j = 1, 2. \tag{2.11.57}$$

In order to obtain a variational formulation of the problem, we express relation (2.11.56) in a superpotential form. From Proposition 1.2.19 a locally Lipschitz (nonconvex) function $j : \mathbb{R} \to \mathbb{R}$ (see (1.2.14)) can be determined up to an additive constant, such that

$$\tilde{\beta}(\xi) = \partial j(\xi). \tag{2.11.58}$$

Moreover, we assume that the following boundary conditions hold on the open subset $\tilde{\Gamma}_j$ of the plate boundaries (cf. (2.7.1) and (2.7.2))

$$M_{nj}(\zeta^{(j)}) \in \beta_j\left(\frac{\partial \zeta^{(j)}}{\partial n}\right) \quad \text{on } \tilde{\Gamma}_j, \quad j = 1, 2, \tag{2.11.59}$$

$$-K_{nj}(\zeta^{(j)}) \in \beta'_j(\zeta^{(j)}) \quad \text{on } \tilde{\Gamma}_j, \quad j = 1, 2 \tag{2.11.60}$$

whereas

$$\zeta^{(j)} = \frac{\partial \zeta^{(j)}}{\partial n} = 0 \quad \text{on } \Gamma_j \backslash \tilde{\Gamma}_j, \quad j = 1, 2. \tag{2.11.61}$$

Here $\beta_j, \beta'_j, j = 1, 2$ are possibly multivalued maximal monotone operators from $\mathbb{R}$ into $2^\mathbb{R}$. According to Proposition 1.3.15, convex, l.s.c., proper functionals $\phi_j, \phi'_j, j = 1, 2$ can be determined such that

$$\beta_j\left(\frac{\partial \zeta^{(j)}}{\partial n}\right) = \partial \phi_j\left(\frac{\partial \zeta^{(j)}}{\partial n}\right), \quad j = 1, 2, \tag{2.11.62}$$

and

$$\beta'_j(\zeta^{(j)}) = \partial \phi'_j(\zeta^{(j)}), \quad j = 1, 2. \tag{2.11.63}$$

For example, with respect to the rotation of the plate Ω_j the friction boundary conditions can be given by the relations

$$\text{if } \mid M_{nj} \mid < M_{0j} \text{ then } \frac{\partial \zeta^{(j)}}{\partial n} = 0, \tag{2.11.64}$$

$$\text{if } \mid M_{nj} \mid = M_{0j} \text{ then there exists } \lambda \geq 0$$
$$\text{such that } M_{nj} = \lambda \frac{\partial \zeta^{(j)}}{\partial n}. \tag{2.11.65}$$

In this case β_j has the form

$$\beta_j \left(\frac{\partial \zeta^{(j)}}{\partial n} \right) = \begin{cases} [-M_{0j}, M_{0j}] & \text{if } \frac{\partial \zeta^{(j)}}{\partial n} = 0 \\ +M_{0j} & \text{if } \frac{\partial \zeta^{(j)}}{\partial n} > 0 \\ -M_{0j} & \text{if } \frac{\partial \zeta^{(j)}}{\partial n} < 0 \end{cases} \tag{2.11.66}$$

Here M_{0j} is a prescribed bending moment. If we assume that a plastic hinge is formed on the boundary of the plate then M_{0j} is the plastic moment of the plate cross-section.

For another example, the boundary conditions of unilateral rotation of the plate Ω_j have a form similar to the Signorini-Fichera boundary conditions, i.e.

$$\frac{\partial \zeta^{(j)}}{\partial n} \leq 0, \, M_{nj} \leq 0, \, M_{nj} \frac{\partial \zeta^{(j)}}{\partial n} = 0.$$

The respective operator β_j is then given by

$$\beta_j \left(\frac{\partial \zeta^{(j)}}{\partial n} \right) = \begin{cases} 0 & \text{if } \frac{\partial \zeta^{(j)}}{\partial n} < 0 \\ \mathbb{R}_- & \text{if } \frac{\partial \zeta^{(j)}}{\partial n} = 0 \\ \emptyset & \text{if } \frac{\partial \zeta^{(j)}}{\partial n} > 0 \end{cases}$$

or also

$$\beta_j \left(\frac{\partial \zeta^{(j)}}{\partial n} \right) = \partial \Psi_{\mathbb{R}_+}(\zeta^{(j)}).$$

Various examples for the unilateral conditions on the shearing forces can also be given. For example, Signorini-Fichera boundary conditions

$$\zeta^{(j)} \geq 0, \, K_{nj} \geq 0, \, K_{nj} \zeta^{(j)} = 0. \tag{2.11.67}$$

The respective operator β'_j is given by

$$\beta'_j(\zeta^{(j)}) = \begin{cases} 0 & \text{if } \zeta^{(j)} > 0 \\ (-\infty, 0] & \text{if } \zeta^{(j)} = 0 \\ \emptyset & \text{if } \zeta^{(j)} > 0 \end{cases} \tag{2.11.68}$$

that is also

$$\beta_j'(\zeta^{(j)}) = \partial\Psi_{\mathbb{R}+}(\zeta^{(j)}).$$

Let us introduce the Sobolev space $H^2(\Omega_j)$ for the deflections $\zeta^{(j)}$ and define the following convex, l.s.c. and proper functionals Φ_j by

$$D(\Phi_j) = \{u \in H^2(\Omega_j) : \phi_j(\gamma_{1\tilde{\Gamma}_j} z^{(j)}) \in L^1(\tilde{\Gamma}_j)$$

$$\text{and } \phi_j'(\gamma_{\tilde{\Gamma}_j} z^{(j)}) \in L^1(\tilde{\Gamma}_j)\}$$

and

$$\Phi_j(z^{(j)}) = \int_{\tilde{\Gamma}_j} \phi_j(\gamma_{1\tilde{\Gamma}_j} z^{(j)})ds + \int_{\tilde{\Gamma}_j} \phi_j'(\gamma_{\tilde{\Gamma}_j} z^{(j)})ds, \qquad (2.11.69)$$

on its domain and $\Phi_j(z^{(j)}) = +\infty$ otherwise. Recall that formally $\gamma_{\tilde{\Gamma}_j}(z) = z_{|\tilde{\Gamma}_j}$ and $\gamma_{1\tilde{\Gamma}_j}(z) = \frac{\partial z}{\partial n}_{|\tilde{\Gamma}_j}$. For the in-plane displacement we assume the boundary conditions

$$\sigma_{\alpha\beta}^{(j)} n_\beta^{(j)} = 0 \quad \text{on } \Gamma_j. \qquad (2.11.70)$$

We can now derive the variational formulation of the problem. From (2.11.51), by assuming sufficiently regular functions, multiplying by $z^{(j)} - \zeta^{(j)}$, integrating and applying the Green-Gauss theorem, we obtain the expressions $(j = 1, 2)$:

$$a_j(\zeta^{(j)}, z^{(j)} - \zeta^{(j)}) + \int_{\Omega_j} h_j \sigma_{\alpha\beta}^{(j)} \zeta_{,\alpha}^{(j)}(z^{(j)} - \zeta^{(j)})_{,\beta}dx$$

$$= \int_{\Gamma_j} h_j \sigma_{\alpha\beta}^{(j)} \zeta_{,\beta}^{(j)} n_\alpha^{(j)}(z^{(j)} - \zeta^{(j)})ds + \int_{\Omega_j} \bar{\bar{f}}^{(j)}(z^{(j)} - \zeta^{(j)})dx$$

$$+ \int_{\Gamma_j} K_{nj}(\zeta^{(j)})(z^{(j)} - \zeta^{(j)})ds$$

$$- \int_{\Gamma_j} M_{nj}(\zeta^{(j)})\frac{\partial(z^{(j)} - \zeta^{(j)})}{\partial n^{(j)}}ds. \qquad (2.11.71)$$

Here $n^{(j)}$ denotes the outward normal unit vector to Γ_j,

$$a_j(\zeta, z) = K_j \int_{\Omega_j} [(1 - \nu_j)\zeta_{,\alpha\beta}^{(j)} z_{,\alpha\beta}^{(j)}$$

$$+ \nu_j \Delta\zeta^{(j)} \Delta z^{(j)}]dx, \alpha, \beta = 1, 2, \qquad (2.11.72)$$

$$M_{nj}(\zeta) = -K_j[\nu_j\,\Delta\zeta^{(j)} + (1-\nu_j)(2n_1^{(j)}n_2^{(j)}\zeta_{,12}^{(j)}$$

$$+n_1^{(j)2}\zeta_{,11}^{(j)} + n_2^{(j)2}\zeta_{,22}^{(j)}], \tag{2.11.73}$$

$$K_{nj}(\zeta) = -K_j\left[\frac{\partial\Delta\zeta^{(j)}}{\partial n} + (1-\nu_j)\frac{\partial}{\partial\tau}[n_1^{(j)}n_2^{(j)}(\zeta_{,22}^{(j)} - \zeta_{,11}^{(j)})\right.$$

$$\left. +(n_1^{(j)2} - n_2^{(j)2})\zeta_{,12}^{(j)}]\right]. \tag{2.11.74}$$

Applying the same technique to (2.11.52), it implies the expression ($j = 1, 2; \alpha, \beta = 1, 2$) :

$$\int_{\Omega_j} \sigma_{\alpha\beta}^{(j)}\varepsilon_{\alpha\beta}^{(j)}(v^{(j)} - u^{(j)})dx = \int_{\Gamma_j} \sigma_{\alpha\beta}^{(j)}n_\beta^{(j)}(v_\alpha^{(j)} - u_\alpha^{(j)})ds. \tag{2.11.75}$$

Further, the following notations are introduced:

$$R^{(j)}(m, k) = \int_\Omega C_{\alpha\beta\gamma\delta}^{(j)}m_{\alpha\beta}k_{\gamma\delta}dx, \quad \alpha, \beta, \gamma, \delta = 1, 2 \tag{2.11.76}$$

and

$$P(\zeta, z) = \{\zeta_{,\alpha}z_{,\beta}\}, \quad P(\zeta, \zeta) = P(\zeta), \tag{2.11.77}$$

where $m = \{m_{\alpha\beta}\}$ and $k = \{k_{\alpha\beta}\}, \alpha, \beta = 1, 2$ are 2×2 tensors.

Let us also introduce a functional framework for the B.V.P. We assume that $u^{(j)}, v^{(j)} \in H^1(\Omega_j; \mathbb{R}^2)$ and that $\zeta^{(j)}, z^{(j)} \in Z_j$,where

$$Z_j = \left\{z \in H^2(\Omega_j) : \gamma_{U_j}(z) = 0 \ \text{ a.e. on }\ U_j, \right.$$

$$\left. \gamma_{1,U_j}(z) = 0 \ \text{ a.e. on}\, U_j\right\}, \tag{2.11.78}$$

with $U_j = \Gamma_j\backslash\tilde{\Gamma}_j$. Taking into account the notations introduced in (2.11.76), (2.11.77) and the variational equalities (2.11.71), (2.11.75), the boundary conditions (2.11.59)-(2.11.61) and (2.11.70), the interface conditions (2.11.56), (2.11.57), the definition (2.11.69) we obtain the following problem : Find $u^{(j)} \in H^1(\Omega_j; \mathbb{R}^2)$ and $\zeta^{(j)} \in Z_j, j = 1, 2$, such as to satisfy the variational-hemivariational inequality

$$\sum_{j=1}^{2} a_j(\zeta^{(j)}, z^{(j)} - \zeta^{(j)})$$

$$+ \sum_{j=1}^{2} h_j R^{(j)}(\varepsilon^{(j)}(u^{(j)}) + \frac{1}{2}P(\zeta^{(j)}), P(\zeta^{(j)}, z^{(j)} - \zeta^{(j)}))$$

$$+ \int_{\Omega'} j^0([\zeta]; [z] - [\zeta])dx + \sum_{j=1}^{2}\{\Phi_j(z^{(j)}) - \Phi_j(\zeta^{(j)})\}$$

$$\geq \sum_{j=1}^{2} \int_{\Omega_j} \overline{\overline{f}}^{(j)} (z^{(j)} - \zeta^{(j)})dx, \forall z^{(1)} \in Z_1, z^{(2)} \in Z_2, \qquad (2.11.79)$$

and the variational equalities $(j = 1, 2)$

$$R^{(j)}(\varepsilon(u^{(j)}) + \frac{1}{2}P(\zeta^{(j)}), \varepsilon^{(j)}(v^{(j)} - u^{(j)})) = 0, \forall v^{(j)} \in H^1(\Omega_j; \mathbb{R}^2).$$
$$(2.11.80)$$

Analogously we may derive the variational formulation for r-laminae. Then in (2.11.79) the summation $\sum_{j=1}^{2}$ is replaced by $\sum_{j=1}^{r}$ and the term $\int_{\Omega'} j^0(\cdot, \cdot)dx$ is replaced by the term $\sum_{m=1}^{m'} \int_{\Omega'_m} j_m^0(\cdot, \cdot)dx$, where m' is the total number of interfaces, i.e. $m' = r - 1$. Then (2.11.79) must hold for every $z^{(j)} \in Z_j$, $j = 1, \ldots, r$. We assume now that the classical boundary conditions of each plate in bending define the kinematically admissible sets Z_j which are assumed to be (linear) subspaces of $H^2(\Omega_j)$. The following problem is formulated: Find $\zeta^{(j)} \in Z_j$ and $u^{(j)} \in H^1(\Omega_j; \mathbb{R}^2), j = 1, \ldots, r$, such as to satisfy the variational-hemivariational inequality $(j = 1, \cdots, r)$:

$$\sum_{j=1}^{r} a_j(\zeta^{(j)}, z^{(j)} - \zeta^{(j)})$$

$$+ \sum_{j=1}^{r} h_j R^{(j)}(\varepsilon^{(j)}(u^{(j)}) + \frac{1}{2}P(\zeta^{(j)}), P(\zeta^{(j)}, z^{(j)} - \zeta^{(j)}))$$

$$+ \sum_{m=1}^{m'} \int_{\Omega'_m} j_m^0([\zeta]^{(m)}; [z]^{(m)} - [\zeta]^{(m)})dx$$

$$+ \sum_{j=1}^{r}\{\Phi_j(z^{(j)}) - \Phi_j(\zeta^{(j)})\}$$

$$\geq \sum_{j=1}^{r} \int_{\Omega_j} \overline{\overline{f}}^{(j)} (z^{(j)} - \zeta^{(j)})dx, \forall z^{(j)} \in Z_j, \qquad (2.11.81)$$

and the variational equalities $(j = 1, \cdots, r)$:

$$R^{(j)}(\varepsilon^{(j)}(u^{(j)}) + \frac{1}{2}P(\zeta^{(j)}), \varepsilon^{(j)}(v^{(j)} - u^{(j)})) = 0,$$

$$\forall \, v^{(j)} \in H^1(\Omega_j; \mathbb{R}^2). \tag{2.11.82}$$

Further we shall eliminate the in-plane displacements of the plate. To this end we note first that $R(\cdot, \cdot)$ as defined in (2.11.76) is a continuous symmetric, coercive bilinear form on $L^2(\Omega_j; \mathbb{R}^4)$, and that P : $H^2(\Omega_j; \mathbb{R}^2) \to L^2(\Omega_j; \mathbb{R}^4) \times L^2(\Omega_j; \mathbb{R}^4)$ of (2.11.77) is a completely continuous operator (cf. e.g. [46], [47] and [336]). For every deflection $\zeta^{(j)} \in Z_j, j = 1, 2, \ldots, r$, there corresponds a plane displacement $u^{(j)}(\zeta^{(j)}) \in H^1(\Omega_j; \mathbb{R}^2)$. Indeed, due to Korn's second inequality $R^{(j)}(\varepsilon^{(j)}(u), \varepsilon^{(j)}(v))$ is a bilinear semicoercive form on the space $H^1(\Omega_j; \mathbb{R}^2)$. Note that the space of in-plane rigid-plate displacements is given by

$$\bar{R}_j = \{\bar{r} \in H^1(\Omega_j; \mathbb{R}^2) : \quad \bar{r}_1 = \alpha_1 + bx_2,$$

$$\bar{r}_2 = \alpha_2 - bx_1, \alpha_1, \alpha_2, b \in \mathbb{R}\}. \tag{2.11.83}$$

From (2.11.82) it results that

$$\varepsilon^{(j)}(u^{(j)}(\zeta^{(j)})) : Z_j \to L^2(\Omega_j; \mathbb{R}^4) \tag{2.11.84}$$

is uniquely determined and is a completely continuous quadratic function of $\zeta^{(j)}, j = 1, 2, \ldots, r$, since $\varepsilon^{(j)}(u^{(j)}(\zeta^{(j)}))$ is a linear continuous function of $P(\zeta^{(j)})$. We also introduce the completely continuous, quadratic functions $G_j : Z_j \to L^2(\Omega_j; \mathbb{R}^4)$ which are defined by

$$\zeta^{(j)} \to G_j(\zeta^{(j)}) = \varepsilon^{(j)}(u^{(j)}(\zeta^{(j)})) + \frac{1}{2}P(\zeta^{(j)}) \tag{2.11.85}$$

and satisfy the equations (cf. (2.11.72))

$$R^{(j)}(G_j(\zeta^{(j)}), \varepsilon^{(j)}(u^{(j)}(\zeta^{(j)}))) = 0. \tag{2.11.86}$$

We now define the operators $A_j : Z_j \to Z_j^*$ and $C_j : Z_j \to Z_j^*$ such that

$$a_j(\zeta^{(j)}, z^{(j)}) = \langle A_j \zeta^{(j)}, z^{(j)} \rangle_j \tag{2.11.87}$$

and

$$h_j R^{(j)}(G_j(\zeta^{(j)}), P(\zeta^{(j)}, z^{(j)})) = \langle C_j(\zeta^{(j)}), z^{(j)} \rangle_j. \tag{2.11.88}$$

Let

$$T_j = A_j + C_j. \tag{2.11.89}$$

The A_j's are continuous symmetric linear operators. Note also that

$$Ker\ A_j = \{q \in Z_j : a_j(q,q) = 0\}.$$

The C_j's are cubic and completely continuous operators; and $\langle \cdot, \cdot \rangle_j$ denotes the duality pairing between Z_j and Z_j^*. Let us now recall the most important properties of these operators (see [83], [361]-[362], [46], [47], [336]). By means of (2.11.86) we obtain that

$$\langle C_j(\zeta^{(j)}), \zeta^{(j)} \rangle_j = h_j R^{(j)}(G_j(\zeta^{(j)}), 2G_j(\zeta^{(j)})) \geq 0,$$

$$\forall\, \zeta^{(j)} \in Z_j, \ j = 1, 2, \ldots, r. \tag{2.11.90}$$

Moreover if $a_j(\zeta^{(j)}, \zeta^{(j)})$ are coercive on Z_j, i.e. if the boundary conditions in bending prevent rigid-plate deflections (e.g. a plate is partially clamped, or has a curved boundary partly fixed with free rotation) then

$$\langle (A_j + C_j)(\zeta^{(j)}), \zeta^{(j)} \rangle_j = \langle T_j(\zeta^{(j)}), \zeta^{(j)} \rangle_j \geq c\|\zeta^{(j)}\|_{2,2;\Omega_j}^2,$$

$$\forall\, \zeta^{(j)} \in Z_j, \ j = 1, 2, \ldots, r, (c\ \text{const} > 0). \tag{2.11.91}$$

It can be proved that $C_j(\zeta^{(j)})$ is the Fréchet derivative of

$$\begin{aligned} F_j(\zeta) &= \frac{1}{4} h_j R^{(j)}(G_j(\zeta^{(j)}), P(\zeta^{(j)})) \\ &= \frac{1}{4}\langle C_j(\zeta^{(j)}), \zeta^{(j)} \rangle_j. \end{aligned}$$

Moreover, we have the estimate

$$\| C_j(\zeta_1^{(j)}) - C_j(\zeta_2^{(j)}) \|_{[H^2(\Omega_j)]^*} \leq c'(\| \zeta_1^{(j)} \|_{2,2;\Omega_j}^2$$

$$+ \| \zeta_2^{(j)} \|_{2,2;\Omega_j}^2)\, \| \zeta_1^{(j)} - \zeta_2^{(j)} \|_{1,4;\Omega_j}\ (c'\ \text{const} > 0),\ \forall\, \zeta^{(j)} \in Z_j \backslash \{0\}.$$

Note that we have used the notation

$$\| \cdot \|_{p,q;\Omega_j} := \| \cdot \|_{W^{p,q}(\Omega_j)} .$$

If the plate Ω_j is partially clamped then

$$\langle C_j(\zeta^{(j)}), \zeta^{(j)} \rangle_j > 0, \ \forall\, \zeta^{(j)} \in Z_j \backslash \{0\}.$$

If $Z_j = H^2(\Omega_j)$ then A_j is semicoercive and $Ker\ A_j$ is the space of all polynomials of degree ≤ 1. We may also show that

$$\langle C_j(\zeta^{(j)}), q^{(j)} \rangle_j = 0, \ \forall\, \zeta^{(j)} \in H^2(\Omega_j),\ q^{(j)} \in Ker\ A_j.$$

Moreover, the functional F_j is $Ker\ A_j$-invariant. Indeed, from (2.11.90), we have

$$\langle C_j(\zeta^{(j)}), \zeta^{(j)} \rangle_j = h_j R^{(j)}(G_j(\zeta^{(j)}), 2G_j(\zeta^{(j)}))$$

so that

$$\langle C_j(\zeta^{(j)} + q^{(j)}), \zeta^{(j)} + q^{(j)} \rangle_j = h_j R^{(j)}(G_j(\zeta^{(j)} + q^{(j)}), 2G_j(\zeta^{(j)} + q^{(j)}))$$

We may prove that $G_j(\zeta^{(j)} + q^{(j)}) = G_j(\zeta^{(j)})$ (see [336] for details) and thus

$$\begin{aligned}
\langle C_j(\zeta^{(j)} + q^{(j)}), \zeta^{(j)} + q^{(j)} \rangle_j &= h_j R^{(j)}(G_j(\zeta^{(j)}), 2G_j(\zeta^{(j)})) \\
&= \langle C_j(\zeta^{(j)}), \zeta^{(j)} \rangle_j.
\end{aligned}$$

Thus (2.11.81)-(2.11.82) yield the following variational problem: Find $\zeta^{(j)} \in Z_j$, $j = 1, 2, \ldots, r$, so as to satisfy the variational-hemivariational inequality

$$\sum_{j=1}^{r} \langle T_j(\zeta^{(j)}), z^{(j)} - \zeta^{(j)} \rangle_j + \sum_{m=1}^{m'} \int_{\Omega'_m} j_m^0([\zeta]^{(m)}; [z]^{(m)} - [\zeta]^{(m)}) dx$$

$$+ \sum_{j=1}^{r} \{\Phi_j(z^{(j)}) - \Phi_j(\zeta^{(j)})\} \geq \sum_{j=1}^{r} \int_{\Omega_j} \overline{\overline{f}}^{(j)} (z^{(j)} - \zeta^{(j)}) dx,$$

$$\forall\ z^{(j)}\ \in\ Z_j. \tag{2.11.92}$$

The last variational-hemivariational inequality describes a large class of nonlinear B.V.Ps for the laminated von Kármán plates.

2.11.5   UNILATERAL BUCKLING OF A PLATE

We consider an elastic von Kármán plate Ω with constant thickness h, which is connected with an adhesive material of negligible thickness to the rigid plane body Ω' on Ω_0, $\Omega_0 \subset \Omega \cap \Omega'$ (Fig. 2.11.3). We suppose that $\bar{\Omega}_0 \cap \Gamma = \emptyset$ where Γ is the boundary of Ω which is assumed to be $C^{1,1}$. The points of the plate are referred to a fixed right-handed Cartesian coordinate system $\{O; x_1; x_2; x_3\}$ and the middle plane of the underformed plate coincides with the $\{O; x_1; x_2\}$-plane. The sets Ω, Ω_0 are nonempty, open, bounded and connected subsets of $\mathbb{R}^2$.

We suppose that the plate has a buckling because of the boundary loading in the plane of the plate. The theory of von Kármán plates leads

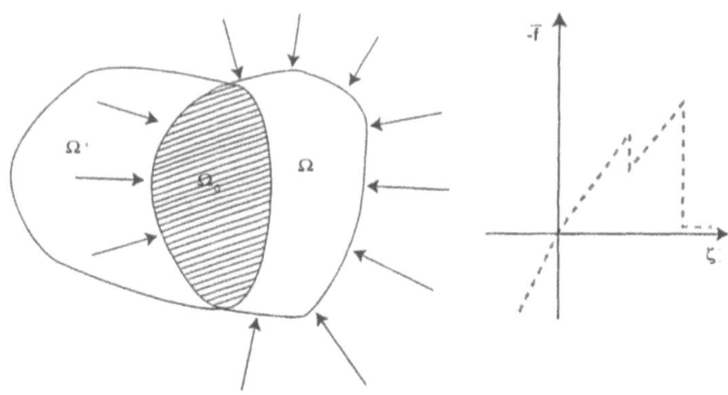

Figure 2.11.3.   On the buckling of a von Kármán plate in adhesive contact.

to the differential equations (2.11.51)-(2.11.53) for $j = 1$. The index "1" from Section 2.11.4 is here useless and therefore removed (e.g. $f^{(1)} = f$, $\Omega_1 = \Omega$). The boundary loading which is responsible for the buckling has the form

$$\sigma_{\alpha\beta}n_\beta = \lambda g_\alpha, \qquad \alpha = 1, 2 \text{ on } \Gamma \qquad (2.11.93)$$

where $g = g(x), x \in \Gamma$ is a self-equilibrating compressive load distribution on the plate boundary Γ, that are

$$\int_\Gamma g_\alpha ds = 0 \text{ and } \int_\Gamma g_1 x_2 - g_2 x_1 ds = 0,$$

and λ is a real number measuring the magnitude of the load distribution. When λ exceeds a certain value called the critical value, the plate leaves its position of equilibrium and buckling occurs. In this case, the small displacement theory is not adequate for precise description of buckling phenomena, and thus the von Kármán theory must be applied.

We assume now that $f = \bar{\bar{f}} + \bar{f}$ where generally $\bar{\bar{f}} \in L^2(\Omega)$ and $\bar{f}$ is a given function of ζ and describes the action of the given support on the plate. Let $\bar{f}$ be a function of ζ in Ω_0 and

$$-\bar{f} \in \tilde{b}(\zeta) \text{ on } \Omega_0 \subset \Omega, \ \bar{f} = 0 \text{ on } \Omega\backslash\Omega_0 \qquad (2.11.94)$$

where $\tilde{b}$ denotes a nonmonotone multivalued function on $\mathbb{R}$ which results from $b \in L^\infty_{\text{loc}}(\mathbb{R})$ by means of the formula

$$\tilde{b}(\zeta) = [\underline{b}(\zeta), \overline{b}(\zeta)]. \qquad (2.11.95)$$

We denote by j the nonconvex superpotential corresponding to $\tilde{b}$ through $\tilde{b} \equiv \partial j$. The expressions for the virtual work for the bending and the streching respectively of a free plate are (see Section 2.11.4)

$$a(\zeta, z - \zeta) + \int_\Omega h\sigma_{\alpha\beta}\zeta_{,\alpha}(z - \zeta)_{,\beta}dx = \int_\Gamma h\sigma_{\alpha\beta}\zeta_{,\alpha}n_\beta(z - \zeta)ds$$

$$+ \int_\Gamma K_n(\zeta)(z - \zeta)ds - \int_\Gamma M_n(\zeta)\frac{\partial(z - \zeta)}{\partial n}ds$$

$$+ \int_\Omega \bar{\bar{f}}\,(z - \zeta)dx + \int_{\Omega_0} \bar{f}(z - \zeta)dx \qquad (2.11.96)$$

and

$$\int_\Omega \sigma_{\alpha\beta}\varepsilon_{\alpha\beta}(v - u)dx = \int_\Gamma \sigma_{\alpha\beta}n_\beta(v_\alpha - u_\alpha)ds \qquad (2.11.97)$$

where $\alpha, \beta = 1, 2$, $n = (n_1, n_2)$ denotes the outward normal unit vector on Γ, $M_n = M_n(\zeta)$, $K_n = K_n(\zeta)$ are supposed to be known functions of ζ expressing the bending moment and the total shearing force at the plate boundary Γ. Now, we define the space

$$Z = \{z \in H^2(\Omega) : \gamma(z) = 0 \text{ a.e. on } \Gamma$$

$$\text{and } \gamma_1(z) = 0 \text{ a.e. on } \Gamma\} \qquad (2.11.98)$$

where $H^2(\Omega)$ is the Sobolev space for the plate deflections. Then, setting in (2.11.96)) $\bar{\bar{f}} = 0$, and assuming that in (2.11.93) $g \in L^2(\Gamma)$, and (2.11.98) yield the inequality

$$a(\zeta, z - \zeta) + \int_\Omega h\sigma_{\alpha\beta}\zeta_{,\alpha}(z - \zeta)_{,\beta}dx + \int_{\Omega_0} j^0(\zeta; z - \zeta)dx$$

$$\geq \lambda \int_\Gamma hg_\alpha\zeta_{,\alpha}(z - \zeta)ds \ \forall\, z \in Z. \qquad (2.11.99)$$

Here, following the method which is presented in Section 2.11.4 and provided that the symbols and the notions are the same as there, we obtain the following problem after eliminating the in-plane displacements: Find $\lambda \in \mathbb{R}$ and $\zeta \in Z$ such as to satisfy the hemivariational inequality eigenvalue problem

$$\langle A(\zeta) + C(\zeta), z - \zeta\rangle + \int_{\Omega_0} j^0(\zeta; z - \zeta)dx$$

$$\geq \lambda\langle B\zeta, z - \zeta\rangle, \ \forall\, z \in Z \qquad (2.11.100)$$

where $B : Z \to Z^*$ is a symmetric, linear and compact operator (see e.g. [336]).

Suppose now that the plate is resting without friction on a plane rigid support, clamped on $\Gamma_1 \subset \Gamma$ ($\zeta = 0, \frac{\partial \zeta}{\partial n} = 0$) and simply supported on $\Gamma_2 = \Gamma \backslash \overline{\Gamma}_1 (\zeta = 0, M_n = 0)$. Then we may also set $f = \overline{\overline{f}} + \overline{f}$ where $\overline{\overline{f}} \in L^2(\Omega)$ is given and $\overline{f}$ describes the action of the support, that is here

$$-\overline{f} \in \partial \Psi_{\mathbb{R}^+}(\zeta) \text{ on } \Omega.$$

Here we set

$$Z = \{z \in H^2(\Omega) : \gamma(z) = 0 \text{ a.e. on } \Gamma \text{ and } \gamma_{1,\Gamma_1}(z) = 0 \text{ a.e. on } \Gamma_1\}$$

and

$$K = \{z \in Z : z \geq 0 \text{ a.e. on } \Omega\}.$$

Here the B.V.P. leads to the variational inequality eigenvalue problem: Find $\lambda \in \mathbb{R}$ and $\zeta \in C$ such that

$$\langle A\zeta + C(\zeta) - \lambda B\zeta, v - \zeta \rangle \geq 0, \ \forall \, v \in K. \tag{2.11.101}$$

If the plate is clamped on Γ_1 and subjected on Γ_2 to the conditions

$$\zeta = 0 \text{ and } M_n(\zeta) \in \partial\varphi(\frac{\partial \zeta}{\partial n}),$$

where φ is some proper convex and l.s.c. function on $\mathbb{R}$, then the B.V.P. leads to the variational inequality eigenvalue problem: Find $\lambda \in \mathbb{R}, \zeta \in Z$ such that

$$\langle A\zeta + C(\zeta) - \lambda B\zeta, v - \zeta \rangle + \Psi(v) - \Psi(\zeta) \geq 0, \ \forall \, v \in Z, \tag{2.11.102}$$

where

$$\Psi(\zeta) = \begin{cases} \int_{\Gamma_2} \varphi(\gamma_{1,\Gamma_2}(\zeta))ds & \text{if} \quad \varphi(\gamma_{1,\Gamma_2}(z)) \in L^1(\Gamma_2) \\ +\infty & \text{if} \quad \text{not.} \end{cases} \tag{2.11.103}$$

2.11.6   UNILATERAL BENDING OF A BEAM

Let us consider a beam $[0, 1]$ simply supported at its ends and compressed by a force P. Moreover, we assume that the beam is subject to a transversal density force $IEf(x)$ ($x \in (0, 1)$) where E is Young's modulus of elasticity and I the bending moment. In the framework of a large scale theory, the bending $x \to u(x)$ of the beam is described by the B.V.P. (see e.g. [99], [380]).

$$a^2 u^{(4)}(x) + u''(x) = a^2 f(x) \text{ on } (0,1) \tag{2.11.104}$$

$$u(0) = u(1) = 0 \tag{2.11.105}$$

$$u''(0) = u''(1) = 0, \tag{2.11.106}$$

with $a^2 = \frac{IE}{P}$. Multiplying (2.11.104) by v, integrating over $(0,1)$, we obtain

$$a^2 \int_0^1 u^{(4)} v \, dx + \int_0^1 u'' v \, dx = a^2 \int_0^1 f \, v \, dx.$$

One has

$$\int_0^1 u'' v \, dx = - \int_0^1 u' v' \, dx + [u' v]_0^1$$

and

$$\begin{aligned}
\int_0^1 u^{(4)} v \, dx &= - \int_0^1 u^{(3)} v' \, dx + [u^{(3)} v]_0^1 \\
&= + \int_0^1 u'' v'' \, dx - [u'' v']_0^1 + [u^{(3)} v]_0^1 \\
&= + \int_0^1 u'' v'' \, dx + [u^{(3)} v]_0^1.
\end{aligned}$$

To get the last inequality we have used the boundary conditions (2.11.106). Setting now

$$V = H^2(\Omega) \cap H_0^1(\Omega),$$

we obtain the weak formulation for $u \in V$:

$$\int_0^1 u'' v'' \, dx - \frac{1}{a^2} \int_0^1 u' v' \, dx = \int_0^1 f \, v \, dx, \ \forall \, v \in V.$$

We assume now that $f = \bar{\bar{f}} + \bar{f}$ where $\bar{\bar{f}} \in L^2(0,1)$ and $\bar{f}$ is a given function of u which can be written by means of a locally Lipschitz superpotential j, that is

$$-\bar{f} \in \partial j(u), \ \forall \, x \in (\theta_1, \theta_2), \tag{2.11.107a}$$

$$-\bar{f} = 0, \ \forall \, x \in (0, \theta_1) \cup (\theta_2, 1), \tag{2.11.107b}$$

where $0 < \theta_1 < \theta_2 < 1$. For example, if the part (θ_1, θ_2) is connected with an adhesive material to a rigid body support then $\bar{f}$ describes the action of the given support on the beam.

Setting $\lambda = \frac{1}{a^2}$, we obtain the eigenvalue problem for the hemivariational inequality: Find $(\lambda, u) \in \mathbb{R}_+ \times V$ such that

$$a(u, v) - \lambda \langle Lu, v \rangle + \int_{\theta_1}^{\theta_2} j^0(u; v) dx \geq \int_0^1 f \, v dx, \ \forall \, v \, \in \, V, \quad (2.11.108)$$

where

$$a(u, v) = \int_0^1 u'' v'' dx, \ \forall \, u, \, v \, \in \, V, \quad (2.11.109)$$

is a bilinear, continuous, symmetric and coercive form on V and

$$\langle Lu, v \rangle = \int_0^1 u' v' dx \quad (2.11.110)$$

is a linear, symmetric and compact operator from V onto V^*.

2.11.7   LOADING AND UNLOADING PROBLEMS

In solid mechanics and especially in structural analysis hysteresis loops appear in loading-unloading processes [72]. In a simple bar submitted to forces acting at the two ends along its axis the loading-unloading process may be easily understood and gives rise to one-dimensional stress-strain $\{\sigma, \varepsilon\}$ diagrams like the one depicted in Fig. 2.11.4. Here we accept that the stress-strain law is generally nonmonotone and multivalued, i.e. it contains filled-in parts parallel to the σ-axis.

A loading $t \rightarrow p(t)$ is given and one has to determine the stress, strain and displacement fields $\sigma = \{\sigma_{ij}\}$, $\varepsilon = \{\varepsilon_{ij}\}$ and $u = \{u_i\}$ at every point of the body. We know the stress-strain law in loading (path OAC) and we also know that in the case of unloading (resp. loading) at A the path AΓBE (resp. AC) will be realized. Analogously, if at a given moment an element has a stress and strain state corresponding to B (resp. to B'), then in the case of loading the path BΔA' (resp. B'Δ'C') will be realized, whereas in the case of unloading the path BE (resp. B'E'). Note that we do not know a priori the points A, A' etc or B, B' etc, where the branching of the solution takes place and we have to determine these points for varying external loading. It is now well-known that a nonmonotone possibly multivalued, (i.e. with vertical parts, but without branching) stress-strain or reaction-displacement law gives rise to a hemivariational inequality with respect to the displacements.

In the case of smooth $\sigma - \varepsilon$ laws the hemivariational inequality reduces to a variational equality which is the variational form of a partial

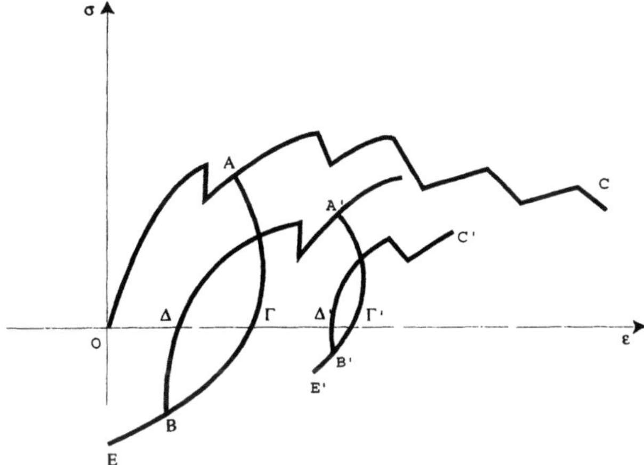

Figure 2.11.4.   Loading-unloading hysteresis loops.

differential equation. Then the branching at A leads (cf. e.g. [72], [413]) to an eigenvalue problem for a variational equality. In the case of a nonsmooth $\sigma - \varepsilon$ law with multivalued parts the branching at A leads analogously to an eigenvalue problem for a hemivariational inequality. The potential energy Π is a functional of the "generalized" displacement u. It also depends on a scalar variable λ which determines the magnitude of the external loading on the system. The variable u must satisfy certain boundary or support conditions of classical (non-unilateral) type, e.g. $u = 0$, on a boundary part Γ_1. Conditions resulting from nonconvex superpotential conditions between the stress vector S and the displacements u or relative displacements $[u]$ are not taken into account on the remaining part of the boundary $\Gamma\backslash\overline{\Gamma}_1$ for the moment, i.e. we assume that the corresponding stress vector (interface forces, or reaction forces) S is given. Since we have a bilateral structure the principle of virtual work reads (δ is the symbol of classical variation, $\langle .,. \rangle$ is a duality pairing – physically a work expression)

$$\delta\Pi(u,\lambda) = \langle \Pi'_u(u,\lambda), \delta u \rangle = 0. \qquad (2.11.111)$$

for every admissible variation δu. Here $\Pi'_u(u,\lambda)$ denotes the Gâteaux differential of Π with respect to u. We assume now that the structure is linear elastic and that a nonmonotone reaction-displacement relation $(-S, u)$ holds on the boundary part $\Gamma_2 = \Gamma\backslash\overline{\Gamma}_1$, which has a graph analogous to the one-dimensional (σ, ε) graph of Fig. 2.11.4. Moreover the branches OAC, EBΓAC, EBΔA$'$ etc. are described by relations of the type

$$-S \in \partial j_\alpha(u), \ \alpha = 1, 2, \ldots, \tag{2.11.112}$$

where j_α is a suitable locally Lipschitz function.

Suppose now that there exists a fundamental solution u_0 varying with λ which corresponds to the OAC curve of Fig. 2.11.4, and let another solution exists, say the one corresponding to the curve AΓB in Fig. 2.11.4 that intersects the fundamental solution at $\lambda = \lambda_c$ and coincides with it for $\lambda < \lambda_c$. Let the other solution be $u = u_0(\lambda) + v(\lambda)$. Then

$$\lim_{\lambda \to \lambda_c} v(\lambda) = 0. \tag{2.11.113}$$

Moreover we have that

$$\langle \Pi'_u(u_0(\lambda); \lambda), \delta u \rangle = 0 \text{ and } \langle \Pi'_u(u_0(\lambda) + v(\lambda); \lambda), \delta u \rangle = 0. \tag{2.11.114}$$

We denote by

$$u_1 = \lim_{\lambda \to \lambda_c} \left(v(\lambda) \| v(\lambda) \|^{-1} \right)$$

the "hysteresis bifurcation mode". We write now the total potential energy functional Π as

$$\Pi(u) = \frac{1}{2} a(u, u) - \langle S(u), u \rangle_{\Gamma_2} - \lambda \langle 1, \theta(u) \rangle_{\Omega_0} \tag{2.11.115}$$

where $a(\cdot, \cdot)$ denotes the quadratic elastic strain energy, S is assumed as given and θ denotes the load-shortening function [72] assumed to be of the type $\theta(u) = \frac{1}{2}u^2 + cu + d$, a fact justified by the physical nonlinearity. Here $\langle \cdot, \cdot \rangle_{\Gamma_2}$ and $\langle \cdot, \cdot \rangle_{\Omega_0}$ denote expressions of work; the first is extended over Γ_2 and the second over the body part Ω_0 , on which the external loading is applied. We have

$$a(u_0(\lambda), \delta u) - \langle S(u_0(\lambda)), \delta u \rangle_{\Gamma_2} - \langle \lambda \theta'(u_0(\lambda)), \delta u \rangle_{\Omega_0} = 0$$

and

$$a(u_0(\lambda) + v(\lambda), \delta u) - \langle S(u_0(\lambda) + v(\lambda)), \delta u \rangle_{\Gamma_2} - \langle \lambda \theta'(u_0(\lambda) + v(\lambda)), \delta u \rangle_{\Omega_0} = 0.$$

Expanding θ' in Taylor series around $u_0(\lambda)$, we get

$$\theta'(u_0(\lambda) + h) = \theta'(u_0(\lambda)) + \theta''(u_0(\lambda))h + \frac{1}{2}\theta^{(3)}(u_0(\lambda))h^2 + \ldots,$$

that is here with our expression for θ:

$$\theta'(u_0(\lambda) + h) = \theta'(u_0(\lambda)) + h.$$

Thus

$$a(v(\lambda), \delta u) - (\langle S(u_0(\lambda) + v(\lambda)), \delta u \rangle_{\Gamma_2} - \langle S(u_0(\lambda)), \delta u \rangle_{\Gamma_2}) - \langle \lambda v(\lambda), \delta u \rangle_{\Omega_0} = 0.$$

Dividing by $\| v(\lambda) \|$ and letting $\lambda \to \lambda_c$, we get the expression

$$0 \leq a(u_1(\lambda_c), \delta u) - \langle \lambda u_1(\lambda_c), \delta u \rangle_{\Omega_0}$$

$$+ \limsup_{\lambda \to \lambda_c} [- \frac{(\langle S(u_0(\lambda) + v(\lambda)), \delta u \rangle_{\Gamma_2} - \langle S(u_0(\lambda)), \delta u \rangle_{\Gamma_2})}{\| v(\lambda) \|}].$$

We suppose that

$$\liminf_{\lambda \to \lambda_c} [\frac{(\langle S(u_0(\lambda) + v(\lambda)), \delta u \rangle_{\Gamma_2} - \langle S(u_0(\lambda)), \delta u \rangle_{\Gamma_2})}{\| v(\lambda) \|}] \geq \langle S(u_1(\lambda)), \delta u \rangle.$$

This last condition holds if for example, we assume that S is a linear mapping of the displacements in a small neighborhood around the bifurcation point λ_c (this assumption is justified by the linear elasticity model holding before the introduction of the nonmonotone possibly multivalued conditions). Finally, we obtain the expression

$$a(u_1(\lambda_c), \delta u) - \langle S(u_1(\lambda_c)), \delta u \rangle_{\Gamma_2} - \langle \lambda_c u_1(\lambda_c), \delta u \rangle_{\Omega_0} = 0. \qquad (2.11.116)$$

Then a boundary condition of the form

$$\int_{\Gamma_2} j^0(u_1(\lambda_c); w) ds \geq -\langle S(u_1(\lambda_c)), w \rangle_{\Gamma_2} \ \forall \, w \in V_{\text{ad}}, \qquad (2.11.117)$$

where V_{ad} denotes the kinematically admissible set yields an eigenvalue problem for a hemivariational inequality of the type: Find λ_c and $u_1(\lambda_c) \in V_{\text{ad}}$ such as to satisfy

$$a(u_1(\lambda_c), w) + \int_{\Gamma_2} j^0(u_1(\lambda_c); w) ds$$

$$\geq \int_{\Omega_0} \lambda_c u_1(\lambda_c) w \, dx, \ \forall \, w \in V_{\text{ad}}. \qquad (2.11.118)$$

If Γ_1 is fixed then a more precise formulation of (2.11.118) is: Find $\lambda_c \in \mathbb{R}$ and $u_1(\lambda_c) \in V_{ad} := \{ u \in H^1(\Omega) : \gamma_{|\Gamma_1} u_1(\lambda_c) = 0 \text{ a.e. on } \Gamma_1 \}$ such that

$$a(u_1(\lambda_c), w) + \int_{\Gamma_2} j^0(\gamma_{|\Gamma_2} u_1(\lambda_c); w) ds$$

$$\geq \int_{\Omega_0} \lambda_c u_1(\lambda_c) w \, dx, \ \forall \, w \in V_{\text{ad}}.$$

2.11.8   OBSTACLE PROBLEM FOR A MEMBRANE

Let $\Omega \subset \mathbb{R}^2$ be a nonempty open bounded and connected subset with boundary $\Gamma \in C^{1,1}$. The stationary behavior of a membrane Ω subject to a body force f is described by the equation

$$-\Delta u = f \text{ in } \Omega \tag{2.11.119}$$

and boundary conditions. Let us now assume that $f = \overline{f} + \overline{\overline{f}}$ where $\overline{\overline{f}}$ is given and $\overline{f}$ is a known function of u that can be put in the general superpotential form

$$-\overline{f} \in \partial j(u) \text{ in } \Omega. \tag{2.11.120}$$

If j is locally Lipschitz then a weak formulation of (2.11.119) (together with suitable assumptions on the data) leads to the problem: Find $u \in U_{ad}$ such that

$$\int_\Omega \nabla u \nabla v dx + \int_\Omega j^0(u; v) dx$$

$$\geq \int_\Omega f v dx + \int_\Gamma \frac{\partial u}{\partial n} v ds, \ \forall v \in U_{ad}, \tag{2.11.121}$$

where U_{ad} denotes the set of kinematically admissible displacements. The expression

$$\int_\Gamma \frac{\partial u}{\partial n} v ds$$

can be estimated as soon as (possibly unilateral) boundary conditions are specified.

If the membrane is constrained to stay on or above an obstacle whose geometry is given by a function Ψ, then we consider the relation

$$-\overline{f} \in N_{C(x)}(u(x)) \text{ a.e. } x \in \Omega \tag{2.11.122}$$

where $C(x)$ is defined by

$$C(x) = \{z \in \mathbb{R}^2 : z \geq \Psi(x)\}.$$

Using (2.11.119) together with (2.11.122) we obtain the variational inequality problem: Find $u \in C$ such that

$$\int_\Omega \nabla u \nabla (v - u) dx \geq \int_\Omega f(v - u) dx$$

$$+ \int_\Gamma \frac{\partial u}{\partial n}(v - u) ds, \ \forall v \in C, \tag{2.11.123}$$

where

$$C = \{u \in U_{ad} : u(x) \geq \Psi(x), \text{ a.e. } x \in \Omega\}. \tag{2.11.124}$$

Note that $C = \{u \in U_{ad} : u(x) \in C(x), \text{ a.e. } x \in \Omega\}$. In the case of a dynamic problem, the evolution of the displacement field is described by the equation

$$\rho\frac{\partial^2 u}{\partial t^2} - \Delta u = \overline{f} + \overline{\overline{f}} \text{ in } \Omega, \tag{2.11.125}$$

where ρ is the density of the membrane. Unilateral condition of the form

$$-\overline{f} \in \partial j(\frac{\partial u}{\partial t}) \text{ in } \Omega \tag{2.11.126}$$

with j locally Lipschitz can be used to describe friction effects on the membrane. The condition (2.11.122) can here again be used to describe the obstacle problem. The weak formulation of these two B.V.P.s lead respectively to the following models:

$$u \in U_{ad}, \text{ a.e. } t \in (0,T), \int_\Omega \rho u'' v + \nabla u \nabla v \, dx + \int_\Omega j^0(u';v)dx$$

$$\geq \int_\Omega \overline{\overline{f}} \, v dx, \forall \, v \in U_{ad}, \text{ a.e. } t \in (0,T) \tag{2.11.127}$$

and

$$u \in C, \text{ a.e. } t \in (0,T), \int_\Omega \rho u''(v-u) + \nabla u \nabla (v-u)dx$$

$$\geq \int_\Omega f(v-u)dx, \forall \, v \in C, \text{ a.e. } t \in (0,T) \tag{2.11.128}$$

where $T > 0$ is fixed. In order to complete the formulation of these dynamic problems we have also to impose initial conditions. For example, we may suppose that the displacement and velocity field at time $t = 0$ are given, i.e.

$$u(0) = u_0, u'(0) = u_1 \text{ on } \Omega. \tag{2.11.129}$$

For some problems it can be interesting to consider periodic conditions, i.e.

$$u(0) = u(T), u'(0) = u'(T) \text{ on } \Omega, \tag{2.11.130}$$

for some $T > 0$.

Various kinds of inequality problems can be formulated as the result of (2.11.119) or (2.11.125), in-plane unilateral conditions like (2.11.120), (2.11.122) or (2.11.126), possibly unilateral boundary conditions and time-initial conditions or time-periodic conditions. In addition the problem is relatively simple to formulate in a suitable framework and for these reasons the unilateral membrane problem (and particularly the obstacle problem for a membrane) constitutes a basic model on which mathematical results are usually tested.

2.11.9   SEMIPERMEABLE MEDIA PROBLEMS

Consider a medium Ω of $\mathbb{R}^n (n = 2$ or 3 in normal applications) with boundary Γ. One assumes that the medium Ω is occupied by a fluid for which the pressure is $u(x,t)$. The normalized diffusion equation is

$$\frac{\partial u}{\partial t} - \Delta u = f \text{ in } \Omega. \tag{2.11.131}$$

Here f denotes a body force field acting on Ω. Suppose now that we apply on the boundary Γ a fluid pressure $h(x)$ and assume that the boundary Γ allows fluid entering Γ to pass freely but prevents fluid to leave. Then the unilateral boundary conditions are

$$u - h \geq 0, \text{ on } \Gamma \tag{2.11.132}$$

$$\frac{\partial u}{\partial n} \geq 0, \text{ on } \Gamma \tag{2.11.133}$$

and

$$(u - h)\frac{\partial u}{\partial n} = 0, \text{ on } \Gamma. \tag{2.11.134}$$

The condition (2.11.134) means that if $u > h$, i.e. if the fluid has tendency to leave Γ then the flux is zero. If $u = h$ then the fluid tends to enter Ω and the wall allowing this, the flux $\frac{\partial u}{\partial n}$ satisfies (2.11.133). Proceeding now as usually, we may write a weak formulation of (2.11.131). Taking into account (2.11.132)-(2.11.134), we obtain the variational inequality: Find $u \geq h$ on Γ, a.e. $t \in (0,T)$ such that

$$\int_\Omega \frac{\partial u}{\partial t}(v - u)dx + \int_\Omega \nabla u \nabla(v - u)dx$$

$$\geq \int_\Omega f(v - u)dx, \forall\, v \geq h \text{ on } \Gamma, \text{ a.e. } t \in (0,T), \tag{2.11.135}$$

where $T > 0$ is fixed.

Suppose now that $f = \overline{f} + \overline{\overline{f}}$ where $\overline{\overline{f}}$ is given and $\overline{f}$ satisfies the superpotential condition

$$-\overline{f} \in \partial j(u) \text{ in } \Omega, \tag{2.11.136}$$

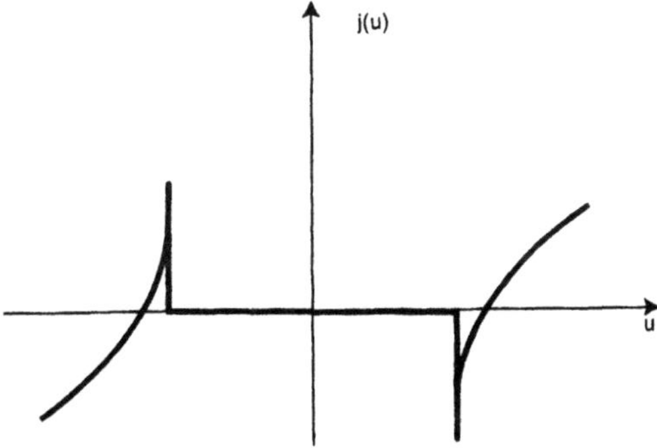

Figure 2.11.5.   Control signal.

where j is a locally Lipschitz function. The unilateral condition (2.11.136) can be used to model a control problem in which the pressure is regulated by some hydraulic control device. The control signal may indeed include some jumps (see Fig. 2.11.5).

In this case the B.V.P. leads to the hemivariational inequality: Find $u \in U_{ad}$, a.e. $t \in (0, T)$ such that

$$\int_\Omega \frac{\partial u}{\partial t} v dx + \int_\Omega j^0(u; v) dx \geq \int_\Omega \overline{\overline{f}} \, v dx$$

$$+ \int_\Gamma \frac{\partial u}{\partial n} v dx, \forall \, v \in U_{ad}, \quad \text{a.e. } t \in (0, T), \tag{2.11.137}$$

where U_{ad} denotes the space of admissible pressure.

Here also to complete the formulation of these models it is necessary to consider time-initial condition like

$$u(0) = u_0 \text{ on } \Omega,$$

where u_0 is given and defined on Ω. For some problems it can be interesting to consider time-periodic conditions, i.e.

$$u(0) = u(T) \text{ on } \Omega.$$

2.11.10   NONSMOOTH OSCILLATOR

Various problems in classical mechanics can be described by the following oscillator model: Find $t \to u(t) \in \mathbb{R}^N$ such that

$$A(t)\ddot{u}(t) + B(u(t)) = 0, \tag{2.11.138}$$

where $A \in L^\infty(T_1, T_2; \mathbb{R}^{N \times N})$ is a time-dependent matrix and $B : \mathbb{R}^N \to \mathbb{R}^N$ a possibly nonlinear mapping. In most systems we may write

$$B(u(t)) \in -\partial J(u(t)) + F(t), \tag{2.11.139}$$

where $F \in L^2(T_1, T_2; \mathbb{R}^N)$ is given and J is a suitable locally Lipschitz superpotential functional. From (2.11.138) and (2.11.139) we deduce the inequality: Find $u \in U_{ad}$

$$-\int_{T_1}^{T_2} A\ddot{u}(t).v(t)dt + \int_{T_1}^{T_2} J^0(u; v)dt$$

$$\geq \int_{T_1}^{T_2} F.vdt, \forall \, v \, \in \, U_{ad}, \tag{2.11.140}$$

where U_{ad} denotes the space of admissible oscillations.

The formula (2.11.140) permits to consider the usual force Q which can be obtained from a smooth potential $J \in C^1(\mathbb{R}^N; \mathbb{R})$, that is

$$Q = -\nabla J(u).$$

In addition, the model (2.11.140) can be used to describe the action of generalized forces which can be obtained from a nonsmooth potential (superpotential). In this case the relation Force-Oscillation may include jumps. For example, the oscillations can be regulated by a nonlinear signal which may include jumps. Together with (2.11.140), we usually consider periodic conditions $(T_1 = 0, T_2 = T > 0)$

$$u(0) = u(T), \, 0 < T < +\infty,$$

or the vanishing condition $(T_1 = -\infty, T_2 = +\infty)$

$$\lim_{t \to -\infty} u(t) = 0, \, \lim_{t \to +\infty} u(t) = 0.$$

The inequality (2.11.140) constitutes here a variational formulation of a system of differential equations involving discontinuous nonlinearities.

2.11.11   FRICTIONLESS CONTACT PROBLEMS IN ENGINEERING STRUCTURES

Usual engineering structures consist in parts which are treated as perfectly rigid bodies. These parts may come into contact but none of them can overlap. That means that the coordinates $u_1, \cdots, u_n$ used to locate every positions of the system need to satisfy some inequality constraints

$$f_\alpha(u_1, \cdots, u_n) \geq 0, \alpha = 1, \cdots, \nu, \qquad (2.11.141)$$

related to these impenetrability properties. Here we suppose that the functions $f_\alpha(\alpha = 1, \cdots, \nu)$ are continuous and convex. One forms the vector $u = (u_1, \cdots, u_n)^T \in \mathbb{R}^n$ and we define the feasible set

$$C = \{u \in \mathbb{R}^n : f_\alpha(u) \geq 0, \alpha = 1, \cdots, \nu\}. \qquad (2.11.142)$$

The formulation of these unilateral constraints has to encompass the associated forces of constraints $r_1, \cdots, r_n$. One forms the vector $r = (r_1, \cdots, r_n)^T \in \mathbb{R}^n$ and in the case of contact without friction, we express that the reaction forces at every possible points of contact are normal to the concerned bodies. So, for the whole system, we write the law

$$u \in C, r \in -N_C(u). \qquad (2.11.143)$$

In general, for a system involving deformable bodies, the equilibrium equations of the system take the form (see e.g. [173], [365], [394], [399])

$$Mu = q + r \qquad (2.11.144)$$

where M is a matrix related to the energy of deformation of the system and q is a vector related to the given forces exerted on the system. Using (2.11.144) together with (2.11.143), we obtain the variational inequality: Find $u \in C$ such that

$$(Mu - q)^T(v - u) \geq 0, \ \forall v \in C. \qquad (2.11.145)$$

Chapter 3

FUNDAMENTAL EXISTENCE THEORY OF INEQUALITY PROBLEMS

The first purpose of this Chapter is to list and prove the fundamental existence theorems applicable to the study of inequality problems. Variational and hemivariational inequalities are studied for several important classes of operators among which monotone and hemicontinuous operators, semicoercive operators, nonlinear perturbations of semicoercive operators, maximal monotone operators and pseudomonotone perturbations of maximal monotone operators. The second purpose of this Chapter is to draw from the aforementioned abstract theorems the basic methods which can be used to study inequality problems. For instance, the monotonicity method (Sections 3.3, 3.11), the projection method (Section 3.2), the Fichera's approach (Section 3.4), the recession approach (Section 3.5), the method of lower and upper solutions (Section 3.6), the method of maximal monotone operators and semigroup of contractions (Sections 3.7, 3.9), the Brézis approach (Section 3.8) are here discussed. The results of this chapter will be used later in Chapters 5, 6 and 7 so as to study various classes of elliptic, parabolic and hyperbolic unilateral problems. In preparing Sections 3.1, 3.2, 3.3, 3.4, 3.6, 3.7, 3.8 and 3.9 we have primarily used the celebrated works of Barbu [39], Brézis [62], [64], Brézis and Haraux [65], Browder [69], [71], Fichera [131], Hartman and Stampacchia [186], Lions and Stampacchia [250], Morosanu [287], Mosco [289], [288] and Zeidler [432]. Sections 3.5, 3.10, 3.11 and 3.12 rely primarily on the recent works of Goeleven [163], [165] and [167] and Naniewicz and Panagiotopoulos [311]. Note that recent related results are also discussed in the works of Dinca, Panagiotopoulos and Pop [107], [108], Fundo [144], [145], Goeleven [168], Miettinen [268], Naniewicz [303], [306], [309], [310], [312] and Verma [422]. The fixed point method discussed in Section 3.13 relies on the approach developed by Awbi, Rochdi and Sofonea [30]. For more details concerning the aspects related to operator theory used through this Chapter we refer the reader to [70]-[71], [64], [119], [358], [359], [421] and [432].

3.1   THE HARTMAN-STAMPACCHIA THEOREM IN $I\!R^N$

Let C be a convex compact subset of $I\!R^n (n \geq 1)$ and let $A : I\!R^n \to I\!R^n$ be a continuous mapping. We consider the variational inequality: Find $u \in C$ such that:

$$\langle Au, v - u \rangle_n \geq 0, \forall\, v \in C. \tag{3.1.1}$$

Recall here that $\langle ., . \rangle_n$ denotes the euclidean scalar product in $I\!R^n$. A direct application of degree theory (see Section 5.1) permits to give a simple proof of the famous Hartman-Stampacchia theorem [186].

Theorem 3.1.1 Let C be a nonempty convex compact subset of $I\!R^n (n \geq 1)$ and let $A : I\!R^n \to I\!R^n$ be a continuous mapping. Then there exists $u \in C$ such that

$$\langle Au, v - u \rangle_n \geq 0, \forall\, v \in C.$$

Proof. Problem (3.1.1) is equivalent to the fixed point problem

$$u = P_C(u - Au).$$

We set

$$H(t, u) = P_C(tu - tAu).$$

It is clear that for $R > 0$ large enough, the problem

$$u = H(t, u)$$

does not have any solution on ∂B_R, and $P_C(0) \in B_R$. Thus

$$\begin{aligned}
deg(u - P_C(u - Au), B_R, 0) &= deg(u - H(1, u), B_R, 0) \\
&= deg(u - H(0, u), B_R, 0) \\
&= 1.
\end{aligned}$$

$\blacksquare$

3.2 THE LIONS-STAMPACCHIA THEOREM AND GENERALIZED PROJECTION MAPPINGS

A particularly useful tool in the field of variational inequalities is given by the famous Lions-Stampacchia theorem [250]. Let X be a real Hilbert space with scalar product denoted by $(.,.)$ and corresponding norm $\| \, . \, \|$. In this case the duality map $J : X \to 2^{X^*}$ defined by $J = \partial \frac{1}{2} \| \, . \, \|^2$ is single-valued and satisfies

$$\langle Ju, v \rangle = (u, v), \ \forall \, u, \, v \, \in \, X.$$

Theorem 3.2.1   Let X be a real Hilbert space, $C \subset X$ a nonempty closed convex set and $A : X \to X^*$ a Lipschitz and strongly monotone operator, that is

$$\| Au - Av \|_* \leq k \| u - v \|, \ \forall \, u, \, v \, \in \, X, \tag{3.2.1}$$

$$\langle Au - Av, u - v \rangle \geq \alpha \| u - v \|^2, \ \forall \, u, \, v \, \in \, X, \tag{3.2.2}$$

for some constants $k, \alpha > 0$. Then for each $f \in X^*$, there exists a unique solution to the variational inequality

$$u \in C : \langle Au - f, v - u \rangle \geq 0, \forall \, v \, \in \, C. \tag{3.2.3}$$

Proof. Problem (3.2.3) is equivalent to the fixed point problem (1.1.49). We claim that for $\rho > 0$ sufficiently small, the map $T : X \to X$ defined by

$$Tu := P_C(u - \rho J^{-1}Au + \rho J^{-1}f)$$

is a strict contraction mapping. Indeed,

$$\begin{aligned} \| Tu - Tv \| \ &\leq \ \| u - \rho J^{-1}Au - v + \rho J^{-1}Av \| \\ &\leq \ \| I - \rho J^{-1}A \| \| u - v \| \end{aligned}$$

where

$$\| I - \rho J^{-1}A \| = \sup_{\substack{x \in X \\ x \neq 0}} \frac{\| x - \rho J^{-1}Ax \|_*}{\| x \|}.$$

We have

$$\| x - \rho J^{-1}Ax \|^2 = \| x \|^2 + \rho^2 \| J^{-1}Ax \|^2 - 2\rho(x, J^{-1}Ax)$$

and since

$$\| J^{-1}Ax \| = \| Ax \|_* \leq \| A \| \| x \|$$

and

$$(x, J^{-1}Ax) = \langle Ax, x \rangle \geq \alpha \| x \|^2,$$

we get

$$\| x - \rho J^{-1}Ax \|_*^2 \leq (1 + \rho^2 \| A \|^2 - 2\alpha\rho) \| x \|^2,$$

so that

$$\| I - \rho J^{-1}A \| \leq (1 + \rho^2 \| A \|^2 - 2\alpha\rho)^{\frac{1}{2}}.$$

If $0 < \rho < 2\alpha \| A \|^{-2}$ then $(1 + \rho^2 \| A \|^2 - 2\alpha\rho) < 1$ and T is a strict contraction mapping. The Banach-Picard theorem on contractions (see e.g. [203]) may be applied to conclude. ∎

The Lions-Stampacchia theorem enables us to define the solution mapping

$$\Pi_{A,C} : X^* \to C, f \to \Pi_{A,C}(f)$$

which associates to $f \in X^*$ the unique solution $\Pi_{A,C}(f)$ of the variational inequality problem (3.2.3). Moreover, the solution mapping is Lipschitz continuous.

Proposition 3.2.2 Let X be a real Hilbert space and $A : X \to X^*$ satisfying (3.2.1) and (3.2.2). Then

$$\| \Pi_{A,C}(f) - \Pi_{A,C}(f') \| \leq \frac{1}{\alpha} \| f - f' \|, \ \forall f, f' \in X^*.$$

Proof. Let $f, f' \in X^*$ be given. We have

$$\langle A\Pi_{A,C}(f) - f, v - \Pi_{A,C}(f) \rangle \geq 0, \forall v \in C \qquad (3.2.4)$$

and

$$\langle A\Pi_{A,C}(f') - f', v' - \Pi_{A,C}(f') \rangle \geq 0, \ \forall v' \in C. \qquad (3.2.5)$$

Setting $v := \Pi_{A,C}(f')$ in (3.2.4) and $v' := \Pi_{A,C}(f')$ in (3.2.5) we obtain

$$-\langle A\Pi_{A,C}(f), \Pi_{A,C}(f') - \Pi_{A,C}(f) \rangle \leq -\langle f, \Pi_{A,C}(f') - \Pi_{A,C}(f) \rangle$$

and

$$\langle A\Pi_{A,C}(f'), \Pi_{A,C}(f') - \Pi_{A,C}(f)\rangle \leq \langle f', \Pi_{A,C}(f') - \Pi_{A,C}(f)\rangle.$$

Adding these two last relations and using (3.2.2), we obtain

$$\alpha \parallel \Pi_{A,C}(f') - \Pi_{A,C}(f) \parallel^2 \leq \parallel f' - f \parallel \parallel \Pi_{A,C}(f') - \Pi_{A,C}(f) \parallel$$

and we may conclude. $\blacksquare$

Let us now consider the more general variational inequality: Find $u \in C$ such that

$$\langle Au - F(u), v - u\rangle \geq 0, \forall v \in C \qquad (3.2.6)$$

where $F : X \to X^*$ is a possibly nonlinear mapping. The inequality problem (3.2.6) is equivalent to the fixed point problem: Find $u \in X$ such that

$$u = \Pi_{A,C}(F(u))$$

provided that A satisfies the assumptions of Theorem 3.2.1

If A is a symmetric, bounded and linear operator satisfying condition (3.2.2) then the bilinear form $((.,.))$ defined by

$$((u,v)) := \langle Au, v\rangle, \ \forall u, v \in X$$

constitutes a scalar product on X, equivalent to the initial one. In this case, problem (3.2.3) is also equivalent to the problem: Find $u \in C$ such that

$$((u - A^{-1}f, v - u)) \geq 0, \forall v \in C,$$

or also

$$u = P_C(A^{-1}f),$$

where the projection operator is defined by (1.1.43) with respect to the norm $\parallel . \parallel := ((.,.))^{\frac{1}{2}}$. That means also that here

$$\Pi_{A,C}(.) = P_C(A^{-1}.)$$

and the corresponding nonlinear variational inequality (3.2.6) is equivalent to the fixed point problem

$$u = P_C(A^{-1}F(u)). \qquad (3.2.7)$$

We usually call the solution mapping $\Pi_{A,C}$ a generalized projection mapping (with respect to A). Indeed, if X is a real Hilbert space and $A = J$ then $\Pi_{A,C} \equiv \Pi_{J,C} \equiv P_C$.

3.3   THE BROWDER THEOREM FOR MONOTONE AND HEMICONTINUOUS VARIATIONAL INEQUALITIES

A variant of the previous discussion can be given via the more general theorem of Browder. Let X be a real Banach space. We recall that an operator $A : X \to X^*$ is said to be monotone if

$$\langle Au - Av, u - v \rangle \geq 0, \ \forall\, u, \ v \ \in \ X,$$

strictly monotone if

$$\langle Au - Av, u - v \rangle > 0, \ \forall\, u, \ v \ \in \ X, \ u \neq v,$$

and hemicontinuous if for every $u, v \in X$ and every sequence $t_n \to 0$ of non-negative real numbers, the sequence $A(u + t_n v)$ converges weakly* to Au in X^*.

Theorem 3.3.1 Let X be a real reflexive Banach space, C a nonempty closed convex subset of X, A a monotone, hemicontinuous and coercive, i.e. there exists $x_0 \in C$ such that

$$\frac{\langle Ax, x - x_0 \rangle}{\| x \|} \to +\infty, \ \text{as} \ \| x \| \to +\infty, \tag{3.3.1}$$

mapping from X onto X^*. Then for each $f \in X^*$, there exists at least one $u \in C$ satisfying

$$\langle Au - f, v - u \rangle \geq 0, \forall\, v \ \in \ C. \tag{3.3.2}$$

Proof. Let us denote by C_n the closed convex set

$$C_n := \{u \in C : \| u \| \leq n\}.$$

If n is large enough, let us say $n \geq n_0$, then C_n is nonempty.

We claim that for each $n \geq n_0$, there exists $u_n \in C_n$ such that

$$\langle Av - f, v - u_n \rangle \geq 0, \ \forall\, v \ \in \ C_n. \tag{3.3.3}$$

Indeed, set

$$g(x, y) := \langle Ax - f, y - x \rangle$$

and

$$f(x, y) := \langle Ay - f, y - x \rangle$$

The monotonicity of A implies that

$$g(x, y) \leq f(x, y).$$

Moreover, $x \to f(x, y)$ is concave and $y \to g(x, y)$ is w-l.s.c. Applying Theorem 1.1.21 with $\Phi \equiv 0$ and $\lambda = 0$, we obtain the existence of $u_n \in C_n$ satisfying (3.3.3). Using Minty's principle (see Section 1.1.8), we obtain also

$$\langle Au_n - f, v - u_n \rangle \geq 0, \ \forall \, v \, \in \, C_n.$$

Let $x_0 \in C$ be given. For n large enough, let us say $n \geq n_1$, we have $x_0 \in C_n$ and thus

$$\langle Au_n - f, u_n - x_0 \rangle \leq 0.$$

The coercivity condition (3.3.1) implies thus that $\| u_n \|$ is bounded. Indeed, if by contradiction we suppose that $\| u_n \| \to +\infty$ then we get the contradiction

$$\| f \|_* + \| f \|_* \frac{\| x_0 \|}{\| u_n \|} \geq \frac{\langle Au_n - f, u_n - x_0 \rangle}{\| u_n \|} \to +\infty \text{ as } n \to +\infty.$$

Thus, along a subsequence, we may assume that $u_n \rightharpoonup u \in C$. Let $v \in C$ be given. There exists $n_2 \in \mathbb{N}$ such that $v \in C_n$, for all $n \geq n_2$. From (3.3.3), we get

$$\langle Av - f, v - u_n \rangle \geq 0, \ \forall \, n \, \geq \, max\{n_0, \, n_2\}.$$

Taking the limit as $n \to +\infty$, we get

$$\langle Av - f, v - u \rangle \geq 0.$$

This last inequality can be checked for any $v \in C$. Using again Minty's principle, we deduce that

$$\langle Au - f, v - u \rangle \geq 0, \forall \, v \, \in \, C.$$

$\blacksquare$

Remark 3.3.2 A consequence of the Minty's principle is that, if A is monotone and hemicontinuous then the solutions set $SOL(A, f, C)$ of problem (3.3.2) satisfies

$$SOL(A, f, C) = \{u \in C : \langle Av - f, v - u \rangle \geq 0, \forall \, v \, \in \, C\}.$$

It results that $SOL(A, f, C)$ is closed and convex. If in addition A satisfies (3.3.1) then we can see by following similar arguments to the ones used in the proof of Theorem 3.3.1 that $SOL(A, f, C)$ is bounded. Note also that if $A : X \to X^*$ is hemicontinuous and satisfies the strong monotonicity condition

$$\langle Au - Av, u - v \rangle \geq \alpha \parallel u - v \parallel^p, \ \forall \, u, \, v \, \in \, X,$$

for some $\alpha > 0$ and $p > 1$ then all assumptions of Theorem 3.3.1 are satisfied. Moreover, in this case, the solution of (3.3.1) is unique and the solution mapping $\Pi_{A,C} : X^* \to C, f \to \Pi_{A,C}(f)$ is here also well-defined and as in Proposition 3.2.2 one proves that $\Pi_{A,C}$ is Lipschitz continuous.

3.4 THE FICHERA'S APPROACH FOR SEMICOERCIVE VARIATIONAL INEQUALITIES

The coercivity condition (3.3.1) appears too much restrictive as soon as we deal with unilateral problems involving insufficiently blocking-up boundary conditions. In this Section, we give a basic existence theorem applicable to the study of a large class of noncoercive variational inequalities. The idea of the following approach goes back to Fichera [131]. More recent works have also been followed, in particular those of Baiocchi, Buttazzo, Gastaldi and Tomarelli [33] and Goeleven [165]. Related results concerning the mathematical study of noncoercive variational inequalities are discussed in [5], [32], [77], [75], [146], [157], [188], [215], [383], [387] and [371].

Let X be a real Hilbert space and C a nonempty closed convex subset of X. Let X_1 be a finite dimensional subspace of X and let $A : X \to X^*$ be an hemicontinuous operator satisfying

$$\langle Au - Av, u - v \rangle \geq \alpha \parallel P_{X_1^\perp} u - P_{X_1^\perp} v \parallel^2, \ \forall \, u, \, v \, \in \, X. \qquad (3.4.1)$$

We obtain the following existence theorem.

Theorem 3.4.1 Suppose that the aforementioned conditions are satisfied. Then for each $f \in X^*$ satisfying

$$\langle f - Ax_0, \phi \rangle < 0, \ \forall \, \phi \, \in \, X_1 \cap C_\infty \backslash \{0\}, \qquad (3.4.2)$$

for some $x_0 \in C$, there exists at least one $u \in C$ satisfying

$$\langle Au - f, v - u \rangle \geq 0, \forall \, v \, \in \, C. \qquad (3.4.3)$$

Proof. Let C_n be the closed convex set defined by

$$C_n = \{u \in C : \| u \| \le n\}.$$

As in the proof of Theorem 3.3.1, we show that for all n large enough, there exists $u_n \in C_n$ such that

$$\langle Av - f, v - u_n \rangle \ge 0, \ \forall v \in C_n, \tag{3.4.4}$$

and thus by using Minty's principle

$$\langle Au_n - f, v - u_n \rangle \ge 0, \ \forall v \in C_n. \tag{3.4.5}$$

We claim that the sequence $\{u_n\}$ is bounded. Suppose on to the contrary that $\| u_n \| \to +\infty$ and set

$$z_n := \frac{u_n}{\| u_n \|}.$$

By considering a subsequence, we may assume that

$$z_n \rightharpoonup z \text{ in } X.$$

Let $h \in C, \alpha > 0$ be given. For n large enough

$$\frac{\alpha}{\| u_n \|} u_n + (1 - \frac{\alpha}{\| u_n \|})h \in C$$

and taking the limit as $n \to +\infty$, we obtain

$$\alpha z + h \in C.$$

This is true for any $h \in C$ and $\alpha > 0$ and therefore

$$z \in C_\infty.$$

For all $n \in \mathbb{N}$, n large enough, we have $x_0 \in C_n$ and thus we may set $v = x_0$ in (3.4.5) to get

$$\langle Au_n, u_n - x_0 \rangle \le \langle f, u_n - x_0 \rangle.$$

We obtain

$$\| P_{X_1^\perp}(u_n - x_0) \|^2 \le \frac{1}{\alpha} \langle Ax_0 - f, x_0 - u_n \rangle \tag{3.4.6}$$

and thus

$$\| P_{X_1^\perp} u_n \|^2 \le \frac{1}{\alpha} \langle Ax_0 - f, x_0 - u_n \rangle - \| P_{X_1^\perp} x_0 \|^2$$

$$+2 \parallel P_{X_1^\perp} u_n \parallel \parallel P_{X_1^\perp} x_0 \parallel . \qquad (3.4.7)$$

Dividing (3.4.7) by $\parallel u_n \parallel^2$ and taking the limit superior as $n \to +\infty$, we obtain

$$P_{X_1^\perp} z_n \to 0.$$

Moreover

$$P_{X_1} z_n \to P_{X_1} z$$

since X_1 is a finite dimensional space. We conclude that

$$z_n = P_{X_1} z_n + P_{X_1^\perp} z_n \to P_{X_1} z \qquad (3.4.8)$$

and

$$z = P_{X_1} z. \qquad (3.4.9)$$

From the strong convergence in (3.4.8) and the fact that $\parallel z_n \parallel = 1$ we obtain $\parallel z \parallel = 1$ and thus

$$z \neq 0.$$

From (3.4.9), we deduce that

$$z \in X_1.$$

In conclusion, we have shown that

$$z \in C_\infty \cap X_1 \backslash \{0\}.$$

From (3.4.6), we deduce also that

$$0 \leq \langle Ax_0 - f, x_0 - u_n \rangle. \qquad (3.4.10)$$

Dividing (3.4.10) by $\parallel u_n \parallel$ and taking the limit as $n \to +\infty$, we obtain

$$\langle Ax_0 - f, z \rangle \leq 0$$

and a contradiction with the assumption (3.4.2). Thus the sequence $\{u_n\}$ is bounded and we may conclude as in the proof of Theorem 3.3.1. ∎

The main consequence of Theorem 3.4.1 results in the following corollary concerning the class of variational inequalities involving semicoercive linear operators. Let X be a real Hilbert space and let $A : X \to X^*$ be a bounded linear operator. We say that A is semicoercive if there exists $\alpha > 0$ such that

$$\langle Au, u \rangle \geq \alpha \parallel P_{(Ker(A+A^*))^\perp} u \parallel^2, \ \forall\, u \in X.$$

One usually sets $P := P_{(Ker(A+A^*))^\perp}$ and $Q := P_{Ker(A+A^*)}$ the orthogonal projector from X onto $Ker(A + A^*)$. Recall also that for a semicoercive operator A, we have

$$Ker(A) = Ker(A^*) \subset Ker(A + A^*)$$

and

$$Ker(A + A^*) = \{u \in X : \langle Au, u \rangle = 0\}.$$

Let us first give some results characterizing this important class of operators. Let H be a second real Hilbert space with norm $| \, . \, |$ and suppose that $X \hookrightarrow H$ with a compact embedding.

Proposition 3.4.2 Suppose that there exist $\lambda \geq 0$ and $c > 0$ such that

$$\langle Au, u \rangle + \lambda \, | \, u \, |^2 \geq c \, \| \, u \, \|^2, \ \forall \, u \in X.$$

If in addition

$$\langle Au, u \rangle \geq 0, \ \forall \, u \in X,$$

then i) $dim\{Ker(A + A^*)\} < +\infty$ and ii) A is semicoercive.

Proof. If $\lambda = 0$ then the result is trivial. Suppose now that $\lambda > 0$. We set

$$Z := Ker(A + A^*).$$

We have

$$\| \, u \, \|^2 \leq \lambda c^{-1} \, | \, u \, |^2, \ \forall \, u \in Z.$$

Let $\{u_n\} \subset Z$ be a sequence satisfying $u_n \rightharpoonup u$ in V. Then $u \in Z$,

$$\| \, u - u_n \, \|^2 \leq \lambda c^{-1} \, | \, u - u_n \, |^2$$

and since $u_n \to u$ in H then the previous relation entails $u_n \to u$ in X. That means that the identity mapping is compact on Z and therefore $dim\{Z\} < +\infty$.

We claim that there exists $\alpha > 0$ such that

$$\langle Au, u \rangle \geq \alpha \, \| \, Pu \, \|^2, \ \forall \, u \in X.$$

It suffices to prove the relation for each $u \in X$ such that $\| \, Pu \, \| \neq 0$. If this is not true then we can find a sequence $\{u_n\} \subset X$ such that $\| \, Pu_n \, \| \neq 0$ and

$$\langle Au_n, u_n \rangle < n^{-1} \, \| \, Pu_n \, \|^2 \, .$$

We set

$$v_n := u_n/ \parallel Pu_n \parallel .$$

We have

$$\langle Av_n, v_n \rangle \leq 1/n$$

and

$$\parallel Pv_n \parallel = 1.$$

We choose v_n' so that $Pv_n' = Pv_n$ and $\parallel Qv_n' \parallel \leq 1$. In this case, we have $\parallel v_n' \parallel^2 = 1 + \parallel Qv_n' \parallel^2$ and thus $1 \leq \parallel v_n' \parallel^2 \leq 2$. We have

$$
\begin{aligned}
\langle Av_n', v_n' \rangle &= \langle A(Pv_n'), Pv_n' \rangle + \langle A(Qv_n'), Qv_n' \rangle \\
&\quad + \langle A(Qv_n'), Pv_n' \rangle + \langle A(Pv_n'), Qv_n' \rangle \\
&= \langle A(Pv_n'), Pv_n' \rangle + \frac{1}{2} \langle (A + A^*)(Qv_n'), Qv_n' \rangle \\
&\quad + \langle (A + A^*)(Qv_n'), Pv_n' \rangle \\
&= \langle A(Pv_n'), Pv_n' \rangle \\
&= \langle A(Pv_n), Pv_n \rangle \\
&= \langle Av_n, v_n \rangle \leq 1/n.
\end{aligned}
$$

Considering a subsequence, we have $v_n' \rightharpoonup v$ in $V, v_n' \to v$ in H and $\limsup_{n \to +\infty} \langle Av_n', v_n' \rangle \leq 0$. Thus $\langle Av, v \rangle \leq 0$ and $v \in Ker(A + A^*)$. We have

$$
\begin{aligned}
\langle A(v_n' - v), v_n' - v \rangle + \lambda \mid v_n' - v \mid^2 &\geq c \parallel v_n' - v \parallel^2 \\
&= c \parallel Pv_n' - Pv \parallel^2 + c \parallel Qv_n' - Qv \parallel^2 \\
&\geq c \parallel Pv_n' - Pv \parallel^2 = c.
\end{aligned}
$$

We have also

$$
\begin{aligned}
\langle A(v_n' - v), v_n' - v \rangle &= \langle Av_n', v_n' \rangle + \langle Av, v \rangle - \langle (A + A^*)v, v_n' \rangle \\
&= \langle Av_n', v_n' \rangle.
\end{aligned}
$$

Thus

$$\langle Av_n', v_n' \rangle + \lambda \mid v_n' - v \mid^2 \geq c.$$

Thus taking the lim sup as $n \to +\infty$, we get $0 \geq c$, which is a contradiction.

∎

Proposition 3.4.3 Let $\{X, (.,.)\}$ be a real Hilbert space and let $A : X \to X$ be a bounded linear operator. Suppose that

$$(Au, u) \geq 0, \ \forall u \in X,$$

and
$$R(A + A^*) \text{ is closed.}$$

Then A is semicoercive.

Proof. The mapping
$$A + A^* : R(A + A^*) \to R(A + A^*)$$

is bijective. Indeed, set $B := A + A^*$ and $E := R(A + A^*)$ and consider the self-adjoint mapping $B : E \to E$. We have here ([67]; Theorem II.18).
$$Ker(A + A^*)^\perp = R(A + A^*),$$

and thus
$$Ker(B) = \{u \in R(A + A^*) : (A + A^*)u = 0\} = \{0\}.$$

The closedness of $R(B)$ together with the previous relation imply that ([67]; Theorem II.19) B is surjective. Moreover, there exists a constant $d > 0$ such that
$$\| u \| \le d \| B^*u \|, \ \forall u \in E.$$

The operator $B = B^*$ is injective and thus the map $B : E \to E$ is bijective. Using the previous relation we see that B^{-1} is continuous. Thus $0 \notin \sigma(B)$. That means that $\inf_{\lambda \in \sigma(B)}\{\lambda\} > 0$ and since
$$\inf_{\lambda \in \sigma(B)} \{\lambda\} \le \inf_{u \in E, \|u\|=1} (Bu, u),$$

we obtain
$$(Bu, u) \ge (\inf_{\lambda \in \sigma(B)} \{\lambda\}) \| u \|^2, \ \forall u \in E.$$

Therefore
$$(Au, u) \ge \frac{1}{2}\alpha \| u \|^2, \ \forall u \in R(A + A^*),$$

with $\alpha := \inf_{\lambda \in \sigma(A+A^*)}\{\lambda\}$. Thus A is semicoercive.

∎

Corollary 3.4.4 Let X be a real Hilbert space, C a nonempty closed convex subset of X and let $A : X \to X^*$ be a bounded linear and semicoercive operator and such that
$$dim\{Ker(A + A^*)\} < +\infty.$$

Then for each $f \in X^*$ satisfying
$$\langle f - Ax_0, \phi \rangle < 0, \ \forall \phi \in Ker(A + A^*) \cap C_\infty \backslash \{0\}, \qquad (3.4.11)$$

for some $x_0 \in C$, there exists at least one $u \in C$ satisfying

$$\langle Au - f, v - u \rangle \geq 0, \forall\, v \in C.$$

Proof. It suffices to apply Theorem 3.4.1 with

$$X_1 = Ker(A + A^*).$$

■

Remark 3.4.5   i) Theorem 3.4.1 can also be used to study nonlinear monotone hemicontinuous perturbations of linear semicoercive variational inequalities. Indeed, if A is a linear operator satisfying the conditions of Corollary 3.4.4 and if B is monotone and hemicontinuous then it is clear that the operator $B' \equiv A + B$ satisfies all the assumptions of Theorem 3.4.1.

ii) Note that if A is symmetric then (3.4.11) is equivalent to

$$\langle f, \phi \rangle < 0, \ \forall\, \phi \in Ker\,(A) \cap C_\infty \backslash \{0\}. \qquad (3.4.12)$$

Indeed, if $\phi \in Ker A$ then

$$\langle Ax_0, \phi \rangle = \langle x_0, A\phi \rangle = 0.$$

iii) The conditions of Theorem 3.4.1 entail also that the solutions set of (3.4.3) is bounded and closed.

The previous result can be refined if A is symmetric. Indeed, in this case, we obtain the following result.

Theorem 3.4.6   Let X be a real Hilbert space, $f \in X^*$, C a nonempty closed convex subset of X and let $A : X \to X^*$ be a bounded linear, symmetric and semicoercive operator such that

$$dim\{Ker(A)\} < +\infty.$$

Then there exists $u \in C$ satisfying

$$\langle Au - f, v - u \rangle \geq 0, \forall\, v \in C \qquad (3.4.13)$$

provided that

$$C - Ker(A) \cap C_\infty \cap Ker(f) \backslash \{0\} \subset C \qquad (3.4.14a)$$

and

$$\langle f, \phi \rangle \leq 0, \ \forall \, \phi \, \in \, Ker(A) \cap C_\infty. \tag{3.4.14b}$$

Proof. Suppose that (3.4.14) is satisfied. Let $\{\varepsilon_n\}$ be a sequence of positive real numbers satisfying $\varepsilon_n \to 0^+$. Using Theorem 3.3.1, there exists $u_n \in C$ such that

$$\langle 2\varepsilon_n J u_n + A u_n - f, v - u_n \rangle \geq 0, \forall \, v \, \in \, C. \tag{3.4.15}$$

Using now the variational principle (Theorem 1.1.26) for the convex functional

$$u \to \varepsilon_n \parallel u \parallel^2 + \frac{1}{2} \langle Au, u \rangle - \langle f, u \rangle,$$

it results that

$$\varepsilon_n \parallel u_n \parallel^2 + \frac{1}{2} \langle A u_n, u_n \rangle - \langle f, u_n \rangle$$

$$\leq \varepsilon_n \parallel v \parallel^2 + \frac{1}{2} \langle Av, v \rangle - \langle f, v \rangle, \forall \, v \, \in \, C. \tag{3.4.16}$$

We claim that the sequence $\{u_n\}$ is bounded. Suppose on to the contrary that $\parallel u_n \parallel \to +\infty$. Then for a subsequence, we may assume that

$$w_n := \frac{u_n}{\parallel u_n \parallel} \rightharpoonup w \in C_\infty.$$

Let $v_0 \in C$ be given. We have

$$\frac{1}{2} \langle A u_n, u_n \rangle - \langle f, u_n \rangle \leq \varepsilon_n \parallel v_0 \parallel^2 + \frac{1}{2} \langle Av_0, v_0 \rangle - \langle f, v_0 \rangle.$$

Dividing this last relation by $\parallel u_n \parallel^2$ and taking the limit as $n \to +\infty$, one obtains

$$\frac{1}{2} \langle Aw, w \rangle \leq 0$$

so that $w \in Ker \, A$. Since A is nonnegative, we have also

$$-\langle f, u_n \rangle \leq \varepsilon_n \parallel v_0 \parallel^2 + \frac{1}{2} \langle Av_0, v_0 \rangle - \langle f, v_0 \rangle.$$

Dividing this last relation by $\parallel u_n \parallel$ and taking the limit as $n \to +\infty$, we get

$$\langle f, w \rangle \geq 0$$

and thus

$$\langle f, w \rangle = 0$$

since (3.4.14b) is satisfied. Moreover

$$\frac{\alpha}{2} \parallel P_{(Ker A)^{\perp}}(u_n) \parallel^2 \le \langle f, u_n \rangle + \varepsilon_n \parallel v_0 \parallel^2 + \frac{1}{2}\langle Av_0, v_0 \rangle - \langle f, v_0 \rangle.$$

Dividing this last inequality by $\parallel u_n \parallel^2$ and taking the limit as $n \to +\infty$, we obtain that

$$P_{(Ker A)^{\perp}} w_n \to 0.$$

Moreover

$$P_{Ker A} w_n \to P_{Ker A} w = w$$

and thus $w_n \to w$. It results that $w \ne 0$ since $\parallel w_n \parallel = 1$, $\forall n \in \mathbb{N}$. We have

$$\langle A(u_n - w), u_n - w \rangle = \langle Au_n, u_n \rangle \tag{3.4.17}$$

and

$$\langle f, u_n - w \rangle = \langle f, u_n \rangle. \tag{3.4.18}$$

Moreover, for n large enough

$$\parallel u_n - w \parallel = \parallel u_n(1 - \frac{1}{\parallel u_n \parallel}) + (w_n - w) \parallel$$

$$\le (1 - \frac{1}{\parallel u_n \parallel}) \parallel u_n \parallel + \parallel w_n - w \parallel$$

$$= \parallel u_n \parallel + (\parallel w_n - w \parallel - 1). \tag{3.4.19}$$

On the other hand, using (3.4.14a), (3.4.16) with $v = u_n - w$, (3.4.17) and (3.4.18), we get

$$\varepsilon_n \parallel u_n \parallel^2 + \frac{1}{2}\langle Au_n, u_n \rangle - \langle f, u_n \rangle$$

$$\le \quad \varepsilon_n \parallel u_n - w \parallel^2 - \langle f, u_n - w \rangle + \frac{1}{2}\langle A(u_n - w), u_n - w \rangle$$

$$= \quad \varepsilon_n \parallel u_n - w \parallel^2 + \frac{1}{2}\langle Au_n, u_n \rangle - \langle f, u_n \rangle.$$

It results that

$$\parallel u_n \parallel \le \parallel u_n - w \parallel$$

and from (3.4.19), we get

$$\parallel u_n \parallel \le \parallel u_n \parallel + (\parallel w_n - w \parallel - 1).$$

However $w_n \to w$ strongly and thus for n large enough, $\parallel w_n - w \parallel < 1$. The contradiction results. Along a subsequence, $u_n \rightharpoonup u \in C$. Passing to

the limit superior in (3.4.15), we obtain that u is a solution of (3.4.13).
∎

Remark 3.4.7   i) If $Ker(A) \cap C_\infty \cap Ker(f)$ is a subspace then

$$-[Ker(A) \cap C_\infty \cap Ker(f)] \subset Ker(A) \cap C_\infty \cap Ker(f) \subset C_\infty$$

and thus

$$C - Ker(A) \cap C_\infty \cap Ker(f) \subset C + C_\infty \subset C$$

so that assumption (3.4.14a) is satisfied in this case.   ii) If $\langle f, \phi \rangle <$ 0, $\forall \phi \in Ker(A) \cap C_\infty \backslash \{0\}$ then conditions (3.4.14a) and (3.4.14b) are also satisfied. Here $Ker(A) \cap C_\infty \cap Ker(f) = \{0\}$.

3.5   THE RECESSION APPROACH FOR NONCOERCIVE VARIATIONAL INEQUALITIES

More general noncoercive variational inequalities are now studied. The following approach called the recession approach is based on the material studied in Section 1.4. In particular we use the concepts of recession cone and recession mapping of Brézis and Nirenberg so as to deal with a large class of noncoercive variational inequalities. Recall here that the same material has already been used in Section 1.4.6 in order to deduce necessary conditions for the existence of solutions of possibly noncoercive variational inequalities. The sufficient conditions are now here discussed. The approach discussed here relies primarily on the works of Adly, Goeleven and Théra [3], [6], Goeleven [165] and Tomarelli [414], [415].

Let X be a real reflexive Banach space and let C be a closed convex subset of X such that

$$0 \in C. \tag{3.5.1}$$

Let $f \in X^*$ be given and let $A : X \to X^*$ be an operator. Let us denote by $\mathcal{S}(A, f, C)$ the set of sequences defined as follows:

$$S(A, f, C) := \left\{ \{v_n\} : v_n = \frac{u_n}{\| u_n \|}, u_n \in C, \| u_n \| \to +\infty \right.$$

$$\left. \text{and } \langle Au_n, u_n \rangle \leq \langle f, u_n \rangle \right\}. \tag{3.5.2}$$

Note that if $\{v_n\} \in S(A, f, C)$ then $\| v_n \| = 1$, $\forall n \in \mathbb{N}$ and any accumulation point of $\{v_n\}$ belongs to C_∞.

Definition 3.5.1   We say that $S(A, f, C)$ is asymptotically compact (shortly a-compact) provided that if $\{v_n\} \in S(A, f, C)$ and $v_n \rightharpoonup w$ then along a subsequence $v_n \to w$.

Let us now introduce the recession set

$$\Lambda_\infty(A, f, C) = \{w \in C_\infty : \exists \{v_n\} \in S(A, f, C), v_n \rightharpoonup w\}. \qquad (3.5.3)$$

The following result yields to a very general approach to deal with noncoercive variational inequalities.

Theorem 3.5.2   Suppose that the aforementioned conditions are satisfied. We assume that A satisfies the following two conditions:

$$S(A, f, C) \text{ is } a-\text{compact} \qquad (3.5.4)$$

and the mapping

$$v \to \langle Av, v - x \rangle \text{ is w.l.s.c., } \forall \, x \in C. \qquad (3.5.5)$$

If there exists a subset $W \subset C_\infty$ such that

$$\Lambda_\infty(A, f, C) \subset W \qquad (3.5.6)$$

and

$$\underline{r}_A(v) > \langle f, v \rangle, \; \forall \, v \in W \backslash \{0\} \qquad (3.5.7)$$

then there exists at least one $u \in C$ satisfying

$$\langle Au - f, v - u \rangle \geq 0, \forall \, v \in C. \qquad (3.5.8)$$

Proof.   Let $C_n \subset X$ be defined by

$$C_n := \{x \in C : \| x \| \leq n\}.$$

The set C_n is nonempty ($0 \in C_n$), closed and convex. Let $G_n : C \to 2^X$ be defined by

$$G_n(x) := \{v \in C_n : \langle Av - f, v - x \rangle \leq 0\}.$$

Assumption (3.5.5) implies that for each $x \in C$, the set $G_n(x)$ is weakly closed. The set C_n being bounded, each set $G_n(x)$ is weakly compact and thus $\cap_{x \in C} G_n(x) \neq \emptyset$ provided that the family of sets $\{G_n(x) : x \in C\}$

has the finite intersection property. From Example 1.1.19, we know that G is a KKM-mapping and we may therefore apply Theorem 1.1.20 to get $u_n \in C_n$ such that

$$\langle Au_n, v - u_n \rangle \geq \langle f, v - u_n \rangle, \ \forall \, v \, \in \, C_n. \tag{3.5.9}$$

Set $v = 0$ in (3.5.9) to obtain

$$\langle Au_n, u_n \rangle \leq \langle f, u_n \rangle. \tag{3.5.10}$$

We claim that the sequence $\{u_n\}$ is bounded. Indeed, if we suppose the contrary then along a subsequence

$$v_n := \frac{u_n}{\| u_n \|} \rightharpoonup w \in C_\infty.$$

Thus $\{v_n\} \in S(A, f, C)$ and $w \in \Lambda_\infty(A, f, C)$.

Assumption (3.5.4) entails that along a subsequence

$$v_n \to w$$

and thus $w \neq 0$. Dividing (3.5.10) by $\| u_n \|$ and taking the limit inferior, we obtain

$$\underline{r}_A(w) \leq \liminf_{n \to +\infty} \frac{\langle Au_n, u_n \rangle}{\| u_n \|} \leq \langle f, w \rangle.$$

This last relation contradicts (3.5.7) since $w \in W \backslash \{0\}$. Thus the sequence $\{u_n\}$ is bounded and along a subsequence we may suppose that $u_n \rightharpoonup u$. If $v \in C$ is fixed then for all n large enough, $v \in C_n$ and

$$\langle Au_n, v - u_n \rangle \geq \langle f, v - u_n \rangle.$$

That is also

$$\langle Au_n, u_n - v \rangle \leq \langle f, u_n - v \rangle$$

and taking the limit inferior as $n \to +\infty$, we obtain

$$\langle Au, u - v \rangle \leq \langle f, u - v \rangle.$$

Since v is arbitrary in C, we finally obtain

$$\langle Au, v - u \rangle \geq \langle f, v - u \rangle, \forall \, v \, \in \, C.$$

$\blacksquare$

The previous result can be extended to more general nonempty sets C provided that we adapt the previous concepts. Let $u_0 \in C$ be given. We set

$$S(A, f, C, u_0) = \{\{v_n\} : v_n = \frac{u_n}{\| u_n \|}, u_n \in C, \| u_n \| \to +\infty$$

$$\text{and } \langle Au_n, u_n - u_0 \rangle \leq \langle f, u_n - u_0 \rangle \}, \tag{3.5.11}$$

and as above, we say that $S(A, f, C, u_0)$ is a-compact provided that if $\{v_n\} \in S(A, f, C, u_0)$ and $v_n \rightharpoonup w$ then along a subsequence $v_n \to w$. We also set

$$\Lambda_\infty(A, f, C, u_0) = \{w \in C_\infty : \exists \{v_n\} \in S(A, f, C, u_0),$$

$$v_n \rightharpoonup w\}. \tag{3.5.12}$$

Theorem 3.5.3 Let X be a real reflexive Banach space and let C be a nonempty closed convex subset of X. Let $A : X \to X^*$ be an operator satisfying condition (3.5.5). We assume that

$$S(A, f, C, u_0) \text{ is } a-\text{compact,}$$

and

$$\Lambda_\infty(A, f, C, u_0) \subset W,$$

for some subset $W \subset C_\infty$. If

$$\underline{r}_{A,u_0}(v) > \langle f, v \rangle, \ \forall \, v \, \in \, W \backslash \{0\}, \tag{3.5.13}$$

then there exists at least one $u \in C$ such that

$$\langle Au - f, v - u \rangle \geq 0, \forall \, v \, \in \, C.$$

Proof. For n large enough, the set

$$C_n := \{x \in C : \| x \| \leq n\}$$

is nonempty, closed and convex. As in the proof of Theorem 3.5.2, we obtain a sequence $\{u_n\}$ such that $u_n \in C_n$ and

$$\langle Au_n, v - u_n \rangle \geq \langle f, v - u_n \rangle, \ \forall \, v \, \in \, C_n.$$

Set $v = u_0$ to obtain, for n sufficiently large,

$$\langle Au_n, u_n - u_0 \rangle \leq \langle f, u_n - u_0 \rangle. \tag{3.5.14}$$

We claim that the sequence $\{u_n\}$ is bounded. Indeed, if we suppose the contrary then along a subsequence,

$$v_n := \frac{u_n}{\| u_n \|} \rightharpoonup w \in C_\infty.$$

We conclude as in the proof of Theorem 3.5.2 that $w \in W \backslash \{0\}$. From (3.5.14), we obtain also that

$$\frac{\langle Au_n, u_n - u_0 \rangle}{\| u_n \|} \leq \frac{\langle f, u_n - u_0 \rangle}{\| u_n \|}$$

and thus

$$\underline{r}_{A,u_0}(w) \leq \langle f, w \rangle$$

which is a contradiction to (3.5.13). The sequence $\{u_n\}$ is therefore bounded and we may conclude as in the proof of Theorem 3.5.2.　　∎

Remark 3.5.4　The conditions of Theorem 3.5.2 (or Theorem 3.5.3) ensure also that the solutions set of (3.5.8) is bounded and weakly closed.

Let us now study some classes of operators satisfying the conditions (3.5.4) and (3.5.6) required by Theorem 3.5.2 and its generalized version given by Theorem 3.5.3. One recalls that an operator $A : X \to X^*$ is said to satisfy the S^+-property provided that whenever $u_n \rightharpoonup u$ and

$$\limsup_{n \to \infty} \langle Au_n, u_n - u \rangle \leq 0$$

then $u_n \to u$. Let us also recall that one says that A is weakly continuous provided that whenever $u_n \rightharpoonup u$ then $Au_n \rightharpoonup Au$.

Proposition 3.5.5　Let C be a nonempty closed convex subset of a real Banach space X and let $u_0 \in C$ be given. If

(i)　　A satisfies the S^+-property;

(ii)　$\langle Ax, x \rangle \geq 0, \forall \, x \in X$;

(iii)　A is weakly continuous;

(iv)　A is positively homogeneous;

(v)　　B is monotone on X,

then

　　　　(1) $\mathcal{S}(A + B, f, C, u_0)$ is a-compact;

and

　　　　(2) $\Lambda_\infty(A + B, f, C, u_0) \subset \{w \in C_\infty : \langle Aw, w \rangle = 0\}$.

Proof. Suppose that $w \in \Lambda_\infty(A + B, f, C, u_0)$ and let $u_n \in C$ be such that $t_n := \| u_n \| \to +\infty, w_n := \frac{u_n}{t_n} \rightharpoonup w$ and

$$\langle Au_n + Bu_n, u_n - u_0 \rangle \leq \langle f, u_n - u_0 \rangle. \tag{3.5.15}$$

Since B is monotone, (3.5.15) implies that

$$\langle Bu_0, u_n - u_0 \rangle + \langle Au_n, u_n - u_0 \rangle \leq \langle f, u_n - u_0 \rangle. \tag{3.5.16}$$

Dividing (3.5.16) by t_n^2, we obtain

$$\langle Aw_n, w_n \rangle \leq \langle Aw_n, u_0 t_n^{-1} \rangle + \langle t_n^{-1} Bu_0, u_0 t_n^{-1} - w_n \rangle$$

$$+ \langle f t_n^{-1}, w_n - u_0 t_n^{-1} \rangle, \tag{3.5.17}$$

and by using assumption (iii),

$$\limsup_{n \to \infty} \langle Aw_n, w_n \rangle \leq 0. \tag{3.5.18}$$

We have

$$\limsup_{n \to \infty} \langle Aw_n, w_n - w \rangle \leq \limsup_{n \to \infty} \langle Aw_n, w_n \rangle + \limsup_{n \to \infty} \langle Aw_n, -w \rangle.$$

Thus, by using assumption (ii) and (iii) we obtain

$$\limsup_{n \to \infty} \langle Aw_n, w_n - w \rangle \leq \limsup_{n \to \infty} \langle Aw_n, w_n \rangle.$$

This together with (3.5.18) imply that

$$\limsup_{n \to \infty} \langle Aw_n, w_n - w \rangle \leq 0,$$

and thus, by assumption (i), the sequence w_n is strongly convergent to w, which proves the a-compacity of $\mathcal{S}(A + B, f, C, u_0)$. Moreover, using (3.5.17) again and taking into account that along a subsequence $w_n \to w$ and $Aw_n \rightharpoonup Aw$, we obtain

$$\langle Aw, w \rangle \leq 0.$$

This together with assumption (ii) implies that

$$\Lambda_\infty(A + B, f, C, u_0) \subset \{w \in C_\infty : \langle Aw, w \rangle = 0\}.$$

$\blacksquare$

Remark 3.5.6 i) It is easy to see that Proposition 3.5.5 remains true if assumption (v) is replaced by the following one

$$(v') \quad \langle Bx, x - u_0 \rangle \geq 0, \forall\, x \in X.$$

ii) Let X be a real Hilbert space. If A is bounded, linear and semicoercive then A satisfies assumptions (i)-(iv).

Proposition 3.5.7   Let C be a nonempty closed convex subset of a real Banach space $X, u_0 \in C$ and let $(T, \mathcal{T}, \mu)$ be a complete positive measure space. Let $\gamma : X \to L^p(T)$ denote a continuous and linear mapping $(1 \le p < \infty)$. Let $h : T \times \mathbb{R} \to \mathbb{R}$ be a Caratheodory function satisfying

$$\mid h(x, u) \mid \le a \mid u \mid^{\frac{p}{q}} + b(x), \ a \in \mathbb{R}, b \in L^q(T)(q^{-1} = 1 - p^{-1})$$

and

$$uh(x, u) \ge -c(x) \mid u \mid - d(x), \ c \in L^q(T), d \in L^1(T).$$

Let $A : X \to X^*$ be an operator satisfying conditions (i)-(iv) of Proposition 3.5.5 and let $B : X \to X^*$ be defined by

$$\langle Bu, v \rangle := \int_T h(x, \gamma(u))\gamma(v) \, d\mu, \ \forall \, u, \, v \, \in X.$$

Then

$$(1) \ \mathcal{S}(A + B, f, C, u_0) \text{ is } a\text{-compact};$$

and

$$(2) \ \Lambda_\infty(A + B, f, C, u_0) \subset \{w \in C_\infty : \langle Aw, w \rangle = 0\}.$$

Proof.   Suppose that $w \in \Lambda_\infty(A + B, f, C, u_0)$ and let $u_n \in C$ be such that $t_n := \parallel u_n \parallel \to +\infty, w_n := \frac{u_n}{t_n} \rightharpoonup w$ and

$$\langle Au_n + Bu_n, u_n - u_0 \rangle \le \langle f, u_n - u_0 \rangle.$$

We obtain

$$\langle Aw_n, w_n \rangle + \langle t_n^{-1} Bu_n, w_n \rangle$$

$$\le \langle Aw_n, u_0 t_n^{-1} \rangle + \langle Bu_n, u_0 t_n^{-2} \rangle + \langle f t_n^{-1}, w_n - t_n^{-1} u_0 \rangle$$

which means that

$$\langle Aw_n, w_n \rangle \ \le \ \int_T c(x) \mid \gamma(u_n) \mid t_n^{-2} d\mu + \int_T d(x) t_n^{-2} d\mu$$

$$+ \ \langle Aw_n, u_0 t_n^{-1} \rangle + a \int_T \mid \gamma(u_n) \mid^{\frac{p}{q}} \mid \gamma(u_0) \mid t_n^{-2} d\mu$$

$$+ \ \int_T b(x) \mid \gamma(u_0) \mid t_n^{-2} d\mu + \langle f t_n^{-1}, w_n - u_0 t_n^{-1} \rangle.$$

The continuity of $\gamma : X \hookrightarrow L^p(\Omega)$ yields a positive constant $c > 0$ such that

$$| \gamma(u) |_{0,p} \leq c \parallel u \parallel, \forall\, u \in X.$$

Then it is easy to find positive constants C_1 and C_2 such that

$$\langle Aw_n, w_n \rangle \;\leq\; \langle Aw_n, u_0 t_n^{-1} \rangle + \langle f t_n^{-1}, w_n - u_0 t_n^{-1} \rangle$$

$$+ C_1 t_n^{-1} + C_2 t_n^{-2}.$$

Thus

$$\limsup_{n \to \infty} \langle Aw_n, w_n \rangle \leq 0,$$

and we may conclude by following the same steps as in the proof of Proposition 3.5.5.   ∎

These last properties can be used together with Theorem 3.5.2 (or Theorem 3.5.3) so as to study precise noncoercive variational inequalities. Such approach will be considered later in Chapter 6 of this book.

3.6   THE MONOTONICITY PRINCIPLE AND THE METHOD OF LOWER AND UPPER SOLUTIONS FOR VARIATIONAL INEQUALITIES

Let Ω be an open bounded subset of $\mathbb{R}^n$ and let X be a real reflexive Banach space. We assume that X is a sublattice of $L^p(\Omega)$ $(1 < p < +\infty)$ for the ordering

$$u \leq v \Leftrightarrow u(x) \leq v(x), \text{ a.e. } x \in \Omega.$$

Moreover, we assume that the embedding $X \hookrightarrow L^p(\Omega)$ is continuous.

Let us recall that an operator $A : X \to X^*$ is said to be T-monotone if

$$\langle Ax - Ay, (x - y)^+ \rangle \geq 0, \; \forall\, x,\, y \in X, \; (x - y)^+ \in X$$

and strictly T-monotone, if

$$\langle Ax - Ay, (x - y)^+ \rangle > 0, \tag{3.6.1}$$

$\forall x, y \in X, (x - y)^+ \in X \backslash \{0\}$.

Let us now consider a strictly T-monotone and hemicontinuous operator $A : X \to X^*$ satisfying the coercivity condition (3.3.1). Let $C \subset X$

be a nonempty closed convex subset of X with the property that

$$x \wedge y \in C \text{ and } x \vee y \in C, \forall x, y \in C. \tag{3.6.2}$$

Note that

$$\langle Ax - Ay, x - y \rangle = \langle Ax - Ay, (x - y)^+ \rangle + \langle Ay - Ax, (y - x)^+ \rangle.$$

It results that A is monotone and using Theorem 3.3.1 we obtain that for each $f \in L^q(\Omega)(\frac{1}{p} + \frac{1}{q} = 1)$, there exists at least one $u \in C$ solution of the inequality problem $VI(A, f, C)$:

$$\langle Au, v - u \rangle \geq \int_\Omega f(x)(v(x) - u(x))dx, \forall v \in C. \tag{3.6.3}$$

We have the following result.

Lemma 3.6.1 Let u_1 and u_2 be arbitrary solutions of $VI(A, f_1, C)$ and $VI(A, f_2, C)$ respectively. Then

$$f_2 \leq f_1 \Rightarrow u_2 \leq u_1.$$

Proof. We have $u_1, u_2 \in C$,

$$\langle Au_1, v - u_1 \rangle \geq \int_\Omega f_1(v - u_1)dx, \forall v \in C \tag{3.6.4}$$

and

$$\langle Au_2, v - u_2 \rangle \geq \int_\Omega f_2(v - u_2)dx, \forall v \in C. \tag{3.6.5}$$

Set $v = u_1 \vee u_2 = u_1 + (u_2 - u_1)^+ \in C$ in (3.6.4). One obtains

$$-\langle Au_1, (u_2 - u_1)^+ \rangle \leq -\int_\Omega f_1(u_2 - u_1)^+ dx. \tag{3.6.6}$$

Set $v = u_1 \wedge u_2 = u_2 - (u_2 - u_1)^+ \in C$ in (3.6.5). One finds

$$\langle Au_2, (u_2 - u_1)^+ \rangle \leq \int_\Omega f_2(u_2 - u_1)^+ dx. \tag{3.6.7}$$

Adding (3.6.6) and (3.6.7), we get

$$\langle Au_2 - Au_1, (u_2 - u_1)^+ \rangle \leq \int_\Omega (f_2 - f_1)(u_2 - u_1)^+ dx.$$

If $f_2 \leq f_1$ then

$$\langle Au_2 - Au_1, (u_2 - u_1)^+ \rangle \leq 0$$

and thus

$$(u_2 - u_1)^+ = 0$$

entailing that $u_2 \leq u_1$ ∎

A consequence of Lemma 3.6.1 is that the solution of (3.6.3) is unique and we may therefore define the solution mapping

$$f \to \Pi_{A,C}(f)$$

as a well-defined mapping from $L^q(\Omega)$ onto $C \subset L^p(\Omega)$. Moreover, Lemma 3.6.1 implies that $\Pi_{A,C}$ is monotone nondecreasing on $L^q(\Omega)$. The following result is known as the Monotonicity Principle.

Theorem 3.6.2   (Monotonicity Principle) a) Let $A : X \to X^*$ be a strictly T-monotone and hemicontinuous operator satisfying condition (3.3.1) and let $C \subset X$ be a nonempty closed convex subset satisfying condition (3.6.2). Then the mapping

$$\Pi_{A,C} : L^q(\Omega) \to C, f \to \Pi_{A,C}(f)$$

is monotone and nondecreasing, i.e.

$$f \leq h \Rightarrow \Pi_{A,C}(f) \leq \Pi_{A,C}(h). \tag{3.6.8}$$

b) If in addition A is strongly monotone in the sense that

$$\langle Au - Av, u - v \rangle \geq \alpha \parallel u - v \parallel^2, \ (\alpha > 0) \ \forall \, u, \, v \, \in \, X,$$

then there exists a constant $k > 0$ such that

$$\mid \Pi_{A,C}(f) - \Pi_{A,C}(h) \mid_{0,p} \leq k \mid f - h \mid_{0,q}, \ \forall \, f, \, h \, \in \, L^q(\Omega). \tag{3.6.9}$$

Proof.   Part a) is a direct consequence of Lemma 3.6.1. Let us now check part b). We have

$$\langle A(\Pi_{A,C}(f)) - A(\Pi_{A,C}(h)), \Pi_{A,C}(f) - \Pi_{A,C}(h) \rangle$$
$$\leq \int_\Omega (f - h)(\Pi_{A,C}(f) - \Pi_{A,C}(h)) dx$$
$$\leq \mid f - h \mid_{0,q} \mid \Pi_{A,C}(f) - \Pi_{A,C}(h) \mid_{0,p}.$$

Using the strong monotonicity of A and the continuous embedding $X \hookrightarrow L^p(\Omega)$, i.e.

$$| v |_{0,p} \leq c \| v \|, \forall v \in X,$$

for some $c > 0$, we get

$$| \Pi_{A,C}(f) - \Pi_{A,C}(h) |_{0,p} \leq \frac{c}{\alpha} | f - h |_{0,q} .$$

$\blacksquare$

A method of lower and upper solutions can now be developed for variational inequalities. Let $f : \Omega \times \mathbb{R} \to \mathbb{R}$ be a function satisfying the Caratheodory conditions:

$$f(.,y) : \Omega \to \mathbb{R} \text{ is measurable for all fixed } y \in \mathbb{R}$$

and

$$f(x,.) : \mathbb{R} \to \mathbb{R} \text{ is continuous for almost all } x \in \Omega.$$

Moreover, we assume that

$$| f(x,u) | \leq a_1(x) + a_2(x) | u |^{\frac{p}{q}}, \forall u \in \mathbb{R}, x \in \Omega, \qquad (3.6.10)$$

where $a_1 \in L^q(\Omega)$ and a_2 is any (non-negative) $L^\infty(\Omega)$-function. Then the Nemyckii operator F defined by

$$F(u)(x) = f(x,u(x))$$

is a well-defined bounded continuous operator from $L^p(\Omega)$ into $L^q(\Omega)$ (see e.g. [432]). We consider the variational inequality: Find $u \in C$ such that

$$\langle Au, v - u \rangle \geq \int_\Omega f(x,u(x))(v(x) - u(x))dx, \forall v \in C. \qquad (3.6.11)$$

One introduces a concept of lower and upper solutions for the inequality problem (3.6.11).

Definition 3.6.3 We say that $\underline{u} \in L^p(\Omega)$ is a lower-solution of (3.6.11) if $\underline{u} \leq \Pi_{A,C}(F(\underline{u}))$.

Definition 3.6.4 We say that $\overline{u} \in L^p(\Omega)$ is an upper-solution of (3.6.11) if $\overline{u} \geq \Pi_{A,C}(F(\overline{u}))$.

Let us now introduce a concept of maximal and minimal solution for the inequality problem (3.6.11).

Definition 3.6.5 We say that u is a minimal (maximal) solution of (3.6.11) in a set $D \subset X$ if $u \leq z$ ($u \geq z$) for every solution z of (3.6.11) in D.

The following results show that if $f(x,.)$ is monotone non-decreasing, for each $x \in \Omega$, then the existence of at least one lower-solution and upper-solution entails the existence of a minimal and a maximal solution of (3.6.11).

Theorem 3.6.6 Let X and C as defined above. Let A be a strictly T-monotone and hemicontinuous operator satisfying condition (3.3.1) and let $f : \Omega \times \mathbb{R} \to \mathbb{R}$ be a Caratheodory function satisfying (3.6.10) and monotone nondecreasing in the second variable, i.e.

$$y_1 \leq y_2 \Rightarrow f(x,y_1) \leq f(x,y_2), \ \forall \, x \, \in \, \Omega. \qquad (3.6.12)$$

If problem (3.6.11) has at least one lower-solution $\underline{u} \in L^p(\Omega)$ and at least one upper-solution $\overline{u} \in L^p(\Omega)$ such that $\underline{u} \leq \overline{u}$ then (3.6.11) has a minimal solution u_* and a maximal solution u^* in $[[\underline{u},\overline{u}]] := \{v \in X : \underline{u} \leq v \leq \overline{u}\}$.

Proof. The mapping

$$\Phi : L^p(\Omega) \to L^p(\Omega); h \to \Phi(h)$$

where Φ is defined by

$$\Phi(h) := \Pi_{A,C}(F(h))$$

is monotone nondecreasing. The space $L^p(\Omega)(1 < p < \infty)$ is a Dedeking complete Riesz space and we may apply the Tarski-Knaster-Kantorovich-Birkhoff theorem (see e.g. [52]) to lead to the existence of a minimal and a maximal fixed point of Φ. The fixed point problem $u = \Phi(u)$ is equivalent to the inequality problem (3.6.11) and we may conclude. ∎

Theorem 3.6.7 Let X and C as defined above. Let $A : X \to X^*$ be a strictly T-monotone, strongly monotone and hemicontinuous operator and let $f : \Omega \times \mathbb{R} \to \mathbb{R}$ be a Caratheodory function satisfying conditions (3.6.10) and (3.6.12). Moreover, we assume that

$$f(x,0) \geq 0, \text{ a.e. } x \in \Omega$$

and
$$f(x, z) \leq m(x), \ \forall \, z \, \in \, \mathbb{R}, \text{ a.e. } x \, \in \, \Omega,$$
for some $m \in L^q(\Omega)$.

If $\Pi_{A,C}(F(0)) \geq 0$ then (3.6.11) has at least one solution.

Proof. Let us consider the sequence $\{u_n\}$ defined by
$$u_0 = 0$$
and
$$u_{n+1} = \Pi_{A,C}(F(u_n)). \tag{3.6.13}$$
The sequence of inequalities
$$0 \leq u_1 \leq \cdots \leq u_n \leq u_{n+1} \leq \cdots$$
is satisfied. Moreover,
$$u_n \leq \Pi_{A,C}(m), \ \forall \, n \, \in \, \mathbb{N}.$$
That means that for almost all $x \in \Omega$, the sequence $\{u_n(x)\}$ is a monotone nondecreasing and upper-bounded sequence. Thus
$$u_n(x) \to u(x) \ \text{ in } \mathbb{R}$$
for some $u(x)$ satisfying $\mid u(x) \mid = u(x) \leq l(x)$ where
$$l = \Pi_{A,C}(m) \in L^p(\Omega).$$
Using Beppo-Levi's convergence theorem (see e.g. [385]) for monotone sequence of nonnegative functions, we obtain
$$u_n \to u \ \text{ in } L^p(\Omega).$$
The corresponding sequence $\{f(x, u_n(x))\}$ is a monotone nondecreasing sequence and
$$f(x, u_n(x)) \to f(x, u(x)) \text{ a.e. } x \in \Omega.$$
Moreover,
$$\mid f(x, u(x)) \mid = f(x, u(x)) \leq m(x)$$
and using again Beppo-Levi's convergence theorem for monotone sequences of nonnegative functions, we obtain
$$F(u_n) \to F(u) \text{ in } L^q(\Omega).$$

Thus
$$\Pi_{A,C}(F(u_n)) \to \Pi_{A,C}(F(u)) \text{ in } L^p(\Omega).$$
Taking the limit as $n \to +\infty$ in (3.6.13), we get
$$u = \Pi_{A,C}(F(u))$$
and thus u is a solution of (3.6.11).   ∎

Another approach of lower and upper solutions can be stated for the class of variational inequalities which can be studied via the ordered complementarity formulation (1.1.70) (see Section 1.1.10).

Let X be a reflexive Banach lattice ordered by a pointed closed convex cone K. The norm in X is denoted by $\| \, . \, \|$. Let $A : X \to X$ be a given operator. We consider the problem (see Section 1.4.4): Find $x \in X$ such that
$$\wedge\{x, Ax\} = 0 \tag{3.6.14}$$
which is equivalent to the fixed point problem: Find $x \in X$ such that
$$x = H(x)$$
with
$$H(x) = \vee\{0, x - Ax\}.$$

Definition 3.6.8   We say that $\underline{x}$ is a lower-solution of (3.6.14) if $\underline{x} \leq H(\underline{x})$.

Definition 3.6.9   We say that $\overline{x}$ is an upper-solution of (3.6.14) if $\overline{x} \geq H(\overline{x})$.

The concept of maximal and minimal solution for problem (4.6.14) are defined as in Definition 3.6.5.

A notion introduced by Isac in [197] is now introduced.

Definition 3.6.10 We say that H is Φ-isotone if there exists a mapping $\Phi : X \to X$ such that
$$I + \Phi \text{ is invertible,}$$
$$0 \leq x \leq y \Rightarrow (H + \Phi)(x) \leq (H + \Phi)(y),$$
and
$$0 \leq x \leq y \Rightarrow (I + \Phi)^{-1}(x) \leq (I + \Phi)^{-1}(y).$$

It is easy to see that x is a fixed point of H if and only if x is a fixed point of $(I + \Phi)^{-1}(H + \Phi)$. The following result holds.

Theorem 3.6.11   Let X, K and H as specified above. Suppose in addition that:

(i) The cone K is normal, i.e.

$$0 \le x \le y \Rightarrow \parallel x \parallel \le \delta \parallel y \parallel,$$

for some $\delta > 0$.

(ii) H is Φ-isotone.

(iii) The application $(I + \Phi)^{-1}(H + \Phi)$ is weakly continuous.

If Problem (3.6.14) has at least one lower-solution $\underline{x}$ and at least one upper-solution $\overline{x}$ such that $\underline{x} \le \overline{x}$, then (3.6.14) has a minimal solution x_* and a maximal solution x^* in $[[\underline{x}, \overline{x}]]$.

Proof. Let $\{x_n\}$ and $\{y_n\}$ be the sequences defined by

$$x_0 = \underline{x},$$

$$x_{n+1} = (I + \Phi)^{-1}(H + \Phi)(x_n) \tag{3.6.15}$$

and

$$y_0 = \overline{x},$$

$$y_{n+1} = (I + \Phi)^{-1}(H + \Phi)(y_n). \tag{3.6.16}$$

We have

$$x_0 = \underline{x} \le H(x_0), \quad y_0 = \overline{y} \ge H(y_0), \quad x_0 \le y_0.$$

Then using assumption (ii), we get

$$x_1 = (I + \Phi)^{-1}(H + \Phi)(x_0) \ge (I + \Phi)^{-1}(I + \Phi)(x_0) = x_0$$

and

$$y_1 = (I + \Phi)^{-1}(H + \Phi)(y_0) \le (I + \Phi)^{-1}(I + \Phi)(y_0) = y_0.$$

Moreover,

$$x_1 = (I + \Phi)^{-1}(H + \Phi)(x_0) \le (I + \Phi)^{-1}(H + \Phi)(y_0) = y_1$$

since $x_0 \leq y_0$. Thus

$$0 \leq x_0 \leq x_1 \leq y_1 \leq y_0.$$

By induction, we check that

$$0 \leq x_0 \leq x_1 \leq ... \leq x_k \leq y_k \leq ... \leq y_1 \leq y_0.$$

From assumption (i), we deduce that the sequences $\{x_n\}$ and $\{y_n\}$ are bounded. We may thus find subsequences again denoted by $\{x_n\}$ and $\{y_n\}$ such that $x_n \rightharpoonup x_*$ and $y_n \rightharpoonup x^*$. Using (3.6.15) and (3.6.16) together with assumption (iii), we see that $x_* = (I+\Phi)^{-1}(H+\Phi)(x_*)$ and $x^* = (I+\Phi)^{-1}(H+\Phi)(x^*)$. It results that $x_* = H(x_*)$ and $x^* = H(x^*)$.

Let z be another solution of problem (3.6.14) in $[[\underline{x}, \overline{x}]]$. Then $z = (I+\Phi)^{-1}(H+\Phi)(z)$. Moreover, by definition of the sequences $\{x_n\}$ and $\{y_n\}$ and by induction, we have that $x_n \leq z \leq y_n$ and thus taking the limit as $n \to \infty$, we get $x_* \leq z \leq x^*$. ∎

Example 3.6.12   Variational inequalities involving Z-matrices

Let $M \in \mathbb{R}^{n \times n}$ be a $Z-$matrix, i.e. $m_{ij} \leq 0, \forall\, i \neq j,\ i,j \in \{1, ..., n\}$. Let q be a given vector in $\mathbb{R}^n$. We denote by Φ the diagonal matrix:

$$\Phi = \mathrm{diag}\{\max\{0, m_{ii}\}\}.$$

It is clear that $\Phi \geq 0$, $\Phi - M \geq 0$, $(I + \Phi)^{-1}$ exists and $(I + \Phi)^{-1} \geq 0$. The inequality problem: Find $x \in \mathbb{R}_+^n$ such that

$$(Mx - q)^T (v - x) \geq 0, \forall v \in \mathbb{R}_+^n \qquad (3.6.17)$$

can also be formulated as in (3.6.14) with $X = \mathbb{R}^n, K = \mathbb{R}_+^n$ and $A(x) = Mx - q$ (see Section 1.1.10). Here

$$H(x) = \vee\{0, x - Mx + q\}.$$

Setting $\Phi(x) = \Phi x$, we see that H is Φ-isotone. Indeed, we have

$$H(x) + \Phi(x) = \vee\{\Phi x, x + (\Phi - M)x + q\}.$$

If $x \leq y$ then $\Phi x \leq \Phi y$, $x + (\Phi - M)x + q \leq y + (\Phi - M)y + q$ and thus

$$(H + \Phi)(x) = \vee\{\Phi x, x + (\Phi - M)x + q\}$$

$$\leq \vee\{\Phi y, y + (\Phi - M)y + q\} = (H + \Phi)(y).$$

On the other hand if $x \leq y$ then $(I+\Phi)^{-1}x \leq (I+\Phi)^{-1}y$. The function H is thus Φ-isotone.

Moreover $0 \leq H(0) = \vee\{0, q\}$ so that 0 is a lower-solution of problem (3.6.17). So, if there exists a vector $y_0 \in \mathbb{R}^n$ satisfying $y_0 \leq H(y_0)$, that is

$$\wedge\{y_0, My_0 - q\} \geq 0$$

or equivalently

$$y_0 \geq 0, My_0 - q \geq 0$$

then we may apply Theorem 3.6.11 to get the existence of a minimal and a maximal solution of problem (3.6.17).

Remark 3.6.13   For further results concerning the study of ordered complementarity problems we refer the reader to the works of Borwein [55], Goeleven [166], Isac and Goeleven [201] and Isac and Kostreva [199].

3.7   THE GENERALIZED HILLE-YOSIDA THEOREM AND THE SEMI-GROUP APPROACH FOR EVOLUTION VARIATIONAL INEQUALITIES

The following approach of maximal monotone operators and nonlinear semigroups is a very efficient tool in the study of evolution variational inequalities.

Let H be a real Hilbert space and $A : D(A) \subset H \to 2^H$ a given set-valued operator. The set

$$D(A) = \{u \in H : Au \neq \emptyset\}$$

is called the effective domain of A. The set

$$R(A) = \cup_{u \in D(A)} Au$$

is called the range of A. The set

$$G(A) = \{(u, v) \in H \times H : u \in D(A), v \in Au\}$$

is called the graph of A. One says that A is maximal monotone if A is monotone on $D(A)$, i.e.

$$(x^* - y^*, x - y) \geq 0, \ \forall \, x, \, y \, \in \, D(A), \ x^* \, \in \, Ax, \ y^* \, \in \, Ay$$

and it follows from $(u, u^*) \in H \times H$ and

$$(u^* - v^*, u - v) \geq 0, \ \forall \, (v, \ v^*) \ \in \ G(A)$$

that $(u, u^*) \in G(A)$. Let us also recall that the graph $G(A)$ of a maximal monotone operator A is demi-closed (see e.g. [64], [358]).

If $(T, \mathcal{T}, \mu)$ denotes a complete positive measure space with $\mu(T) < +\infty$ then one can define the operator $\mathcal{A}$ on $L^2(0, T; H)$ extension of A to $L^2(0, T; H)$ by setting $f \in \mathcal{A}(u)$ if and only if $f(x) \in A(u(x)) \ \mu - a.e. \ x \in T$, which is maximal monotone provided that A is maximal monotone (see e.g. [287]).

The following theorem constitutes a generalized version of the Hille-Yosida theorem.

Theorem 3.7.1   Let H be a real Hilbert space and $A : D(A) \subset H \to 2^H$ a maximal monotone operator.   Let $u_0 \in D(A)$ and $f \in W^{1,1}(0, T; H)(0 \leq T < +\infty)$. Then there exists a unique $u \in W^{1,\infty}(0, T; H)$ satisfying

$$u(t) \in D(A), \ \forall \, t \ \in \ (0, T]; \tag{3.7.1}$$

$$u(0) = u_0 \tag{3.7.2}$$

and

$$\frac{du}{dt} + A(u) \ni f(t), \ \text{a.e. } t \in (0, T). \tag{3.7.3}$$

In addition, if u and v are the solutions corresponding to $(u_0, f), (v_0, g) \in D(A) \times W^{1,1}(0, T; H)$, then

$$\| \, u(t) - v(t) \, \| \leq \| \, u_0 - v_0 \, \| + \int_0^t \| \, f(s) - g(s) \, \| \, ds, \ \forall \, t \ \in \ [0, T]. \tag{3.7.4}$$

Proof. We start the proof with property (3.7.4). We have

$$f(t) - \frac{du}{dt} \in A(u)$$

and

$$g(t) - \frac{dv}{dt} \in A(v).$$

We know that

$$(f(t) - g(t), u - v) \geq (\frac{du}{dt} - \frac{dv}{dt}, u - v),$$

so that

$$\frac{1}{2}\frac{d}{dt}\parallel u(t)-v(t)\parallel^2 \le \parallel f(t)-g(t)\parallel\parallel u(t)-v(t)\parallel, \text{ a.e. } t\in(0,T).$$

Thus

$$\frac{1}{2}\parallel u(t)-v(t)\parallel^2 \le \frac{1}{2}\parallel u_0-v_0\parallel^2$$

$$+\int_0^t\parallel f(s)-g(s)\parallel\parallel u(s)-v(s)\parallel ds, \forall 0\le t\le T. \qquad (3.7.5)$$

Let us here recall that if $\Psi\in L^1(a,b)$ $(-\infty<a<b<+\infty)$ with $\Psi\ge 0$ a.e. on (a,b) and $c\in\mathbb{R}$ then if $h\in C^0([a,b])$ verifies

$$\frac{1}{2}h^2(t)\le\frac{1}{2}c^2+\int_a^t\Psi(s)h(s)ds, \ \forall\, t\in[a,b], \qquad (3.7.6a)$$

then

$$\mid h(t)\mid\le\mid c\mid+\int_a^t\Psi(s)ds, \ \forall\, t\in[a,b]. \qquad (3.7.6b)$$

See for instance [287] for the proof of this type of Gronwall's inequality.

Setting $a=0, b=T, h(t)=\parallel u(t)-v(t)\parallel$, $c=\parallel u_0-v_0\parallel$ and $\Psi(s)=\parallel f(s)-g(s)\parallel$, the inequality (3.7.5) takes the form of (3.7.6a) and thus (3.7.6b) holds, i.e.

$$\parallel u(t)-v(t)\parallel\le\parallel u_0-v_0\parallel+\int_0^t\parallel f(s)-g(s)\parallel ds, \ \forall\, t\in[0,T].$$

From this inequality, we deduce also that the solution of (3.7.1)-(3.7.3) is unique provided that it exists. Let us denote by A_λ the Yosida approximation of A [64], i.e.

$$A_\lambda=\lambda^{-1}(I-(I+\lambda A)^{-1}).$$

The mapping A_λ is Lipschitz continuous and there exists a unique $u_\lambda\in C^1([0,T];H)$ satisfying

$$\frac{du_\lambda}{dt}+A_\lambda(u_\lambda(t))=f(t), 0\le t\le T,$$

$$u_\lambda(0)=u_0.$$

As above and using the fact that A_λ is monotone, we obtain

$$\frac{1}{2}\frac{d}{dt}\parallel u_\lambda(t+h)-u_\lambda(t)\parallel^2\le\parallel f(t+h)-f(t)\parallel\parallel u_\lambda(t+h)-u_\lambda(t)\parallel,$$

$$\forall\, t,\ t\,+\,h\,\in\,[0,T],$$

and thus

$$\frac{1}{2}\parallel u_\lambda(t+h)-u_\lambda(t)\parallel^2\,\le\,\frac{1}{2}\parallel u_\lambda(h)-u_0\parallel^2$$

$$+\int_0^t\parallel f(s+h)-f(s)\parallel\parallel u_\lambda(s+h)-u_\lambda(s)\parallel ds,$$

$$\forall\, t,\ t\,+\,h\,\in\,[0,T]. \tag{3.7.7}$$

Set $a=0,b=T,h(t)=\parallel u_\lambda(t+h)-u_\lambda(t)\parallel,\ c=\parallel u_\lambda(h)-u_0\parallel$ and $\Psi(s)=\parallel f(s+h)-f(s)\parallel$. The inequality (3.7.7) takes the form (3.7.6a) and we obtain thus that (3.7.6b) is satisfied, that is here

$$\parallel u_\lambda(t+h)-u_\lambda(t)\parallel\le\parallel u_\lambda(h)-u_0\parallel+\int_0^t\parallel f(s+h)-f(s)\parallel ds,$$

$$\forall\, t,\ t\,+\,h\,\in\,[0,T]. \tag{3.7.8}$$

Dividing (3.7.8) by $h>0$ and taking the limit as $h\to 0^+$, we obtain for any $0\le t\le T$ the estimation

$$
\begin{aligned}
\parallel\frac{du_\lambda}{dt}(t)\parallel\ &\le\ \parallel\frac{du_\lambda}{dt}(0)\parallel+\int_0^t\parallel\frac{d}{ds}f(s)\parallel ds\\[2mm]
&=\ \parallel f(0)-A_\lambda(u_0)\parallel+\int_0^t\parallel\frac{d}{ds}f(s)\parallel ds\\[2mm]
&\le\ \parallel f(0)\parallel+\parallel A^0(u_0)\parallel\\[2mm]
&\quad+\int_0^T\parallel\frac{d}{ds}f(s)\parallel ds,\ 0\le t\le T,
\end{aligned}
\tag{3.7.9}
$$

where $A^0(u_0)$ denotes the element of $A(u_0)$ for which $\parallel A(u_0)\parallel=\inf\{\parallel z\parallel:z\in A(u_0)\}$. To prove (3.7.9) we have used the property that $\parallel A_\lambda(x)\parallel\le\parallel g\parallel,\ \forall g\in A(x)$ [287] so that in particular $\parallel A_\lambda(u_0)\parallel\le\parallel A^0(u_0)\parallel$. Thus

$$\parallel A_\lambda(u_\lambda(t))\parallel\le\parallel f(t)\parallel+\parallel f(0)\parallel+\parallel A^0(u_0)\parallel+\int_0^T\parallel\frac{d}{ds}f(s)\parallel ds$$

$$\le C,\ \forall\, t\,\in\,[0,T],\ \forall\,\lambda\,>\,0, \tag{3.7.10}$$

for some positive constant C. Let $\lambda,\mu>0$. We have

$$\frac{du_\lambda}{dt}(t)-\frac{du_\mu}{dt}(t)+A_\lambda(u_\lambda(t))-A_\mu(u_\mu(t))=0$$

and thus

$$(\frac{du_\lambda}{dt}(t) - \frac{du_\mu}{dt}(t), u_\lambda(t) - u_\mu(t)) = -(A_\lambda(u_\lambda(t)) - A_\mu(u_\mu(t)), u_\lambda(t) - u_\mu(t)).$$

We have

$$\lambda A_\lambda(u_\lambda) = u_\lambda - (I + \lambda A)^{-1} u_\lambda$$

and

$$\mu A_\mu(u_\mu) = u_\mu - (I + \mu A)^{-1} u_\mu.$$

Therefore

$$(\frac{du_\lambda}{dt}(t) - \frac{du_\mu}{dt}(t), u_\lambda(t) - u_\mu(t))$$
$$= -(A_\lambda(u_\lambda(t)) - A_\mu(u_\mu(t)), \lambda A_\lambda(u_\lambda(t)) - \mu A_\mu(u_\mu(t)))$$
$$-(A_\lambda(u_\lambda(t)) - A_\mu(u_\mu(t)), (I + \lambda A)^{-1} u_\lambda - (I + \mu A)^{-1} u_\mu).$$

Taking into account that

$$A_\alpha(u_\alpha(t)) \in A(I + \alpha A)^{-1}(u_\alpha(t)) \ (\alpha = \lambda, \mu),$$

we obtain that

$$(\frac{du_\lambda}{dt}(t) - \frac{du_\mu}{dt}(t), u_\lambda(t) - u_\mu(t))$$
$$\leq -(A_\lambda(u_\lambda(t)) - A_\mu(u_\mu(t)), \lambda A_\lambda(u_\lambda(t)) - \mu A_\mu(u_\mu(t)))$$
$$\leq \lambda \parallel A_\lambda(u_\lambda(t)) \parallel [\parallel A_\lambda(u_\lambda(t)) \parallel + \parallel A_\mu(u_\mu(t)) \parallel]$$
$$+\mu \parallel A_\mu(u_\mu(t)) \parallel [\parallel A_\lambda(u_\lambda(t)) \parallel + \parallel A_\mu(u_\mu(t)) \parallel].$$

Using now (3.7.10), we get

$$(\frac{du_\lambda}{dt}(t) - \frac{du_\mu}{dt}(t), u_\lambda(t) - u_\mu(t)) \leq (\lambda + \mu)K$$

for some positive constant K. Thus

$$\frac{d}{dt} \parallel u_\lambda(t) - u_\mu(t) \parallel^2 \leq 2K(\lambda + \mu)$$

and integrating over $[0, t]$, we get

$$\parallel u_\lambda(t) - u_\mu(t) \parallel^2 \leq \parallel u_\lambda(0) - u_\mu(0) \parallel + 2K(\lambda + \mu)t.$$

However $u_\lambda(0) = u_\mu(0) = u_0$ and thus

$$\parallel u_\lambda(t) - u_\mu(t) \parallel \leq k(\lambda + \mu)^{\frac{1}{2}} t^{\frac{1}{2}}, \ \forall t \in [0, T], \tag{3.7.11}$$

where $k := \sqrt{2K}$.

From (3.7.11), we deduce that there exists $u \in C^0([0,T]; H)$ such that

$$u_\lambda \to u \text{ in } C^0([0,T]; H), \text{ as } \lambda \to 0^+. \qquad (3.7.12)$$

We know from (3.7.9) that

$$\| \frac{du_\lambda}{dt}(t) \| \le c, \ 0 \le t \le T, \lambda > 0,$$

for some constant $c > 0$. Thus

$$\frac{du_\lambda}{dt} \xrightarrow{*} \frac{du}{dt} \text{ in } L^\infty(0,T; H), \text{ as } \lambda \to 0^+ \qquad (3.7.13)$$

and

$$u \in W^{1,\infty}(0,T; H).$$

From (3.7.12), we deduce also that $u(0) = u_0$. We have

$$A_\lambda(u_\lambda) = f - \frac{du_\lambda}{dt} \xrightarrow{*} f - \frac{du}{dt} \text{ in } L^\infty(0,T; H),$$

as $\lambda \to 0^+$. Setting $J_\lambda = (I + \lambda A)^{-1}$ we obtain

$$J_\lambda u_\lambda = u_\lambda - \lambda A_\lambda(u_\lambda)$$

and

$$\| J_\lambda u_\lambda - u \|_\infty \le \| u_\lambda - u \|_\infty + \lambda \| A_\lambda(u_\lambda) \|_\infty.$$

Using (3.7.12) and (3.7.13), we obtain

$$J_\lambda u_\lambda \to u \text{ in } C^0([0,T]; H) \text{ as } \lambda \to 0^+.$$

Let $t \in [0,T]$ be given. We have

$$\begin{aligned}
\| J_\lambda(u(t)) - u(t) \| &\le \| J_\lambda(u_\lambda(t)) - u(t) \| + \| J_\lambda(u_\lambda(t)) - J_\lambda(u(t)) \| \\
&\le \| J_\lambda(u_\lambda(t)) - u(t) \| + \| u_\lambda(t) - u(t) \|.
\end{aligned}$$

To get this last inequality we have used the property that J_λ is non-expansive [287]. Thus

$$J_\lambda(u(t)) \to u(t).$$

We know that

$$J_\lambda(u_\lambda(t)) \in D(A),$$

and since $\| A_\lambda(u_\lambda(t)) \|$ remains in a bounded set, we may conclude that

$$u(t) \in D(A)$$

since $G(A)$ is demi-closed.

Denote by $\mathcal{A} : L^2(0, T; H) \to 2^{L^2(0,T;H)}$ the operator defined by

$$(\mathcal{A}u)(t) := A(u(t))$$

and

$$D(\mathcal{A}) := \{u \in L^2(0, T; H) : u(t) \in D(A) \text{ a.e. } t \in (0, T)\}.$$

It is known [287] that $\mathcal{A}$ is a maximal monotone operator (see the beginning of this Section). From (3.7.12) and (3.7.13) we deduce that (recall that $J_\lambda u_\lambda \to u$ in $C^0([0, T]; H)$ as $\lambda \to 0^+$).

$$
\begin{aligned}
J_\lambda u_\lambda &\to u \text{ in } L^2([0, T]; H), \\
f - \frac{du_\lambda}{dt} &\rightharpoonup f - \frac{du}{dt} \text{ in } L^2([0, T]; H)
\end{aligned}
$$

and we have

$$f - \frac{du_\lambda}{dt} \in \mathcal{A}(J_\lambda u_\lambda).$$

Thus

$$f - \frac{du}{dt} \in \mathcal{A}(u)$$

since $G(\mathcal{A})$ is demi-closed. That is

$$f(t) - \frac{du}{dt}(t) \in A(u)(t) \text{ a.e. } t \in (0, T).$$

$\blacksquare$

Let $C \subset H$ be a nonempty closed convex subset of H and let $B : H \to H$ be a monotone and Lipschitz continuous operator. Then the operator

$$A := B + \partial \Psi_C$$

is maximal monotone [64] and we may apply Theorem 3.7.1 to get the following result for variational inequalities.

Corollary 3.7.2 Let H be a real Hilbert space, C a nonempty closed convex subset of H and $B : H \to H$ a monotone and Lipschitz continuous operator. Let $u_0 \in C$ and $f \in W^{1,1}(0, T; H)(0 < T < +\infty)$. Then there exists a unique $u \in W^{1,\infty}(0, T; H)$ such that

$$u(t) \in C, \ \forall t \in (0, T]; \tag{3.7.14}$$

$$u(0) = u_0 \tag{3.7.15}$$

and

$$\left(\frac{du}{dt}(t), v - u(t)\right) + (Bu(t) - f(t), v - u(t)) \geq 0,$$

$$\forall v \in C, \text{ a.e. } t \in (0, T). \tag{3.7.16}$$

From Corollary 3.7.2, we know that for any $u_0 \in C$ there exists a unique solution $u(t), t \in [0, T]$ of the evolution problem (3.7.14)-(3.7.16). One sets

$$S(t)u_0 = u(t), 0 \leq t \leq T.$$

We see that for each $x \in C$,

$$S(t + s)x = S(t)S(s)x, \ \forall \, x \, \in \, C, \ \forall \, t, \ s \, \in \, \mathbb{R}, 0 \leq t + s \leq T,$$

$$S(0)x = x$$

and the mapping

$$t \to S(t)x$$

is continuous. We see that the family of operators $\{S(t) : C \to C, T \geq t \geq 0\}$ constitutes a continuous semigroup [359] whose infinitesimal generator is given by

$$G(x) = \lim_{h \to 0^+} \frac{S(h)x - x}{h}.$$

Moreover, from (3.7.4) we obtain

$$\| \, S(t)x - S(t)y \, \| \leq \| \, x - y \, \|, \ \forall \, x, \, y \, \in \, C, \, 0 \, \leq \, t \, \leq \, T$$

so that the operator $S(t) : C \to C$ is nonexpansive. Let now $0 < T < +\infty$ be fixed. If C is a nonempty bounded and closed convex subset of the Hilbert space H, we may use the Browder-Petryshin (see e.g. [203]) fixed point theorem for nonexpansive mappings to get the existence of a fixed point $u_0 \in C$ of $S(T)$, i.e.

$$u_0 = S(T)u_0.$$

That means that

$$u(t) = S(t)u_0$$

solves (3.7.14)-(3.7.16) and satisfies

$$\begin{aligned} u(T) &= S(T)u_0 \\ &= u_0 \\ &= u(0), \end{aligned}$$

that is $S(t)u_0$ in a periodic solution of the evolution problem (3.7.14)-(3.7.16). So, we have the following result.

Corollary 3.7.3 Let H be a real Hilbert space, C a nonempty closed convex and bounded subset of H and $B : H \to H$ a monotone and Lipschitz continuous operator. Let $f \in W^{1,1}(0,T;H)(0 < T < +\infty)$. Then there exists at least one $u \in W^{1,\infty}(0,T;H)$ such that

$$u(t) \in C, \ \forall \, t \in (0,T]; \tag{3.7.17}$$

$$u(0) = u(T) \tag{3.7.18}$$

and

$$(\frac{du}{dt}(t), v - u(t)) + (Bu(t) - f(t), v - u(t)) \geq 0,$$

$$\forall \, v \in C, \text{ a.e. } t \in (0,T). \tag{3.7.19}$$

If B is strongly monotone then further results can be obtained.

Corollary 3.7.4 Let H be a real Hilbert space, C a nonempty closed convex subset of H and $B : H \to H$ a strongly monotone, i.e.

$$(Bx - Bx', x - x') \geq \beta \parallel x - x' \parallel^2, (\beta > 0), \ \forall \, x, \, x' \in H,$$

and Lipschitz continuous operator. Let $u_0 \in C$ and $f \in W^{1,1}(0,T;H)(0 < T < +\infty)$. Then there exists a unique $u \in W^{1,\infty}(0,T;H)$ solution of (3.7.14)-(3.7.16). Moreover

$$\parallel u(t) - \Pi_{B,C}(0) \parallel \leq e^{-\beta t} \parallel u_0 - \Pi_{B,C}(0) \parallel$$

$$+ \int_0^t e^{\beta(s-t)} \parallel f(s) \parallel ds, \ \forall \, 0 \leq t \leq T. \tag{3.7.20}$$

If in addition $f \in W^{1,1}(0,\infty;H)$ and

$$\lim_{t \to +\infty} \frac{\int_0^t e^{\beta s} \parallel f(s) \parallel ds}{e^{\beta t}} = 0$$

then

$$u(t) \to \Pi_{B,C}(0) \text{ as } t \to +\infty, \text{ strongly in } H.$$

Proof. The existence and uniqueness of the solution follows from Corollary 3.7.2. We know that

$$\Pi_{B,C}(0) \in C$$

and thus

$$(\frac{d}{dt}u(t), u(t) - \Pi_{B,C}(0)) + (Bu(t), u(t) - \Pi_{B,C}(0))$$

$$\leq (f(t), u(t) - \Pi_{B,C}(0)), a.e.t \in (0, T).$$

We obtain

$$(\frac{d}{dt}u(t) - \frac{d}{dt}\Pi_{B,C}(0), u(t) - \Pi_{B,C}(0))$$

$$+(Bu(t) - B(\Pi_{B,C}(0)), u(t) - \Pi_{B,C}(0))$$

$$\leq (f(t), u(t) - \Pi_{B,C}(0)), a.e.t \in (0, T)$$

since $\frac{d}{dt}\Pi_{B,C}(0) = 0$ and $-(B(\Pi_{B,C}(0)), u(t) - \Pi_{B,C}(0)) \leq 0$. Thus

$$\frac{d}{dt}(\frac{1}{2} \| u(t) - \Pi_{B,C}(0) \|^2) + \beta \| u(t) - \Pi_{B,C}(0) \|^2$$

$$\leq \| f(t) \| \| u(t) - \Pi_{B,C}(0) \|$$

a.e. $t > 0$. Hence, multiplying this last relation by $e^{2\beta t}$ and integrating over $[0, r]$, $(r \in [0, T])$ we obtain

$$\frac{1}{2}e^{2\beta r} \| u(r) - \Pi_{B,C}(0) \|^2 \leq \frac{1}{2} \| u_0 - \Pi_{B,C}(0) \|^2$$

$$+ \int_0^r e^{2\beta t} \| u(t) - \Pi_{B,C}(0) \| \| f(t) \| dt.$$

Using Gronwall inequality (see (3.7.6a) and (3.7.6b)), we obtain

$$e^{\beta r} \| u(r) - \Pi_{B,C}(0) \| \leq \| u_0 - \Pi_{B,C}(0) \| + \int_0^r e^{\beta t} \| f(t) \| dt,$$

and thus

$$\| u(r) - \Pi_{B,C}(0) \| \leq e^{-\beta r} \| u_0 - \Pi_{B,C}(0) \|$$

$$+ \int_0^r e^{\beta(t-r)} \| f(t) \| dt, \forall r \in [0, T].$$

If $f \in W^{1,1}(0, \infty; H)$ then the previous inequality holds for any $r \geq 0$ and $u(t) \to \Pi_{B,C}(0)$ as $t \to +\infty$ strongly in H provided that $\lim_{t\to+\infty} \int_0^t e^{\beta(s-t)} \| f(s) \| ds = 0$. ∎

Corollary 3.7.5 Let H be a real Hilbert space, C a nonempty closed convex subset of H and $B : H \to H$ a strongly monotone and Lipschitz continuous operator. Let $f \in W^{1,1}(0,T;H)(0 < T < +\infty)$. Then there exists at least one $u \in W^{1,\infty}(0,T;H)$ such that

$$u(t) \in C, \ \forall \, t \in (0,T], \tag{3.7.21}$$

$$u(0) = u(T) \tag{3.7.22}$$

and

$$(\frac{du}{dt} + Bu - f, v - u) \geq 0, \forall \, v \in C, \text{ a.e. } t \in (0,T). \tag{3.7.23}$$

Proof. Let u_0 be fixed in C and consider a sequence of positive numbers $\mu_n \in (0,1), \mu_n \to 1$. We know that the assumptions of Corollary 3.7.4 ensure among other things that the mapping $x \to S(t)x$ is a well-defined nonexpansive mapping. The mapping

$$x \to \mu_n S(T)x + (1 - \mu_n)u_0$$

is a contraction mapping from C onto C and thus there exists $x_n \in C$ such that

$$\mu_n S(T)x_n + (1 - \mu_n)u_0 = x_n.$$

Thus

$$S(T)x_n = \frac{x_n + (\mu_n - 1)u_0}{\mu_n}.$$

We know that the mapping $t \to u_n(t) = S(t)x_n$ satisfies

$$u_n(t) \in C, \ \forall \, t \in (0,T],$$

$$u_n(0) = x_n$$

and

$$(\frac{du_n(t)}{dt} + Bu_n(t) - f(t), v - u_n(t)) \geq 0, \forall \, v \in C, \text{ a.e. } t \in (0,T).$$

Moreover

$$u_n(T) = S(T)x_n = \frac{x_n + (\mu_n - 1)u_0}{\mu_n}.$$

From Corollary 3.7.4, we deduce that

$$\| \frac{x_n + (\mu_n - 1)u_0}{\mu_n} - \Pi_{B,C}(0) \|$$

$$\leq \ e^{-\beta T} \| x_n - \Pi_{B,C}(0) \| + \int_0^T e^{\beta(s-T)} \| f(s) \| \, ds.$$

Thus

$$\frac{\parallel x_n \parallel}{\mu_n} \;\leq\; \frac{(1-\mu_n)}{\mu_n}\parallel u_0 \parallel + \parallel \Pi_{B,C}(0) \parallel + e^{-\beta T}\parallel x_n \parallel$$

$$+ e^{-\beta T}\parallel \Pi_{B,C}(0) \parallel + \int_0^T \parallel f(s) \parallel ds$$

$$\leq \frac{(1-\mu_n)}{\mu_n}\parallel u_0 \parallel + e^{-\beta T}\parallel x_n \parallel + k$$

for some positive constant k. We obtain

$$\parallel x_n \parallel \;\leq\; (1-\mu_n)\parallel u_0 \parallel + \mu_n e^{-\beta T}\parallel x_n \parallel + \mu_n k. \qquad (3.7.24)$$

We claim that the sequence $\{x_n\}$ is bounded in H. Indeed, if $\parallel x_n \parallel \to +\infty$, then we may divide (3.7.24) by $\parallel x_n \parallel$ to get

$$1 \leq \frac{(1-\mu_n)\parallel u_0 \parallel}{\parallel x_n \parallel} + \mu_n e^{-\beta T} + \frac{\mu_n k}{\parallel x_n \parallel}$$

and taking the limit as $n \to +\infty$, we obtain

$$1 \leq e^{-\beta T} < 1$$

which is a contradiction. Along a subsequence we may assume that $x_n \rightharpoonup x_0 \in C$. If x is arbitrary in H then

$$\parallel x_n - x \parallel^2 = \parallel (x_n - x_0) + (x_0 - x) \parallel^2$$

$$= \parallel x_n - x_0 \parallel^2 + \parallel x_0 - x \parallel^2 + 2(x_n - x_0, x_0 - x). \qquad (3.7.25)$$

We have

$$S(T)x_n - x_n \;=\; \mu_n S(T)x_n + (1-\mu_n)u_0 - x_n + (1-\mu_n)(S(T)x_n - u_0)$$

$$=\; (1-\mu_n)(S(T)x_n - u_0)$$

Thus

$$S(T)x_n - x_n \to 0 \text{ as } n \to +\infty. \qquad (3.7.26)$$

From (3.7.25) with $x = S(T)x_0$, we obtain

$$\parallel x_n - S(T)x_0 \parallel^2 - \parallel x_n - x_0 \parallel^2$$

$$= \parallel x_0 - S(T)x_0 \parallel^2 + 2(x_n - x_0, x_0 - S(T)x_0)$$

and thus

$$\lim_{n\to\infty}(\parallel x_n - S(T)x_0 \parallel^2 - \parallel x_n - x_0 \parallel^2)$$

$$= \parallel x_0 - S(T)x_0 \parallel^2 . \qquad (3.7.27)$$

We have

$$\| x_n - S(T)x_0 \| \ \leq \ \| x_n - S(T)x_n \| + \| S(T)x_n - S(T)x_0 \|$$
$$\leq \ \| x_n - S(T)x_n \| + \| x_n - x_0 \|,$$

and thus

$$\| x_n - S(T)x_0 \|^2 - \| x_n - x_0 \|^2$$
$$\leq \ \| x_n - S(T)x_n \|^2 + 2 \| x_n - S(T)x_n \| \| x_n - x_0 \|$$

and we obtain from (3.7.26) that

$$\lim_{n\to\infty} (\| x_n - S(T)x_0 \|^2 - \| x_n - x_0 \|^2) = 0.$$

From (3.7.27), we deduce that

$$x_0 = S(T)x_0.$$

We have found a fixed point of the mapping $x \to S(T)x$ and thus the mapping defined by

$$u(t) = S(t)x_0$$

satisfies (3.7.21)-(3.7.23).   ∎

3.8   THE BREZIS APPROACH FOR EVOLUTION VARIATIONAL INEQUALITIES

Let X be a real reflexive Banach space. Let $D(L)$ be a vector subspace of X and $L : D(L) \to X^*$ a linear and monotone operator. Let $C \subset X$ be a closed convex subset of X such that $C \cap D(L) \neq \emptyset$.

Let us first note that the concept of maximal monotonicity introduced in Section 3.7 can be extended in the framework of operators $A : D(A) \subset X \to 2^{X^*}$ acting from X into its dual X^*, the domain, range and graph of which being respectively defined by

$$D(A) = \{u \in X : Au \neq \emptyset\},$$

$$R(A) = \cup_{x \in D(A)} Ax$$

and

$$G(A) = \{(u, v) \in X \times X^* : u \in D(A), v \in Au\}.$$

One says that A is maximal monotone if

$$\langle x^* - y^*, x - y \rangle \geq 0, \ \forall \, x, \, y \, \in \, D(A), \ x^* \, \in \, Ax, \ y^* \, \in \, Ay$$

and it follows from $(u, u^*) \in X \times X^*$ and

$$\langle u^* - v^*, u - v \rangle \geq 0, \ \forall \, (v, \, v^*) \, \in \, G(A)$$

that $(u, u^*) \in G(A)$. The graph $G(A)$ of a maximal monotone operator A is demi-closed. If X is a real reflexive Banach space such that both X and X^* are strictly convex then the monotone operator $A : X \to 2^{X^*}$ is maximal monotone if and only if $R(A + J) = X^*$. In this case the duality mapping $J : X \to X^*$ is single-valued, strictly monotone, bijective, odd, demi-continuous (i.e. $u_n \to u \Rightarrow Ju_n \rightharpoonup Ju$), bounded, positively homogeneous and

$$\langle Ju, u \rangle = \| u \|^2, \ \ \| Ju \|_* = \| u \|,$$

for all $u \in X$. Note that the inverse operator $J^{-1} : X^* \to X$ is the duality map of X^* provided that one identifies X^{**} with X. Let us also recall that if X is a real reflexive Banach space with both X and X^* strictly convex then for each $\lambda > 0$, the inverse operator

$$(A + \lambda J)^{-1} : X^* \to X$$

is single-valued, demi-continuous and maximal monotone. Note that in every reflexive Banach space X, an equivalent norm can be introduced so that X and X^* are strictly convex with respect to the new norms on X and X^* (see e.g. Proposition 32.23 in [432]). Finally, we recall that for a linear operator $L : D(L) \subset X \to X^*$ on the real reflexive Banach space, the maximal monotonicity of L holds if and only if $D(L)$ is dense in X, L and L^* are monotone and $G(L)$ is closed (see e.g. [432] for the details).

Let us now consider the inequality problem: Find $u \in C$ such that

$$\langle Lv + Au - f, v - u \rangle \geq 0, \ \forall \, v \, \in \, C \cap D(L) \qquad (3.8.1)$$

where $A : X \to X^*$ is an operator and $f \in X^*$ is given.

Lemma 3.8.1 Suppose that the aforementioned conditions are satisfied. Suppose in addition that

$$v \to \langle Av, v - x \rangle \text{ is w.l.s.c.}, \ \forall \, x \, \in \, C \qquad (3.8.2)$$

and that C is bounded. Then for each $f \in X^*$, problem (3.8.1) has at least one solution.

Proof. Let F be an arbitrary finite dimensional subspace of X. Let us denote by $i : F \to X$ the canonical injection and let $i^* : X^* \to F^*$ be its adjoint. The set $D(L) \cap F$ is a finite dimensional vector space and the set

$$C \cap D(L) \cap F$$

is thus a compact subset of F.

If $C \cap D(L) \cap F \neq \emptyset$ then there exists $u_F \in C \cap D(L) \cap F$ such that

$$\langle i^* Li(v) + i^* Ai(u_F) - i^* f, v - u_F \rangle_{F^*, F} \geq 0, \; \forall \, v \; \in \; C \cap D(L) \cap F. \quad (3.8.3)$$

Indeed, let

$$G : C \cap D(L) \cap F \to 2^F$$

be defined by

$$G(x) := \{ v \in C \cap D(L) \cap F : \langle i^* Ai(v) + i^* Li(x) - i^* f, v - x \rangle_{F^*, F} \leq 0 \}.$$

Using assumption (3.8.2), we obtain that $G(x)$ is closed in F and thus compact in F. We claim that G is a KKM-mapping. Indeed, suppose by contradiction that there exists $y_0 \in conv\{x_1, \cdots, x_n\}$ such that $y_0 \notin \bigcup_{i=1}^{n} G(x_i)$ for some $x_1, \cdots, x_n \in C \cap D(L) \cap F$.

Then

$$\langle i^* Ai(y_0) + i^* Li(x_j) - i^* f, y_0 - x_j \rangle_{F^*, F} > 0, \; \forall \, j = 1, \cdots, n.$$

Therefore

$$x_j \in \{ x \in F : \langle i^* Ai(y_0) + i^* Li(x) - f, y_0 - x \rangle_{F^*, F} > 0 \}$$

and thus since L is monotone

$$x_j \in \Lambda := \{ x \in F : \langle i^* Ai(y_0) + i^* Li(y_0) - f, y_0 - x \rangle_{F^*, F} > 0 \}.$$

The set Λ is convex and therefore $y_0 \in \Lambda$. This gives an obvious contradiction. From Theorem 1.1.20 we deduce that the family $\{ G(x) : x \in C \cap D(L) \cap F \}$ has the finite intersection property and thus since the sets $G(x)$ are compact subsets of F, we obtain the existence of

$$u_F \in \bigcap_{x \in C \cap D(L) \cap F} G(x),$$

that is u_F solves (3.8.3).

For every $v \in C \cap D(L)$, we define

$$M(v) := \{u \in C : \langle Au + Lv - f, v - u \rangle \geq 0\}.$$

The set $M(v)$ is weakly compact and it follows that

$$\bigcap_{v \in C \cap D(L)} M(v) \neq \emptyset$$

provided that

$$\bigcap_{j=1}^{n} M(v_j) \neq \emptyset$$

for any $v_1, \cdots, v_n \in C \cap D(L)$. Let F be the finite dimensional linear space spanned by $\{v_1, \cdots, v_n\}$. We know that there exists $u_F \in C \cap F$ such that

$$\langle i^* A i u_F + i^* L i v - i^* f, v - u_F \rangle_{F^*, F} \geq 0, \ \forall\, v \in C \cap D(L) \cap F$$

and thus

$$\langle A u_F + L v_i - f, v_i - u_F \rangle \geq 0, \ \forall\, i = 1, \cdots, n.$$

That means that

$$u_F \in \bigcap_{j=1}^{n} M(v_j).$$

The result follows. $\blacksquare$

Theorem 3.8.2 Suppose that the aforementioned conditions are satisfied. We assume that

$$v \to \langle Av, v - x \rangle \text{ is w.l.s.c.}, \ \forall\, x \in C \tag{3.8.4}$$

and

$$\frac{\langle Av, v - x_0 \rangle}{\| v \|} \to +\infty \text{ as } \| v \| \to +\infty, v \in C \tag{3.8.5}$$

for some $x_0 \in D(L) \cap C$. Then for each $f \in X^*$, problem (3.8.1) has at least one solution.

Proof. Let C_n be the closed convex set defined by

$$C_n := \{x \in C : \| x \| \leq n\}.$$

There exists $n_0 \in \mathbb{N}$ such that $x_0 \in C_n, \ \forall n \geq n_0$. Using Lemma 3.8.1, we obtain the existence of $u_n \in C_n$ such that

$$\langle A u_n, v - u_n \rangle + \langle L v, v - u_n \rangle \geq \langle f, v - u_n \rangle, \ \forall\, v \in C_n \cap D(L).$$

In particular

$$\langle Au_n, u_n - x_0 \rangle \le \langle f, u_n - x_0 \rangle + \langle Lx_0, x_0 - u_n \rangle. \tag{3.8.6}$$

We claim that the sequence $\{u_n\}$ is bounded. If we suppose the contrary then $\| u_n \| \to +\infty$, and along a subsequence $v_n := \frac{u_n}{\|u_n\|} \rightharpoonup v$. From (3.8.6) we deduce that

$$\frac{\langle Au_n, u_n - x_0 \rangle}{\| u_n \|} \le \langle f, v_n - x_0 \| u_n \|^{-1} \rangle + \langle Lx_0, x_0 \| u_n \|^{-1} - v_n \rangle.$$

Taking the limit as $n \to +\infty$ and using assumption (3.8.5), we obtain the contradiction $\infty \le \langle f - Lx_0, v \rangle$. Thus the sequence $\{u_n\}$ is bounded and along a subsequence we may suppose that $u_n \rightharpoonup u$. Let $v \in C \cap D(L)$ be given. We see that $v \in C_n \cap D(L)$ for all n large enough and

$$\langle Au_n, u_n - v \rangle + \langle Lv, u_n - v \rangle - \langle f, u_n - v \rangle \le 0.$$

Taking the limit inferior as $n \to \infty$ and using assumption (3.8.4), we obtain

$$\langle Au, u - v \rangle + \langle Lv, u - v \rangle - \langle f, u - v \rangle \le 0.$$

This last inequality holds for any $v \in C \cap D(L)$ and we may conclude that u is a solution of (3.8.1). $\blacksquare$

Remark 3.8.3   i) Note that the original approach of Brézis [62] deals with bounded pseudomonotone operators instead of operators satisfying the condition (3.8.4).

ii) More generally, we see from the proof of Theorem 3.8.2 that the coercivity assumption may be replaced by the more general one: There exists $x_0 \in D(L) \cap C$ such that

$$\langle Av, v - x_0 \rangle + \langle Lx_0, v - x_0 \rangle - \langle f, v - x_0 \rangle \to +\infty$$

as $\| v \| \to +\infty, v \in C$.

Let us now consider the problem: Find $u \in C \cap D(L)$ such that

$$\langle Lu + Au - f, v - u \rangle \ge 0, \ \forall v \in C \cap D(L). \tag{3.8.7}$$

If u is a solution of (3.8.7) then the monotonicity of L entails that u is a solution of problem (3.8.1). The converse is true provided that in addition to the previous assumptions the operator L is maximal monotone and the invariance property

$$(J + \varepsilon L)^{-1} J(C) \subset C$$

is satisfied for $\varepsilon > 0$ small. More precisely, we have the following result.

Theorem 3.8.4   Let X be a real Hilbert space, C a nonempty closed convex subset of X, $L : D(L) \to X$ a maximal monotone linear operator satisfying

$$C \cap D(L) \neq \emptyset \qquad (3.8.8)$$

and

$$(J + \varepsilon L)^{-1} J(C) \subset C, \ \forall \, \varepsilon > 0, \ \varepsilon \text{ small.} \qquad (3.8.9)$$

Let $A : X \to X^*$ be a bounded operator satisfying conditions (3.8.4) and (3.8.5) then there exists at least one $u \in D(L) \cap C$ such that

$$\langle Lu + Au - f, v - u \rangle \geq 0, \ \forall \, v \, \in \, C \cap D(L).$$

Proof.   From Theorem 3.8.2, there exists $u \in C$ such that

$$\langle Lv + Au - f, v - u \rangle \geq 0, \ \forall \, v \, \in \, C \cap D(L). \qquad (3.8.10)$$

Let $u_n \in D(L)$ be the unique solution of

$$J u_n + \frac{1}{n} L u_n = J u,$$

that is

$$u_n = (J + \frac{1}{n} L)^{-1} J u.$$

We have (see e.g. [64], [358])

$$u_n \to P_{\overline{D(L)}} u,$$

where $P_{\overline{D(L)}}$ denotes the projection operator from X onto $\overline{D(L)}$. However, since L is linear and maximal monotone

$$\overline{D(L)} = X$$

and thus

$$u_n \to u \text{ as } n \to +\infty.$$

Assumption (3.8.9) implies that $u_n \in C$ and thus setting $v = u_n \in C \cap D(L)$ in (3.8.10) we obtain

$$\langle L u_n, u_n - u \rangle \geq -\langle Au - f, u_n - u \rangle$$

or also

$$\langle Lu_n, u \rangle \leq \langle Au - f, u_n - u \rangle + \langle Lu_n, u_n \rangle. \tag{3.8.11}$$

Recalling that here $J : X \to X^*$ is linear since H is a Hilbert space, we obtain

$$\langle Lu_n, u \rangle = \langle Lu_n, u_n + \frac{1}{n} J^{-1} Lu_n \rangle$$

$$= \langle Lu_n, u_n \rangle + \frac{1}{n} \parallel Lu_n \parallel_*^2 \tag{3.8.12}$$

and using (3.8.11) together with (3.8.12), we get

$$\begin{aligned} \parallel Lu_n \parallel_*^2 \ &\leq \ n\langle Au - f, u_n - u \rangle \\ &= \ \langle Au - f, J^{-1} Lu_n \rangle \\ &\leq \ \parallel Au - f \parallel_* \parallel J^{-1} \parallel \parallel Lu_n \parallel_* \end{aligned}$$

so that $\{Lu_n\}$ is bounded in X^*. Thus, for a subsequence

$$u_n \in D(L) \cap C,$$

$$u_n \to u \text{ in } X$$

and

$$Lu_n \rightharpoonup l \text{ in } X^*.$$

Any maximal monotone operator is graph demi-closed and thus

$$u \in D(L)$$

and

$$l = Lu.$$

Let $h \in C \cap D(L)$ be given. If $0 < \theta \leq 1$ then

$$\theta h + (1 - \theta)u \in C \cap D(L)$$

and setting $v = \theta h + (1 - \theta)u$ in (3.8.10), we obtain

$$\theta\langle L(\theta h + (1 - \theta)u) + Au - f, h - u \rangle \geq 0. \tag{3.8.13}$$

Dividing (3.8.13) by θ and letting $\theta \to 0^+$, we get

$$\langle Lu + Au - f, h - u \rangle \geq 0.$$

Since h is arbitrary in $C \cap D(L)$, we obtain the result.   ∎

3.9   THE MAXIMAL MONOTONE APPROACH

We have already seen in Section 3.6 and 3.7 that maximal monotone operators possess nice properties which can be used to develop theoretical results applicable to the study of some classes of variational inequalities. In this section we discuss some additional results that are of particular interest to study variational inequalities.

3.9.1   THE MAIN THEOREM ON PSEUDOMONOTONE PERTURBATIONS OF MAXIMAL MONOTONE MAPPINGS

In this Section we recall the main theorem on pseudomonotone perturbations of maximal monotone mappings due to Browder [71]. We do not present its proof since it is very involved. The reader can found detailed proof of the result in Browder [71] or Zeidler [432]. Our aim in this Section is to point out the consequences of this result in the field of variational inequalities. Let us first recall that an operator $B : X \to X^*$ is called pseudomonotone if whenever $u_n \rightharpoonup u$ and $\limsup_{n \to \infty} \langle Bu_n, u_n - u \rangle \leq 0$ then

$$\langle Bu, u - x \rangle \leq \liminf_{n \to \infty} \langle Bu_n, u_n - x \rangle, \ \forall x \in X.$$

Recall also that one says that the operator $B : X \to X^*$ is bounded if it maps bounded sets of X into bounded sets of X^*.

Theorem 3.9.1 Let X be a real reflexive Banach space, $A : D(A) \subset X \to 2^{X^*}$ a maximal monotone operator, $B : X \to X^*$ a pseudomonotone, bounded, demi-continuous operator. We assume that there exists $u_0 \in D(A)$ such that

$$\frac{\langle Bu, u - u_0 \rangle}{\| u \|} \to +\infty \text{ as } \| u \| \to +\infty.$$

Then

$$R(A + B) = X^*.$$

Let V be a real reflexive Banach space and H a real Hilbert space. We suppose that

$$V \hookrightarrow H \hookrightarrow V^*,$$

with dense and continuous embeddings. We define

$$Lu = u',$$

and

$$D(L) = \{u \in L^2(0, T; V) : u' \in L^2(0, T; H) \text{ and } u(0) = u_0\}.$$

Set

$$X = L^2(0; T; V).$$

It is known that

$$L : D(L) \subset X \to X^*$$

is maximal monotone ([432]; Proposition 32.10). Let $B : X \to X^*$ be a bounded monotone and hemicontinuous operator and $\varphi : X \to \mathbb{R} \cup \{+\infty\}$ a proper convex and l.s.c. function. Recall here that a monotone and hemicontinuous operator is pseudomonotone and demi-continuous ([432]; figure 27.1). We suppose that

$$D(L) \cap int\{D(\partial\varphi)\} \neq \emptyset$$

and

$$\frac{\langle Bu, u - u_0 \rangle}{\parallel u \parallel} \to +\infty \text{ as } \parallel u \parallel \to +\infty,$$

for some $u_0 \in D(L) \cap D(\partial\varphi)$. Then $A := L + \partial\varphi$ is maximal monotone ([432]; Theorem 32.I) and from Theorem 3.9.1, we may deduce that for each $f \in X^*$, there exists at least one $u \in L^2(0, T; V)$ such that

$$u' \in L^2(0, T; H)$$

$$u(0) = u_0$$

and

$$\langle u' + Bu - f, v - u \rangle + \varphi(v) - \varphi(u) \geq 0,$$

for each $v \in X$.

The operator $L : D(L) \subset X \to X^*$ defined by

$$Lu = u'$$

and

$$D(L) = \{u \in L^2(0, T, V) : u' \in L^2(0, T; H), u(0) = u(T)\}$$

is maximal monotone ([432]; Proposition 32.10). So, with the same conditions on φ and B specified above, we may deduce that for each $f \in X^*$, there exists at least one $u \in L^2(0, T; V)$ such that

$$u' \in L^2(0, T; H)$$

$$u(0) = u(T)$$

and

$$\langle u' + Bu - f, v - u \rangle + \varphi(v) - \varphi(u) \geq 0,$$

for each $v \in X$.

3.9.2   THE BRÉZIS-HARAUX THEOREM

In this Section, we recall a fundamental result due to Brézis and Haraux [65], which characterizes the range of the sum operator $L + \partial\varphi$, where L denotes a maximal monotone operator and φ a convex functional. The result will be used later to study noncoercive evolution unilateral problems involving periodic conditions.

Theorem 3.9.2 Let H be a real Hilbert space, $L : D(L) \subset H \to 2^H$ a maximal monotone operator and $\varphi : H \to \mathbb{R} \cup \{+\infty\}$ a proper convex and lower semicontinuous function such that

$$\varphi((id_H - \lambda L)^{-1}u) \leq \varphi(u),$$

for all $u \in H$ and $\lambda > 0$, λ small. Then

$$\overline{R(L + \partial\varphi)} = \overline{R(L) + R(\partial\varphi)}$$

and

$$int\{R(L + \partial\varphi)\} = int\{R(L) + R(\partial\varphi)\}.$$

Remark 3.9.3 If L is linear, maximal monotone and such that $R(L)$ is closed then $f \in int\{R(L) + R(\partial\varphi)\}$ if and only if there exist $\sigma > 0$ and $M \in \mathbb{R}$ such that [65]

$$\varphi(v) - (f,v)_H \geq \sigma \parallel v \parallel_H -M, \ \forall \, v \in Ker \, L. \tag{3.9.1}$$

Let $L : H \to H$ be a bounded, linear, symmetric and nonnegative operator such that $dim\{Ker L\} < +\infty$ and $R(L)$ is closed. We claim that condition (3.9.1) is satisfied if and only if

$$\varphi_\infty(v) > (f,v)_H, \ \forall \, v \in Ker \, L \backslash \{0\}. \tag{3.9.2}$$

Indeed, suppose that (3.9.1) is satisfied. Then

$$\frac{\varphi(tv)}{t} \geq (f,v)_H + \sigma \parallel v \parallel_H - \frac{M}{t}, \ \forall t > 0, \ v \in Ker \ L$$

and it results that

$$\varphi_\infty(v) \geq (f,v)_H + \sigma \parallel v \parallel_H, \ \forall \, v \in Ker \ L$$

and thus

$$\varphi_\infty(v) > (f,v)_H, \ \forall \, v \in Ker \ L\backslash\{0\}.$$

Conversely, suppose that (3.9.2) is satisfied. We claim that (3.9.1) holds. Suppose by contradiction that (3.9.1) is not satisfied. Then, we may found a sequence $v_n \in KerL$ such that

$$\varphi(v_n) - (f,v_n)_H < \frac{1}{n} \parallel v_n \parallel_H, \ \forall \, n \in \mathbb{N}\backslash\{0\}. \tag{3.9.3}$$

Set $t_n = \parallel v_n \parallel_H$ and $u_n = \frac{v_n}{t_n}$. We may assume that along a subsequence $u_n \rightharpoonup u$ in H. However, $u_n \in KerL$ and thus $P_{R(L)}(u_n) = 0, \ \forall n \in \mathbb{N}$ since the conditions assumed on L entail among other things that $(Ker \ L)^\perp = R(L)$. It results that

$$P_{R(L)}(u_n) \to 0 \text{ as } n \to +\infty$$

and obviously

$$P_{Ker(L)}(u_n) \to u \text{ as } n \to +\infty.$$

Thus $u_n \to u \in Ker(L)\backslash\{0\}$. From (3.9.3), we have

$$\frac{\varphi(t_n u_n)}{t_n} < (f,u_n)_H + \frac{1}{n} \parallel u_n \parallel_H, \ \forall \, n \in \mathbb{N}.$$

Taking the limit inferior as $n \to +\infty$, we obtain

$$\varphi_\infty(u) \leq (f,u)_H,$$

which is a contradiction since $u \in KerL\backslash\{0\}$. It results the following corollary.

Corollary 3.9.4 Let H be a real Hilbert space, $L : H \to H$ a bounded linear and symmetric nonnegative operator such that $dim\{KerL\} < +\infty$ and $R(L)$ is closed, and $\varphi : H \to \mathbb{R}\cup\{+\infty\}$ a proper convex and l.s.c. function such that

$$\varphi((id_H - \lambda L)^{-1})u) \leq \varphi(u),$$

for all $u \in H$ and $\lambda > 0, \lambda$ small. Then for each $f \in H$ satisfying

$$\varphi_\infty(v) > (f, v)_H, \ \forall \, v \in Ker \, L \backslash \{0\}, \tag{3.9.4}$$

there exists $u \in H$ such that

$$(Lu, v - u)_H + \varphi(v) - \varphi(u) \geq (f, v - u)_H, \ \forall \, v \in H.$$

Proof. We have seen that the condition (3.9.4) is equivalent to condition (3.9.1) which implies that

$$f \in int\{R(L) + R(\partial\varphi)\}.$$

From Theorem 3.9.2, we deduce that

$$f \in int\{R(L + \partial\varphi)\} \subset R(L + \partial\varphi).$$

Thus there exists $u \in H$ such that $f \in Lu + \partial\varphi(u)$ and the result follows from the definition of $\partial\varphi$. $\blacksquare$

Note that if L satisfies the condition of Corollary 3.9.4 then it is semicoercive (see Proposition 3.4.3). Another class of operators can now been considered. Let $L : D(L) \subset H \to H$ be a linear maximal monotone operator such that the set

$$\{x \in D(L) : \| x \|_H \leq 1 \text{ and } \| Lx \|_H \leq 1\}$$

is compact. Then condition (3.9.2) is also equivalent to condition (3.9.1). The fact that (3.9.1) implies (3.9.2) is proved as above. Assume that (3.9.2) is satisfied and suppose by contradiction that we may find a sequence $v_n \in Ker L$ satisfying (3.9.3). Setting $u_n = \frac{v_n}{t_n}$ with $t_n = \| v_n \|_H$, we obtain that $\{u_n\}$ is bounded. We have also $Lu_n = 0, \ \forall n \in \mathbb{N}$ and there exists a subsequence such that $u_n \to u \neq 0$ in H. Dividing (3.9.3) by t_n and taking the limit as $n \to +\infty$, we obtain

$$\varphi_\infty(u) \leq (f, u)_H,$$

which is a contradiction. We obtain the following result.

Corollary 3.9.5 Let H be a real Hilbert space and $L : D(L) \subset H \to H$ a linear and maximal monotone operator such that the set

$$\{x \in D(L) : \| x \|_H \leq 1 \text{ and } \| Lx \|_H \leq 1\}$$

is compact. Let $\varphi : H \to \mathbb{R} \cup \{+\infty\}$ be a proper convex and l.s.c. function such that

$$\varphi((id_H - \lambda L)^{-1} u) \leq \varphi(u),$$

for all $u \in H$ and $\lambda > 0$, λ small. Then for any $f \in H$ satisfying

$$\varphi_\infty(v) > (f, v)_H, \ \forall\, v \in Ker\, L \backslash \{0\}, \qquad (3.9.5)$$

there exists $u \in D(L)$ such that

$$(Lu, v - u)_H + \varphi(v) - \varphi(u) \geq (f, v - u)_H, \ \forall\, v \in H.$$

3.10 THE GENERALIZED HARTMAN-STAMPACCHIA THEOREM FOR VARIATIONAL-HEMIVARIATIONAL INEQUALITIES

Let X be a real reflexive Banach space and let $(T, \mathcal{T}, \mu)$ be a complete positive measure space such that $0 < \mu(T) < +\infty$. Let

$$\gamma : X \to L^p(T; \mathbb{R}^m), (m \in \mathbb{N} \backslash \{0\}, p \in [1, +\infty)),$$

be a given linear and compact operator. Note here that at this point the space X and the operator γ are abstractions of respectively the usual Sobolev spaces and the embeddings or trace operators we will use later in the applications. Let $j : T \times \mathbb{R}^m \to \mathbb{R}$ be a function such that

$$x \to j(x, y) : T \to \mathbb{R} \text{ is measurable}, \forall y \in \mathbb{R}^m; j(., e) \in L^1(T), \quad (3.10.1)$$

for some $e \in L^p(T; \mathbb{R}^m)$ and either a)

$$\mid j(x, y_1) - j(x, y_2) \mid \leq k(x) \mid y_1 - y_2 \mid, \forall x \in T, y_1, y_2 \in \mathbb{R}^m, \quad (3.10.2a)$$

for a function $k \in L^q(T; \mathbb{R}^m)(\frac{1}{q} + \frac{1}{p} = 1)$, or b)

$$y \to j(x, y) : \mathbb{R}^m \to \mathbb{R} \text{ is locally Lipschitz}, \ \forall\, x \in T, \qquad (3.10.2b)$$

and there is a constant $c > 0$ such that

$$\mid z \mid \leq c(1 + \mid y \mid^{p-1}), \ \forall\, x \in T, \ y \in \mathbb{R}^m, \ z \in \partial_y j(x, y) \qquad (3.10.2c)$$

Let C be a nonempty closed convex subset of X, $A : X \to X^*$ a (possibly nonlinear) operator, $f \in X^*$ and $\Phi : X \to \mathbb{R} \cup \{+\infty\}$ a convex and l.s.c. function such that

$$D(\Phi) \cap C \neq \emptyset. \tag{3.10.3}$$

We consider the inequality problem: Find $u \in C$ such that

$$\langle Au - f, v - u \rangle + \Phi(v) - \Phi(u)$$

$$+ \int_T j_y^0(x, \gamma u(x); \gamma v(x) - \gamma u(x)) d\mu \geq 0, \forall\, v \,\in\, C. \tag{3.10.4}$$

A direct application of the KKM-principle leads to the following basic lemma.

Lemma 3.10.1 Let X be a real reflexive Banach space, C a nonempty bounded, closed, convex subset of X, $\Phi : X \to \mathbb{R} \cup \{+\infty\}$ a convex and l.s.c. function such that $D(\Phi) \cap C \neq \emptyset$, $A : X \to X^*$ an operator, $Y \subset X$ a Banach space, $L : X \to Y$ a compact and linear mapping and $J : Y \to \mathbb{R}$ a locally Lipschitz function. Suppose in addition that the mapping

$$v \to \langle Av, v - x \rangle,$$

is weakly lower semicontinuous on C, for each $x \in C$. Then for each $f \in X^*$, there exists at least one $u \in C \cap D(\Phi)$ such that

$$\langle Au - f, v - u \rangle + \Phi(v) - \Phi(u) + J^0(L(u); L(v - u)) \geq 0, \forall\, v \,\in\, C. \tag{3.10.5}$$

Proof. Let us define the set-valued mapping $G : C \cap D(\Phi) \to 2^X$ by

$$G(x) = \{ v \in C \cap D(\Phi) : \langle Av - f, v - x \rangle - J^0(L(v); L(x - v))$$

$$+ \Phi(v) - \Phi(x) \leq 0 \}.$$

The set $G(x)$ is weakly closed. Indeed, if $v_n \rightharpoonup v$ then

$$\langle Av, v - x \rangle \leq \liminf_{n \to \infty} \langle Av_n, v_n - x \rangle$$

and

$$\Phi(v) \leq \liminf_{n \to \infty} \Phi(v_n).$$

Moreover, $L(v_n) \to L(v)$ and thus using the upper-semicontinuity of $J^0(x, v)$ as a function of $(x, v) \in Y \times Y$, we obtain also

$$\limsup_{n \to \infty} J^0(L(v_n); L(x - v_n)) \leq J^0(L(v); L(x - v)).$$

Thus

$$-J^0(L(v); L(x - v)) \leq \liminf_{n \to \infty} -J^0(L(v_n); L(x - v_n)).$$

So, if $v_n \in G(x)$ and $v_n \rightharpoonup v$ then

$$\langle Av - f, v - x \rangle - J^0(L(v); L(x - v)) + \Phi(v) - \Phi(x)$$
$$\leq \liminf_{n \to \infty} \{ \langle Av_n - f, v_n - x \rangle - J^0(L(v_n); L(x - v_n))$$
$$+ \Phi(v_n) - \Phi(x) \} \leq 0,$$

so that $v \in G(x)$. Then $G(x)$ is a weakly closed subset of C and thus $G(x)$ is weakly compact since C is bounded. It results that if the family of sets $\{G(x) : x \in C \cap D(\Phi)\}$ has the finite intersection property then

$$\bigcap_{x \in C \cap D(\Phi)} G(x) \neq \emptyset.$$

We could conclude by using KKM-principle provided that we could show that G is a KKM-mapping. Suppose by contradiction that there exists a finite subset $\{x_1, \cdots, x_n\} \subset C \cap D(\Phi)$ and $y_0 \in conv\{x_1, \cdots, x_n\}$ such that $y_0 \notin \bigcup_{i=1}^n G(x_i)$. Then

$$\langle Ay_0 - f, y_0 - x_i \rangle + \Phi(y_0) - \Phi(x_i) - J^0(L(y_0); L(x_i - y_0)) > 0, \ \forall \, i = 1, \cdots, n.$$

Therefore

$$x_i \in \Lambda := \{ x \in X : \langle Ay_0 - f, y_0 - x \rangle - J^0(L(y_0); L(x - y_0))$$
$$+ \Phi(y_0) - \Phi(x) > 0 \},$$

for all $i \in \{1, \cdots, n\}$. The set Λ is convex and thus $y_0 \in \Lambda$, leading to an obvious contradiction. So,

$$\bigcap_{x \in C \cap D(\Phi)} G(x) \neq \emptyset.$$

This yields an element $u \in C \cap D(\Phi)$ such that

$$\langle Au - f, v - u \rangle + \Phi(v) - \Phi(u) + J^0(L(u); L(v - u)) \geq 0, \forall \, v \in C \cap D(\Phi).$$

The inequality is trivially satisfied if $v \notin D(\Phi)$ and the conclusion follows. $\blacksquare$

We may now derive a result applicable to the inequality problem (3.10.4). Indeed, suppose that the aforementioned conditions are satisfied and set

$$Y = L^p(T; \mathbb{R}^m), L \equiv \gamma.$$

Let $J : Y \to \mathbb{R}$ be the function defined by

$$J(u) = \int_T j(x, u(x))d\mu. \tag{3.10.6}$$

The conditions (3.10.1, 2a) or (3.10.1, 2b, 2c) on j ensure that J is locally Lipschitz on Y and

$$\int_T j_y^0(x, u(x), v(x))d\mu \geq J^0(u, v), \ \forall \, u, \, v \, \in Y.$$

It results that

$$\int_T j_y^0(x, \gamma u(x); \gamma v(x))d\mu \geq J^0(\gamma u; \gamma v), \ \forall \, u, \, v \, \in X. \tag{3.10.7}$$

Thus if $u \in C$ solves (3.10.5) then u solves the inequality problem (3.10.4) too. The following result follows.

Theorem 3.10.2 Suppose that the aforementioned conditions are satisfied. Suppose in addition that C is bounded and the mapping

$$v \to \langle Av, v - x \rangle$$

is w.l.s.c. on C for all $x \in C$. Then there exists at least one $u \in C$ solution of (3.10.4).

A variant of Lemma 3.10.1 for monotone and hemicontinuous operators is now obtained.

Lemma 3.10.3 Let X be a real reflexive Banach space, C a nonempty, bounded, closed, convex subset of $X, \Phi : X \to \mathbb{R} \cup \{+\infty\}$ a convex and l.s.c. function such that $D(\Phi) \cap C \neq \emptyset$, $A : X \to X^*$ an operator, $Y \subset X^*$ a Banach space, $L : X \to Y$ a compact and linear mapping and $J : Y \to \mathbb{R}$ a locally Lipschitz function. Suppose in addition that A is monotone and hemicontinuous. Then for each $f \in X^*$, there exists at least one $u \in C \cap D(\Phi)$ satisfying (3.10.5).

Proof. Set

$$g(x, y) = \langle Ax - f, y - x \rangle - J^0(L(y); L(x) - L(y))$$

and

$$f(x, y) = \langle Ay - f, y - x \rangle - J^0(L(y); L(x) - L(y)).$$

The monotonicity of A implies that

$$g(x,y) \le f(x,y), \ \forall \, x, \, y \, \in \, X.$$

The map $x \to f(x,y)$ is concave and the map $y \to g(x,y)$ is w.l.s.c. Applying Theorem 1.1.21 with $\lambda = 0$, we obtain the existence of $u \in C$ satisfying

$$g(v,u) + \Phi(u) - \Phi(v) \le 0, \forall \, v \, \in \, C,$$

that is

$$\langle Av - f, v - u \rangle + \Phi(v) - \Phi(u) + J^0(L(u); L(v-u)) \ge 0,$$

for all $v \in C$. Using Theorem 1.1.23 with $H(u,v) = J^0(L(u); L(v))$, we obtain also that

$$\langle Au - f, v - u \rangle + \Phi(v) - \Phi(u) + J^0(L(u); L(v-u)) \ge 0, \forall \, v \, \in \, C.$$

$$\blacksquare$$

The analogue of Theorem 3.10.2 for monotone hemicontinuous operators can now be given.

Theorem 3.10.4   Suppose that the aforementioned conditions are satisfied. Suppose in addition that C is bounded and A is monotone and hemicontinuous. Then there exists at least one $u \in C$ solution of (3.10.4).

3.11   COERCIVE VARIATIONAL-HEMIVARIATIONAL INEQUALITIES

The notations of Section 3.10 are here again used. Note that if j satisfies conditions (3.10.1) and (3.10.2a) then

$$\mid \int_T j_y^0(x, \gamma u(x); \gamma v(x)) d\mu \mid \le \int_T k(x) \mid \gamma v(x) \mid d\mu$$

$$\le \mid k \mid_{0,q} \mid \gamma v \mid_{0,p}$$

$$\le c \mid k \mid_{0,q} \parallel v \parallel, \tag{3.11.1}$$

where c denotes the constant of continuity of the linear operator γ. On the other hand if j satisfies conditions (3.10.1), (3.10.2b) and (3.10.2c) then

$$\mid j_y^0(x, u(x); v(x)) \mid \le c(1+ \mid \gamma u(x) \mid^{p-1}) \mid \gamma v(x) \mid$$
$$= c \mid \gamma v(x) \mid + c \mid \gamma u(x) \mid^{p-1} \mid \gamma v(x) \mid$$

and thus

$$\mid \int_T j_y^0(x, \gamma u(x); \gamma v(x)) d\mu \mid \le C_1 \parallel v \parallel + C_2 \parallel u \parallel^{p-1} \parallel v \parallel \qquad (3.11.2)$$

for some suitable constants $C_1, C_2 > 0$. Let us now discuss the solvability of coercive variational-hemivariational inequalities.

Theorem 3.11.1 Let C be a nonempty closed convex subset of X, $\Phi : X \to \mathbb{R} \cup \{+\infty\}$ a proper convex and l.s.c. function such that $C \cap D(\Phi) \ne \emptyset$ and $A : X \to X^*$ an operator such that the mapping

$$v \to \langle Av, v - x \rangle$$

is w.l.s.c. on C, for all $x \in C$. Then the following two results hold.

A) If j satisfies conditions (3.10.1) and (3.10.2a), and if there exists $x_0 \in C \cap D(\Phi)$ such that

$$\frac{\langle Au, u - x_0 \rangle + \Phi(u)}{\parallel u \parallel} \to +\infty, \text{ as } \parallel u \parallel \to +\infty \qquad (3.11.3)$$

then for each $f \in X^*$, there exists $u \in C$ such that

$$\langle Au - f, v - u \rangle + \Phi(v) - \Phi(u)$$

$$+ \int_T j_y^0(x, \gamma u(x); \gamma v(x) - \gamma u(x)) d\mu \ge 0, \forall v \in C. \qquad (3.11.4)$$

B) If j satisfies conditions (3.10.1), (3.10.2b) and (3.10.2c) and if there exists $x_0 \in C \cap D(\Phi)$ and $\theta \ge p$ such that

$$\frac{\langle Au, u - x_0 \rangle}{\parallel u \parallel^\theta} \to +\infty, \text{ as } \parallel u \parallel \to +\infty, u \in C \qquad (3.11.5)$$

then for each $f \in X^*$, there exists $u \in C$ satisfying (3.11.4).

Proof. There exists $n_0 \in \mathbb{N}$ such that

$$x_0 \in C_n := \{x \in C : \parallel x \parallel \le n\}, \forall n \ge n_0.$$

Using Lemma 3.10.1 with J defined in (3.10.6),we get the existence of $u_n \in C_n$ such that $(\forall n \geq n_0)$

$$\langle Au_n - f, v - u_n \rangle + \Phi(v) - \Phi(u_n)$$

$$+ J^0(\gamma(u_n); \gamma(v) - \gamma(u_n)) \geq 0, \ \forall \, v \ \in \ C_n. \tag{3.11.6}$$

We claim that the sequence $\{u_n\}$ is bounded. Suppose by contradiction that $\| u_n \| \to +\infty$. Then along a subsequence, we may assume that

$$z_n := \frac{u_n}{\| u_n \|} \rightharpoonup z.$$

Setting $v = x_0$ in (3.11.6) for $n \geq n_0$ and using (3.10.7), we obtain

$$\langle Au_n, u_n - x_0 \rangle + \Phi(u_n) \leq \Phi(x_0) + \langle f, u_n - x_0 \rangle + J^0(\gamma(u_n); \gamma(x_0 - u_n))$$

$$\leq \Phi(x_0) + \langle f, u_n - x_0 \rangle + \left| \int_T j^0(x, \gamma(u_n); \gamma(x_0 - u_n)) d\mu \right| . \tag{3.11.7}$$

Case A. Using (3.11.1) we obtain

$$\langle Au_n, u_n - x_0 \rangle + \Phi(u_n) \leq \Phi(x_0) + \langle f, u_n - x_0 \rangle + c \mid k \mid_{0,q} \| u_n - x_0 \|$$

and thus

$$\frac{\langle Au_n, u_n - x_0 \rangle + \Phi(u_n)}{\| u_n \|} \leq \frac{\Phi(x_0)}{\| u_n \|} + \langle f, z_n - x_0 \| u_n \|^{-1} \rangle$$

$$+ c \mid k \mid_{0,q} \| z_n - x_0 \| u_n \|^{-1} \| .$$

Taking the limit as $n \to +\infty$, the left-hand term tends to $+\infty$ while the right-hand term remains bounded. It results a contradiction.

Case B. The function Φ being convex and l.s.c., we may apply Theorem 1.1.11 to get

$$\Phi(x) \geq \langle \alpha, x \rangle + \beta, \forall \, x \ \in \ X,$$

for some $\alpha \in X^*, \beta \in \mathbb{R}$. That means that

$$\Phi(x) \geq - \| \alpha \|_* \| x \| + \beta, \forall \, x \ \in \ X.$$

From (3.10.7) and (3.11.2), we deduce that

$$\langle Au_n, u_n - x_0 \rangle \leq \Phi(x_0) + \| \alpha \|_* \| u_n \| - \beta + \langle f, u_n - x_0 \rangle$$
$$+ C_1 \| u_n - x_0 \| + C_2 \| u_n \|^{p-1} \| u_n - x_0 \| .$$

Thus

$$\frac{\langle Au_n, u_n - x_0 \rangle}{\| u_n \|^\theta} \leq \| \alpha \|_* \| z_n \| \| u_n \|^{1-\theta} + (\Phi(x_0) - \beta) \| u_n \|^{-\theta}$$

$$+ \langle f, z_n \| u_n \|^{1-\theta} - x_0 \| u_n \|^{-\theta} \rangle$$
$$+ C_1 \| z_n \| u_n \|^{1-\theta} - x_0 \| u_n \|^{-\theta} \|$$
$$+ C_2 \| u_n \|^{p-\theta} \| v_n - x_0 \| u_n \|^{-1} \|$$

and taking the limit as $n \to +\infty$, we obtain a contradiction since $\theta \geq p \geq 1$.

Thus, in case A as well as in case B, the sequence $\{u_n\}$ is bounded. Along a subsequence, we may therefore assume that $u_n \rightharpoonup u \in C$ and

$$\langle Au_n - f, v - u_n \rangle + \Phi(v) - \Phi(u_n) + J^0(\gamma(u_n); \gamma(v) - \gamma(u_n)) \geq 0,$$

for all $v \in C_n$. Let $v \in C$ be given. For n large enough, $v \in C_n$ and thus for n large enough, we have

$$\langle Au_n - f, u_n - v \rangle + \Phi(u_n) - \Phi(v) - J^0(\gamma(u_n); \gamma(v) - \gamma(u_n)) \leq 0. \quad (3.11.8)$$

We have

$$\langle Au, u - v \rangle \leq \liminf_{n\to\infty} \langle Au_n - f, u_n - v \rangle$$
$$\Phi(u) \leq \liminf_{n\to\infty} \Phi(u_n)$$
$$\gamma(u) = \lim_{n\to\infty} \gamma(u_n)$$

and

$$-J^0(\gamma(u); \gamma(v) - \gamma(u)) \leq \liminf_{n\to\infty} -J^0(\gamma(u_n); \gamma(v) - \gamma(u_n)).$$

Taking the limit inferior in (3.11.8) as $n \to +\infty$, we obtain

$$\langle Au - f, u - v \rangle + \Phi(u) - \Phi(v) - J^0(\gamma(u); \gamma(v) - \gamma(u)) \leq 0.$$

Since v has been chosen arbitrarily, we get finally that

$$\langle Au - f, v - u \rangle + \Phi(v) - \Phi(u) + J^0(\gamma(u); \gamma(v) - \gamma(u)) \geq 0, \forall v \in C.$$

Using now property (3.10.7) again, we get finally that u solves (3.11.4).

$\blacksquare$

A similar result can be obtained for monotone hemicontinuous operators. The following result generalizes Theorem 3.3.1 in several directions.

Theorem 3.11.2   Let C be a nonempty closed convex subset of X, $\Phi : X \to \mathbb{R} \cup \{+\infty\}$ a convex and l.s.c. function such that $C \cap D(\Phi) \neq \emptyset$ and $A : X \to X^*$ a monotone and hemicontinuous operator. Then parts A) and B) of Theorem 3.11.1 hold true.

Proof.   Using Lemma 3.10.3, we get a sequence $u_n \in C_n := \{x \in C : \| x \| \leq n\}$ (C_n being nonempty for n large enough) such that

$$\langle Au_n - f, v - u_n \rangle + \Phi(v) - \Phi(u_n)$$

$$+ J^0(\gamma(u_n); \gamma(v) - \gamma(u_n)) \geq 0, \ \forall \, v \in C_n. \tag{3.11.9}$$

As in the proof of Theorem 3.11.1, we see that $\{u_n\}$ is bounded and thus along a subsequence, we may assume that $u_n \rightharpoonup u$. From (3.11.9) and the monotonicity of A, we deduce that

$$\langle Av - f, v - u_n \rangle + \Phi(v) - \Phi(u_n) + J^0(\gamma(u_n); \gamma(v) - \gamma(u_n)) \geq 0, \ \forall \, v \in C_n.$$

Let $v \in C$ be given. For n large enough

$$\langle Av - f, u_n - v \rangle + \Phi(u_n) - \Phi(v) - J^0(\gamma(u_n); \gamma(v) - \gamma(u_n)) \leq 0$$

and taking the limit inferior as $n \to +\infty$, we obtain

$$\langle Av - f, u - v \rangle + \Phi(u) - \Phi(v) - J^0(\gamma(u); \gamma(v) - \gamma(u)) \leq 0.$$

Since v has been chosen arbitrarily, it results that

$$\langle Av - f, v - u \rangle + \Phi(v) - \Phi(u) + J^0(\gamma(u); \gamma(v) - \gamma(u)) \geq 0, \ \forall \, v \in C.$$

Using now Minty's principle (Theorem 1.1.23), we obtain that

$$\langle Au - f, v - u \rangle + \Phi(v) - \Phi(u) + J^0(\gamma(u); \gamma(v) - \gamma(u)) \geq 0, \ \forall \, v \in C$$

and the result follows from property (3.10.7).   ∎

3.12   NONCOERCIVE VARIATIONAL-HEMIVARIATIONAL INEQUALITIES

The approach developed in Section 3.5 is now revisited for more general variational-hemivariational inequalities. Let X be a real reflexive Banach space and let $(T, \mathcal{T}, \mu)$ be a complete positive measure space such that $0 < \mu(T) < +\infty$. Let $m \in \mathbb{N}\setminus\{0\}, p \in [1,+\infty)$, we assume that

$$\gamma : X \to L^p(T; \mathbb{R}^m)$$

denotes a linear and compact operator. Let C be a nonempty closed convex subset of $X, \Phi : X \to \mathbb{R}\cup\{+\infty\}$ a proper convex and l.s.c. function, $f \in X^*$, $A : X \to X^*$ an operator and $j : T \times \mathbb{R}^m \to \mathbb{R}$ a function satisfying (3.10.1) and either condition (3.10.2a) or conditions (3.10.2b)-(3.10.2c). Moreover, we assume that

$$0 \in C \cap D(\Phi). \tag{3.12.1}$$

We consider the inequality problem: Find $u \in C$ such that

$$\langle Au - f, v - u \rangle + \Phi(v) - \Phi(u)$$

$$+ \int_T j_y^0(x, \gamma u(x); \gamma v(x) - \gamma u(x))d\mu \geq 0, \forall\, v \in C. \tag{3.12.2}$$

We have shown in Section 3.11 that if j satisfies conditions (3.10.1) and (3.10.2a) then

$$\mid \int_T j_y^0(x, \gamma u(x); \gamma v(x))d\mu \mid \leq C \parallel v \parallel, \forall\, u,\, v \in X,$$

for some constant $C > 0$. On the other hand, if j satisfies conditions (3.10.1), (3.10.2b) and (3.10.2c) then

$$\mid \int_T j_y^0(x, \gamma u(x); \gamma v(x))d\mu \mid \leq C_1 \parallel v \parallel + C_2 \parallel u \parallel^{p-1} \parallel v \parallel, \forall\, u,\, v \in X,$$

for some constants $C_1, C_2 > 0$. In both cases, there exist $C_1 \geq 0, C_2 > 0$ and $\sigma > 0$ such that

$$\mid \int_T j_y^0(x, \gamma u(x); \pm\gamma u(x))d\mu \mid \leq C_1 + C_2 \parallel u \parallel^p, \forall\, u \in X. \tag{3.12.3}$$

The concepts of Section 3.5 are now adapted. We set

$$\mathcal{S}(A, f, C, \Phi, j) = \{\{v_n\} : v_n = \frac{u_n}{\parallel u_n \parallel}, u_n \in C, \parallel u_n \parallel \to +\infty$$

and

$$\langle Au_n, u_n \rangle + \Phi(u_n) \leq \Phi(0) + \langle f, u_n \rangle$$

$$+ \int_T j_y^0(x, \gamma(u_n); \gamma(-u_n))d\mu\} \tag{3.12.4}$$

and we say that $\mathcal{S}(A, f, C, \Phi, j)$ is a-compact provided that if $\{v_n\} \in \mathcal{S}(A, f, C, \Phi, j)$ and $v_n \rightharpoonup w$ then along a subsequence $v_n \to w$. We also set

$$\Lambda_\infty(A, f, C, \Phi, j) = \{w \in C_\infty : \exists \{v_n\} \in \mathcal{S}(A, f, C, \Phi, j),$$

$$v_n \rightharpoonup w\}. \tag{3.12.5}$$

We obtain the following abstract theorem.

Theorem 3.12.1 Suppose that the aforementioned conditions are satisfied. We assume that

$$\mathcal{S}(A, f, C, \Phi, j) \text{ is } a-\text{compact} \tag{3.12.6}$$

and the mapping

$$v \to \langle Av, v - x \rangle \text{ is w.l.s.c., } \forall\, x \in C. \tag{3.12.7}$$

If there exists a subset $W \subset C_\infty$ such that

$$\Lambda_\infty(A, f, C, \Phi, j) \subset W \tag{3.12.8}$$

and

$$\underline{r}_A(v) + \underline{\gamma}_j(\gamma(v)) + \Phi_\infty(v) > \langle f, v \rangle, \,\forall\, v \in W \backslash \{0\} \tag{3.12.9}$$

then there exists at least one $u \in C$ satisfying (3.12.2).

Proof. Let $C_n \subset X$ be defined by

$$C_n = \{x \in C : \| x \| \le n\}.$$

Using Theorem 3.10.2, we obtain $u_n \in C_n$ such that

$$\langle Au_n - f, v - u_n \rangle + \Phi(v) - \Phi(u_n)$$

$$+ J^0(\gamma(u_n); \gamma(v - u_n))d\mu \ge 0, \,\forall\, v \in C_n, \tag{3.12.10}$$

where J is defined in (3.10.6) and using (3.10.7) we obtain in consequence that

$$\langle Au_n - f, v - u_n \rangle + \Phi(v) - \Phi(u_n)$$

$$+ \int_T j_y^0(x, \gamma(u_n); \gamma(v - u_n))d\mu \ge 0, \,\forall\, v \in C_n. \tag{3.12.11}$$

If we set $v = 0$ in (3.12.11) then we get

$$\langle Au_n, u_n \rangle + \Phi(u_n) \le \Phi(0) + \langle f, u_n \rangle + \int_T j^0(x, \gamma(u_n); -\gamma(u_n))d\mu.$$

We claim that the sequence $\{u_n\}$ is bounded. Indeed, suppose by contradiction that $\| u_n \| \to +\infty$. Then along a subsequence, we may assume that

$$v_n := \frac{u_n}{\| u_n \|} \rightharpoonup w \in C_\infty.$$

Thus $\{v_n\} \in \mathcal{S}(A, f, C, \Phi, j)$ and $w \in \Lambda_\infty(A, f, C, \Phi, j)$. Assumption (3.12.6) entails that along a subsequence

$$v_n \to w$$

and thus $w \neq 0$.

We have

$$\langle Au_n, u_n \rangle + \Phi(u_n) - \int_T \max_{z \in \partial_y j(x, \gamma u_n(x))} \langle z(x), -\gamma u_n(x) \rangle_m d\mu$$

$$\leq \Phi(0) + \langle f, u_n \rangle. \tag{3.12.12}$$

Note that

$$- \max_{z(x) \in \partial_y j(x,u)} \langle z, -v \rangle_m = \min_{z \in \partial_y j(x,u)} \langle z, v \rangle_m. \tag{3.12.13}$$

Dividing (3.12.12) by $t_n := \| u_n \|$ and using (3.12.13), we obtain

$$\langle A(t_n v_n), v_n \rangle + \frac{\Phi(t_n v_n)}{t_n} + \int_T \min_{z(x) \in \partial_y j(x, t_n \gamma v_n(x))} \langle z(x), \gamma v_n(x) \rangle_m d\mu$$

$$\leq \frac{\Phi(0)}{t_n} + \langle f, v_n \rangle.$$

It results that

$$\underline{r}_A(w) + \Phi_\infty(w) + \underline{\gamma}_j(\gamma(w)) \leq \langle f, w \rangle,$$

which is a contradiction to condition (3.12.9). Thus the sequence $\{u_n\}$ is bounded and along a subsequence we may suppose that $u_n \rightharpoonup u$. We may conclude that u solves (3.12.2) by following the same argument as the one used in the end of the proof of Theorem 3.11.1. ∎

Remark 3.12.2 i) The result given by Theorem 3.12.1 remains true if we replace assumption (3.12.7) by the monotonicity and hemicontinuity of A. In this case, the proof uses Theorem 3.10.4 in place of Theorem 3.10.2 and to complete the proof by passing to the limit in (3.12.11), we use as usually the Minty's principle (see Theorem 1.1.23).
ii) The conditions of Theorem 3.12.1 ensure also that the solution set

of (3.12.2) is bounded. Let us now discuss assumptions (3.12.6) and (3.12.8) for some classes of operators and functionals.

Proposition 3.12.3 Let C be a nonempty closed convex subset of X. Suppose that

(i) A satisfies the S^+-property;

(ii) $\langle Ax, x \rangle \geq 0, \forall\, x \in X$;

(iii) A is weakly continuous;

(iv) A is positively homogeneous of degree $\theta > 0$.

Let $\Phi : X \to \mathbb{R} \cup \{+\infty\}$ be a convex and l.s.c. function such that $\Phi(0) < +\infty$ and let $j : T \times \mathbb{R}^m \to \mathbb{R}$ be a function satisfying condition (3.12.3) with $0 \leq \sigma < \theta + 1$. Then

and
$$(1)\ \mathcal{S}(A, f, C, \Phi, j) \text{ is } a-\text{compact}$$

$$(2)\ \Lambda_\infty(A, f, C, \Phi, j) \subset \{w \in C_\infty : \langle Aw, w \rangle = 0\}.$$

Proof. Suppose that $w \in \Lambda_\infty(A, f, C, \Phi, j)$ and let $u_n \in C$ such that $t_n := \| u_n \| \to +\infty$, $w_n = \frac{u_n}{t_n} \rightharpoonup w \in C_\infty$ and

$$\langle Au_n, u_n \rangle + \Phi(u_n) \leq \Phi(0) + \langle f, u_n \rangle + \int_T j_y^0(x, \gamma(u_n); \gamma(-u_n)) d\mu.$$

It results that

$$\langle Au_n, u_n \rangle + \Phi(u_n) \leq \Phi(0) + \langle f, u_n \rangle + C_1 + C_2 \| u_n \|^\sigma .$$

From Theorem 1.1.11, there exist constants $C_1' \geq 0$ and $C_2' \in \mathbb{R}$ such that
$$\Phi(u_n) \geq -C_1' \| u_n \| + C_2'.$$

Thus

$$\langle Au_n, u_n \rangle \leq \Phi(0) + C_1' \| u_n \| - C_2' + \langle f, u_n \rangle + C_1 + C_2 \| u_n \|^\sigma .$$

Dividing this last relation by $t_n^{\theta+1}$, we get

$$\langle Aw_n, w_n \rangle \leq \Phi(0)t_n^{-1-\theta} + C_1' t_n^{-\theta} - C_2' t_n^{-\theta-1} + \langle f, w_n \rangle t_n^{-\theta}$$

$$+C_1 t_n^{-\theta-1} + C_2 t_n^{\sigma-\theta-1}. \tag{3.12.14}$$

Taking the limit superior as $n \to +\infty$, we get

$$\limsup_{n\to\infty}\langle Aw_n, w_n\rangle \le 0.$$

As in the proof of 3.5.5, we deduce that $w_n \to w$ and taking again the limit as $n \to +\infty$ in (3.12.14), we obtain $\langle Aw, w\rangle \le 0$. This together with assumption (ii) implies that $\langle Aw, w\rangle = 0$. $\blacksquare$

Remark 3.12.4   i) Note that the previous result remains true if we consider as in Section 3.5 the operator $A + B$, where B is a nonlinear perturbation of one of the types considered in Section 3.5. ii) It is worthwhile to note the key role played by the "energy functional" J as defined by (3.10.6) and the intermediate abstract problem

$$\langle Au-f, v-u\rangle+\Phi(v)-\Phi(u)+J^0(\gamma(u);\gamma(v-u)) \ge 0, \ \forall\, v \in C. \tag{3.12.15}$$

The use of (3.12.15) is of great interest to apply Lemma 3.10.1 as well as to pass to limits. It is now easy to see that the result of Lemma 3.10.1 can be without any difficulty extended to the following problem: Find $u \in C$ such that

$$\langle Au - f, v - u\rangle + \Phi(v) - \Phi(u)$$

$$+\Sigma_{\alpha=1}^l J_\alpha^0(L_\alpha(u); L_\alpha(v - u)) \ge 0, \ \forall\, v \in C.$$

where the data J_α and $L_\alpha(\alpha = 1,\cdots,l; \ l \in \mathbb{N}\setminus\{0\})$ satisfy the conditions required respectively on J and L in the statement of the lemma. It results that the existence theorems of Sections 3.11 and 3.12 can also be easily generalized to the problem: Find $u \in C$ such that

$$\langle Au - f, v - u\rangle + \Phi(v) - \Phi(u)$$

$$+\Sigma_{\alpha=1}^l \int_{T_\alpha} j_{\alpha,y}^0(x, \gamma_\alpha u(x); \gamma_\alpha v(x) - \gamma_\alpha u(x))d\mu \ge 0, \ \forall\, v \in C,$$

where the data T_α, j_α and γ_α $(\alpha = 1,\cdots,l)$ are defined as T, j and γ above. In this case the use of the "energy functionals" $J_\alpha, (\alpha = 1,\cdots,l)$ defined by

$$J_\alpha(u) = \int_{T_\alpha} j_\alpha(x, u(x))d\mu, u \in L^p(T_\alpha; \mathbb{R}^m)$$

and the intermediate problem

$$\langle Au - f, v - u\rangle + \Phi(v) - \Phi(u)$$

$$+\Sigma_{\alpha=1}^{l} J_{\alpha}^{0}(\gamma_{\alpha}(u); \gamma_{\alpha}(v - u)) \geq 0, \ \forall v \in C.$$

is then appropriate. Many variants of practical interest for the study of special problems in Mechanics (see Chapter 2) can be so deduced.

3.13 A FIXED POINT APPROACH FOR A CLASS OF EVOLUTION VARIATIONAL INEQUALITIES

In this Section we consider the problem: Find $x \in C^1(0, T; H)$ such that

$$f(t) \in A\dot{x}(t) + Gx(t) + \partial\Phi(\dot{x}(t)), \ \forall t \in [0, T], \tag{3.13.1}$$

and

$$x(0) - x_0, \tag{3.13.2}$$

where H is a real Hilbert space, $x_0 \in H, T > 0, f \in C^0(0, T; H),$ $\Phi :$ $H \to \mathbb{R}$ is a proper convex and l.s.c. functional and $A, G : H \to H$ are possibly nonlinear operators satisfying the following conditions:

$$\mid Ax - Ay \mid \leq L \mid x - y \mid, \ \forall x, y \in H \tag{3.13.3}$$

$$(Ax - Ay, x - y) \geq M \mid x - y \mid^2, \ \forall x, y \in H \tag{3.13.4}$$

$$\mid Gx - Gy \mid \leq K \mid x - y \mid, \ \forall x, y \in H, \tag{3.13.5}$$

for some positive constants $L, M, K > 0$.

Theorem 3.13.1 Suppose that the aforementioned conditions are satisfied. Then problem (3.13.1)-(3.13.2) admits a unique solution.

Proof. For each function $\varepsilon \in C^0(0, T; H)$, we consider the problem: Find $x_\varepsilon \in C^1(0, T; H)$ such that

$$f(t) \in A\dot{x}_\varepsilon(t) + \varepsilon(t) + \partial\Phi(\dot{x}_\varepsilon(t)), \ \forall t \in [0, T], \tag{3.13.6}$$

and

$$x_\varepsilon(0) = x_0. \tag{3.13.7}$$

We claim that problem (3.13.6)-(3.13.7) has a unique solution. Indeed, applying Theorem 3.2.1, we get for each $t \in [0, T]$ the existence of a unique $z_\varepsilon(t) \in H$ such that

$$(Az_\varepsilon(t), v - z_\varepsilon(t)) + \Phi(v) - \Phi(z_\varepsilon(t))$$
$$\geq \ (f(t), v - z_\varepsilon(t)) - (\varepsilon(t), v - z_\varepsilon(t)), \ \forall v \in H.$$

Let $t_1, t_2 \in [0, T]$, we have

$$(Az_\varepsilon(t_1) - Az_\varepsilon(t_2), z_\varepsilon(t_1) - z_\varepsilon(t_2)) \leq (f(t_1) - f(t_2), z_\varepsilon(t_1) - z_\varepsilon(t_2))$$

$$-(\varepsilon(t_1) - \varepsilon(t_2), z_\varepsilon(t_1) - z_\varepsilon(t_2)),$$

so that

$$\mid z_\varepsilon(t_1) - z_\varepsilon(t_2) \mid \leq \frac{1}{M}(\mid f(t_1) - f(t_2) \mid + \mid \varepsilon(t_1) - \varepsilon(t_2) \mid)$$

and thus

$$z_\varepsilon \in C^0(0, T; H).$$

It results that the function $x_\varepsilon \in C^1(0, T; H)$ given by

$$x_\varepsilon(t) = x_0 + \int_0^t z_\varepsilon(s)ds, \ t \in [0, T]$$

is the unique solution of the variational inequality

$$(A\dot{x}_\varepsilon(t), v - x_\varepsilon(t)) + \Phi(v) - \Phi(\dot{x}_\varepsilon(t))$$

$$\geq (f(t), v - x_\varepsilon(t)) - (\varepsilon(t), v - x_\varepsilon(t)), \ \forall v \in H$$

and satisfies

$$x_\varepsilon(0) = x_0,$$

i.e. x_ε is the unique solution of problem (3.13.6)-(3.13.7).

Let us now consider the operator $\Lambda : C^0(0, T; H) \to C^0(0, T; H)$ defined by

$$\Lambda\varepsilon = Gx_\varepsilon.$$

We claim that Λ has a unique fixed point $\bar{\varepsilon} \in C^0(0, T; H)$. Indeed, let $\varepsilon_1, \varepsilon_2 \in C^0(0, T; H)$ be given. Setting $z_1 = x_{\varepsilon_1}, z_2 = x_{\varepsilon_2}$, we get

$$M \mid z_1(t) - z_2(t) \mid^2 \leq \mid \varepsilon_1(t) - \varepsilon_2(t) \mid \ \mid z_1(t) - z_2(t) \mid$$

so that

$$\mid z_1(t) - z_2(t) \mid \leq \frac{1}{M} \mid \varepsilon_1(t) - \varepsilon_2(t) \mid.$$

Thus

$$
\begin{aligned}
\mid \Lambda\varepsilon_1(t) - \Lambda\varepsilon_2(t) \mid \ &= \ \mid Gx_{\varepsilon_1}(t) - Gx_{\varepsilon_2}(t) \mid \\
&\leq \ K \mid x_{\varepsilon_1}(t) - x_{\varepsilon_2}(t) \mid \\
&\leq \ K \int_0^t \mid z_{\varepsilon_1}(s) - z_{\varepsilon_2}(s) \mid ds \\
&\leq \ \frac{K}{M} \int_0^t \mid \varepsilon_1(s) - \varepsilon_2(s) \mid ds.
\end{aligned}
$$

It results that

$$\mid \Lambda^n \varepsilon_1(t) - \Lambda^n \varepsilon_2(t) \mid \, \leq \frac{(KTM^{-1})^n}{n!} \parallel \varepsilon_1 - \varepsilon_2 \parallel_{C^0(0,T;H)}$$

and thus

$$\parallel \Lambda^n \varepsilon_1 - \Lambda^n \varepsilon_2 \parallel_{C^0(0,T;H)} \leq \frac{(KTM^{-1})^n}{n!} \parallel \varepsilon_1 - \varepsilon_2 \parallel_{C^0(0,T;H)} \, \cdot$$

The last inequality shows that for a sufficiently large n, the operator Λ^n is a contraction on $C^0(0,T;H)$. Thus, there exists a unique $\bar{\varepsilon} \in C^0(0,T;H)$ such that $\Lambda\bar{\varepsilon} = \bar{\varepsilon}$.

We have

$$f(t) \in A\dot{\bar{x}}_{\bar{\varepsilon}}(t) + \bar{\varepsilon}(t) + \partial\Phi(\dot{x}_{\bar{\varepsilon}(t)}), \ \forall \, t \, \in [0,T] \tag{3.13.8}$$

and

$$\bar{x}_\varepsilon(0) = x_0. \tag{3.13.9}$$

However, $\bar{\varepsilon}(t) = \Lambda\bar{\varepsilon}(t) = Gx_{\bar{\varepsilon}}(t)$ and thus (3.13.8) and (3.13.9) imply that $\bar{x}_\varepsilon$ is a solution of Problem (3.13.1)-(3.13.2).

It remains to prove that the solution is unique. Suppose that x is another solution of Problem (3.13.1)-(3.13.2). Let $\varepsilon := Gx$ and denote by x_ε the unique solution of Problem (3.13.6)-(3.13.7). It is clear that x is solution of Problem (3.13.6)-(3.13.7) with $\varepsilon = Gx$. Thus $x_\varepsilon = x$. On the other hand $\Lambda\varepsilon = Gx_\varepsilon = Gx = \varepsilon$. The uniqueness of the fixed point of Λ yields $\varepsilon = \bar{\varepsilon}$ and thus $x = x_\varepsilon = x_{\bar{\varepsilon}}$. $\blacksquare$

Chapter 4

MINIMAX METHODS FOR INEQUALITY PROBLEMS

We know from Chapter 2 that, if we intend to consider concrete problems in unilateral Mechanics involving both monotone and nonmonotone unilateral boundary (or interior) conditions, then we have in general to deal with a nonsmooth and nonconvex energy functional - expressed as the sum of a locally Lipschitz function $\Phi : X \to \mathbb{R}$ and a proper, convex and lower semi-continuous function $\Psi : X \to \mathbb{R} \cup \{+\infty\}$ - whose critical points are defined as the solutions of the variational-hemivariational inequality

$$u \in X : \quad \Phi^0(u; v - u) + \Psi(v) - \Psi(u) \geq 0, \ \forall\, v \in X. \qquad (4.1)$$

If $\Phi \in C^1(X; \mathbb{R})$ then problem (4.1) reduces to a classical variational inequality:

$$u \in X : \langle \Phi^{'}(u), v - u \rangle + \Psi(v) - \Psi(u) \geq 0, \forall v \in X. \qquad (4.2)$$

A list of typical examples for the "energy" functional Ψ has been given in Section 2.10.8. If $\Psi = 0$ then the problem (4.1) reduces to the hemivariational inequality:

$$u \in X : \Phi^0(u; v - u) \geq 0, \forall\, v \in X. \qquad (4.3)$$

If $\Phi = \Phi_1 + \Phi_2$ with $\Phi_1 \in C^1(X; \mathbb{R})$ and $\Phi_2 \in L_{loc}(X; \mathbb{R})$ then (4.3) reduces to

$$\langle \Phi_1^{'}(u), v - u \rangle + \Phi_2^0(u; v - u) \geq 0, \forall\, v \in X.$$

A list of typical examples for the "energy" functional Φ_2 can be found in Section 2.10.7. If $\Phi \in C^1(X; \mathbb{R})$ and $\Psi = 0$ then the problem reduces to the equation $\Phi'(u) = 0$ which has been the subject of intensive developments. Here we derive a general linking theorem, which is then used to obtain nonsmooth versions of the Mountain-Pass theorem, the Saddle-point theorem and the main results for even functionals. Besides its practical interest in the theory of variational-hemivariational inequalities considered in this book, the present theory unifies the three basic ones, i.e. the one for smooth functionals (see Rabinowitz [369]), the one of Szulkin [405] for variational inequalities and the one of Chang [80] for locally Lipschitz functionals. Moreover, it applies to the sum of locally Lipschitz functionals and proper l.s.c. convex functionals and, to the best of our knowledge, it is the first theory in this direction.

This Chapter is primarily based on the works of Chang [80], Goeleven and Motreanu [170], Motreanu and Panagiotopoulos [296], Lefter [241] and Szulkin [405]. In preparing this Chapter, a number of well-known works on critical point theory (cfr. [11], [15], [87], [92], [111], [123], [126], [193], [218], [253], [265], [300], [369], [390], [416]) have also been used.

4.1   THE GENERAL SETTING

Let us firstly set up the framework of our critical point theory. Namely, we deal with lower semi-continuous functionals $I : X \to (-\infty, +\infty]$ on a real Banach space X satisfying the following structure hypothesis

(H) $I = \Phi + \Psi$, with $\Phi : X \to \mathbb{R}$ locally Lipschitz and $\Psi : X \to (-\infty, +\infty]$ proper (i.e., $\not\equiv +\infty$), convex and l.s.c..

In order to develop a critical point theory for functionals of type (H) we need to introduce the basic concepts of critical point and Palais-Smale condition. The following definition unifies the concepts introduced in Definition 1.2.24 and Definition 1.3.23.

Definition 4.1.1   An element $u \in X$ is called a critical point of the functional I in (H) if

$$\Phi^0(u; v - u) + \Psi(v) - \Psi(u) \geq 0, \ \forall v \in X. \qquad (4.1.1)$$

Definition 4.1.2   The functional $I : X \to (-\infty, +\infty]$ in (H) is said to satisfy the generalized Palais - Smale condition at the level $c \in \mathbb{R}$ (in short, $(PS)_c$) if every sequence $\{u_n\} \subset X$ satisfying $I(u_n) \to c$ and

$$\Phi^0(u_n; v - u_n) + \Psi(v) - \Psi(u_n) \geq -\varepsilon_n \parallel v - u_n \parallel, \ \forall \, v \, \in \, X \qquad (4.1.2)$$

for a sequence $\{\varepsilon_n\} \subset \mathbb{R}_+$ with $\varepsilon_n \to 0$, contains a convergent subsequence. If $(PS)_c$ is verified for all $c \in \mathbb{R}$, I is said to satisfy the Palais - Smale condition (in short, (PS)).

For the functional $I : X \to (-\infty, +\infty]$ in (H) we denote by $K(I)$ the set of critical points of I in the sense of Definition 4.1.1. Given any number $c \in \mathbb{R}$ let us set

$$K_c(I) := K(I) \cap I^{-1}(c). \qquad (4.1.3)$$

For a latter use we put

$$I_a := \{v \in X : \ I(v) \leq a\}, \ \forall \, a \, \in \, \mathbb{R}.$$

The result below points out a relevant aspect of Definition 4.1.1.

Proposition 4.1.3 Any limit point u of a (generalized Palais - Smale) sequence $\{u_n\}$ entering Definition 4.1.1 belongs to $K_c(I)$.

Proof. Without loss of generality we may admit that $u_n \to u$ in X. Then, letting $n \to \infty$, the upper semicontinuity of Φ^0 and the l.s.c. of Ψ allow us to derive (4.1.1) from (4.1.2), so $u \in K(I)$. Taking $v = u$ in (4.1.2) we obtain that

$$\begin{aligned}
\Psi(u) \ &\leq \ \liminf_{n \to \infty} \Psi(u_n) \leq \limsup_{n \to \infty} \Psi(u_n) \\
&\leq \ \Phi^0(u; u - u) + \Psi(u) = \Psi(u).
\end{aligned}$$

This ensures that $I(u) = \lim_{n \to \infty} I(u_n) = c$, so $u \in K_c(I)$. $\blacksquare$

It results from Proposition 4.1.3 that if I satisfies the $(PS)_c$ condition then $K_c(I)$ is compact. Let us now discuss the previous concepts through two particular but important subcases of the structure hypothesis (H).

Let X be a real Banach space and let I be a function on X satisfying the following hypothesis:

(H_1) $I = \Phi + J$ with $\Phi \in C^1(X; \mathbb{R})$ and $J : X \to \mathbb{R}$ a locally Lipschitz function.

Note that the function I is locally Lipschitz as the sum of two locally Lipschitz functions. Recall that a point $u \in X$ is said to be critical for I if

$$\langle \Phi'(u), v \rangle + J^0(u; v) \geq 0, \ \forall \, v \, \in \, X, \qquad (4.1.4)$$

or equivalently if

$$-\Phi'(u) \in \partial J(u).$$

Note that here (by definition of the generalized derivative) we have

$$I^0(u;v) = \langle \Phi'(u), v \rangle + J^0(u;v), \ \forall \, u, \, v \, \in \, X,$$

and thus

$$\partial I(u) = \Phi'(u) + \partial J(u).$$

A number $c \in \mathbb{R}$ such that $I^{-1}(c)$ contains a critical point will be called a critical value. One says that I satisfies the Palais-Smale condition in the sense of Chang [80] if:

(PSC) every sequence $\{u_n\}$ such that $\mid I(u_n) \mid$ is bounded in $\mathbb{R}$ and

$$\mu(u_n) = \min\{\| z \|_* \colon z \in \partial I(u_n)\} \to 0, \tag{4.1.5}$$

contains a convergent subsequence in X.

From Definition 4.1.2, we say that I satisfies the generalized Palais-Smale condition [4] if:

(HPS) every sequence $\{u_n\}$ such that $I(u_n) \to c \in \mathbb{R}$ and

$$\langle \Phi'(u_n), v \rangle + J^0(u_n;v) \geq -\varepsilon_n \| v \|, \ \forall \, v \, \in \, X, \tag{4.1.6}$$

where $\varepsilon_n \to 0$, possesses a convergent subsequence. Condition (HPS) can also be formulated as follows:

(HPS$'$) every sequence $\{u_n\}$ such that $I(u_n) \to c \in \mathbb{R}$ and

$$\langle \Phi'(u_n), v \rangle + J^0(u_n;v) \geq \langle z_n, v \rangle, \ \forall \, v \, \in \, X, \tag{4.1.7}$$

where $z_n \to 0$ in X^*, possesses a convergent subsequence. Condition (4.1.7) expresses the fact that $z_n \in \Phi'(u_n) + \partial J(u_n)$ so that if $J = 0$ then (HPS$'$) is nothing else that the usual Palais-Smale condition (see e.g. [265]).

Using the Szulkin Lemma, we prove the equivalence between these three compactness conditions for functionals satisfying (H_1).

Proposition 4.1.4   Let $I : X \to \mathbb{R}$ be a function satisfying (H_1). Conditions (HPS) and (HPS$'$) are equivalent.

Proof. It is clear that (4.1.7) implies (4.1.6). Suppose now that (4.1.6) is satisfied. If $\varepsilon_n \leq 0$ there we take $z_n = 0$. If $\varepsilon_n > 0$ then we set

$$\mathcal{X}_n(x) = (\langle \Phi'(u_n), x \rangle + J^0(u_n; x))/\varepsilon_n, \ \forall \, x \in X,$$

and (4.1.6) reads

$$\mathcal{X}_n(x) \geq - \| x \|, \ \forall \, x \in X.$$

Moreover $\mathcal{X}_n(0) = 0$, the map $x \to \mathcal{X}_n(x)$ is convex and continuous. Thus all assumptions of Lemma 1.1.12 are satisfied and we may find $w_n \in X^*$ such that $\| w_n \|_* \leq 1$ and

$$\mathcal{X}_n(x) \geq \langle w_n, x \rangle, \ \forall \, x \in X.$$

Setting $z_n = \varepsilon_n w_n$ we see that

$$\langle \Phi'(u_n), v \rangle + J^0(u_n; v) \geq \langle z_n, v \rangle, \ \forall \, v \in X,$$

and $z_n \to 0$ in X^*. Hence (4.1.7) is satisfied.   ∎

Proposition 4.1.5   Let $I : X \to \mathbb{R}$ be a function satisfying (H_1). Then conditions (HPS') and (PSC) are equivalent.

Proof. Suppose that I satisfies (HPS). Let $\{u_n\}$ be a sequence such that $| I(u_n) |$ is bounded and $\mu(u_n) \to 0$. Then there exists a subsequence (denoted again by $\{u_n\}$) and a sequence $\{z_n\}$ in X^* such that $I(u_n) \to c \in \mathbb{R}$, $z_n \in \Phi'(u_n) + \partial J(u_n)$ and $z_n \to 0$. Thus using (HPS) we obtain a convergent subsequence. Suppose now that I satisfies (PSC). Let $\{u_n\}$ be a sequence such that $I(u_n) \to c \in \mathbb{R}$ and satisfying (4.1.7). Then $| I(u_n) |$ is bounded and

$$\begin{aligned} \mu(u_n) \ &= \ \min\{\| z \|_*: z \in \partial I(u_n)\} \\ &= \ \min\{\| z \|_*: z \in \Phi'(u_n) + \partial J(u_n)\} \leq \| z_n \|. \end{aligned}$$

Thus $\mu(u_n) \to 0$ and using (PSC) we conclude to the existence of a convergent subsequence.   ∎

The generalized Palais-Smale condition together with the variational principle of Ekeland can be used to provide a sufficient condition for the existence of a critical point.

Theorem 4.1.6   If I is bounded below and satisfies (H_1) and (HPS), then

$$c = \inf_{u \in X} I(u)$$

is a critical value.

Proof. Let $\{w_n\}$ be a sequence satisfying

$$I(w_n) \leq c + \frac{1}{n}.$$

By Theorem 1.1.13 with $\delta = \frac{1}{n}$ and $\lambda = 1$, we find a sequence $\{u_n\}$ such that

$$I(u_n) \leq I(w_n) \leq c + \frac{1}{n} \tag{4.1.8}$$

and

$$I(w) - I(u_n) \geq -\frac{1}{n} \parallel w - u_n \parallel, \ \forall \ w \ \in \ X.$$

Let $\{u_j^n\} \subset X$ and $\{t_j\} \subset [0,1]$ be sequences such that $u_j^n \to u_n$ and $t_j \downarrow 0$ as $j \to +\infty$. We set $w = u_j^n + tv$, $v \in X$ to obtain

$$I(u_j^n + t_j v) - I(u_j^n) \geq -\frac{t_j}{n} \parallel v \parallel.$$

Dividing by t_j we get

$$\frac{I(u_j^n + t_j v) - I(u_j^n)}{t_j} \geq -\frac{1}{n} \parallel v \parallel$$

and thus

$$\langle \Phi'(u_n), v \rangle + J^0(u_n; v) \geq -\frac{1}{n} \parallel v \parallel, \ \forall \ v \ \in \ X. \tag{4.1.9}$$

So by (HPS), $\{u_n\}$ possesses a subsequence (denoted again by $\{u_n\}$) converging to some $u \in X$. We have

$$\Phi'(u_n) \to \Phi'(u)$$

and

$$\limsup_{n \to \infty} J^0(u_n; v) \leq J^0(u; v).$$

Thus taking the limit superior in (4.1.9) we obtain

$$\langle \Phi'(u), v \rangle + J^0(u; v) \geq 0$$

so that u is a critical point for I. Taking the limit in (4.1.8) we obtain

$$I(u) \leq c$$

and thus

$$I(u) = c$$

since $c \leq I(x)$, $\forall x \in X$. ∎

The structure hypothesis (H_1) being related to the hemivariational inequality (4.1.4), let us now consider a structure hypothesis related to the variational inequality problem. Let $I : X \to \mathbb{R} \cup \{+\infty\}$ be a function satisfying the following hypothesis:

(H_2) $I = \Phi + \Psi$ with $\Phi \in C^1(X; \mathbb{R})$ and $\Psi : X \to (-\infty, +\infty]$ proper, convex and lower semi-continuous.

A point $u \in X$ is said to be critical for I if u satisfies the variational inequality:

$$\langle \Phi'(u), v - u \rangle + \Psi(v) - \Psi(u) \geq 0, \ \forall v \in X. \tag{4.1.10}$$

Here, we say that I satisfies the generalized Palais-Smale condition (in the sense of Szulkin [405]) if

(VPS) every sequence $\{u_n\}$ such that $I(u_n) \to c \in \mathbb{R}$ and

$$\langle \Phi'(u_n), v - u_n \rangle + \Psi(v) - \Psi(u_n) \geq -\varepsilon_n \parallel v - u_n \parallel, \ \forall v \in X, \tag{4.1.11}$$

where $\varepsilon_n \to 0$, possesses a convergent subsequence. Condition (VPS) can also be formulated as follows:

(VPS') every sequence $\{u_n\}$ such that $I(u_n) \to c \in \mathbb{R}$ and

$$\langle \Phi'(u_n), v - u_n \rangle + \Psi(v) - \Psi(u_n) \geq \langle z_n, v - u_n \rangle, \ \forall v \in X, \tag{4.1.12}$$

where $z_n \to 0$ in X^*, possesses a convergent subsequence. We can now state a result corresponding to Proposition 4.1.4.

Proposition 4.1.7   Suppose that I satisfies (H_2). Conditions (VPS) and (VPS') are equivalent

Proof. Inequality (4.1.12) clearly implies (4.1.11). Suppose now that (4.1.11) is satisfied. If $\varepsilon_n \leq 0$ then we take $z_n = 0$. Suppose now that $\varepsilon_n > 0$. We set

$$\mathcal{X}_n(x) = (\langle \Phi'(u_n), x \rangle + \Psi(x + u_n) - \Psi(u_n))/\varepsilon_n, \forall x \in X.$$

Then (4.1.11) reads

$$\mathcal{X}_n(x) \geq - \parallel x \parallel, \forall x \in X.$$

Moreover $X_n(0) = 0$ and the map $x \to X_n(x)$ is convex and l.s.c. We may apply Lemma 1.1.12 in order to find $w_n \in X^*$ such that $\| w_n \|_* \le 1$ and

$$X_n(x) \ge \langle w_n, x \rangle, \forall \, x \in X.$$

We set $z_n = \varepsilon_n w_n$ and see that

$$(\langle \Phi'(u_n), x - u_n \rangle + \Psi(x) - \Psi(u_n))/\varepsilon_n \ge \langle z_n/\varepsilon_n, x - u_n \rangle, \forall \, x \in X.$$

Thus (4.1.12) is satisfied. ∎

We have also a result corresponding to Theorem 4.1.6.

Theorem 4.1.8   If I is bounded below and satisfies (H$_2$) and (VPS), then

$$c = \inf_{u \in X} I(u)$$

is a critical value.

Proof.   Let $\{w_n\}$ be a sequence satisfying

$$I(w_n) \le c + \frac{1}{n}.$$

By Theorem 1.1.13 with $\delta = \frac{1}{n}$ and $\lambda = 1$, we find another sequence $\{u_n\}$ such that

$$I(u_n) \le I(w_n) \le c + \frac{1}{n} \tag{4.1.13}$$

and

$$I(w) - I(u_n) \ge -\frac{1}{n} \| w - u_n \|, \forall \, w \in X.$$

Set $w = (1 - t)u_n + tv, t \in (0, 1)$ for an arbitrary $v \in X$. Using the convexity of Ψ, we obtain

$$\Phi(u_n + t(v - u_n)) - \Phi(u_n) + t(\Psi(v) - \Psi(u_n)) \ge -\frac{1}{n} t \| v - u_n \|.$$

Dividing by t and letting $t \to 0$, we obtain

$$\langle \Phi'(u_n), v - u_n \rangle + \Psi(v) - \Psi(u_n) \ge -\frac{1}{n} \| v - u_n \|. \tag{4.1.14}$$

Using (VPS), we obtain a converging subsequence $u_n \to u$. Taking the limit in (4.1.13) and (4.1.14), we obtain that u is a critical point of I satisfying $I(u) \le c$ and thus the result since $I(u) \ge c$ by   definition of c. ∎

We close this Section with a discussion of some important classes of functions I arising in applications. Let X be a real Hilbert space with inner product $(\cdot, \cdot)$. We assume that X is dense in a Banach space Y, and that the embedding mapping $\gamma : X \to Y$ is compact. Let $H : Y \to \mathbb{R}$ be a locally Lipschitz function and let $A : X \to X$ be a bounded linear and self-adjoint operator. We consider the function $I : X \to \mathbb{R}$ satisfying (H_1),

$$I = \Phi + J, \tag{4.1.15}$$

with

$$\Phi(u) = \frac{1}{2}(Au, u), \ \forall u \in X \tag{4.1.16}$$

and

$$J(u) = H \mid_X (u), \ \forall u \in X. \tag{4.1.17}$$

Proposition 4.1.9   Suppose that A is coercive, i.e. there exists $\alpha > 0$ such that

$$(Au, u) \geq \alpha \parallel u \parallel^2, \ \forall u \in X,$$

and $J : X \to \mathbb{R}$ satisfies

$$\mid J(u) \mid \leq c_0 \parallel u \parallel^2 + c_1, \ \forall u \in X$$

where $c_0 \in (0, \frac{\alpha}{2})$, $c_1 \in \mathbb{R}$. Then the function I defined in (4.1.15) satisfies (HPS) condition.

Proof.   Let $\{u_n\}$ be a sequence such that

$$I(u_n) \to c, \tag{4.1.18}$$

$$(Au_n, v) + J^0(u_n; v) \geq -\varepsilon_n \parallel v \parallel, \ \forall v \in X, \tag{4.1.19}$$

with $\varepsilon_n \to 0$. Then (4.1.18) entails

$$\frac{1}{2}(Au_n, u_n) + J(u_n) \leq c + \frac{1}{n}$$

along a relabelled subsequence. Thus

$$(\frac{\alpha}{2} - c_0) \parallel u_n \parallel^2 \leq c + \frac{1}{n} + c_1,$$

along a relabelled subsequence. We conclude that the subsequence $\{u_n\}$ is bounded in X. Thus by considering eventually a subsequence we may assume that $u_n \rightharpoonup u$ in X and $\gamma(u_n) \to \gamma(u)$ in Y. Inequality (4.1.19) and the fact that $H^0(\gamma(u); \gamma(v)) \geq J^0(u; v)$, $\forall u, v \in X$ (see Theorem 1.2.13 and (1.2.8)), imply

$$(Au_n, v) + H^0(\gamma(u_n); \gamma(v)) \geq -\varepsilon_n \parallel v \parallel, \ \forall v \in X. \qquad (4.1.20)$$

We set $v = u - u_n$ in (4.1.20) to get

$$(Au_n, u_n - u) - H^0(\gamma(u_n); \gamma(u - u_n)) \leq \varepsilon_n \parallel u - u_n \parallel$$

and thus

$$\alpha \parallel u_n - u \parallel^2 - H^0(\gamma(u_n); \gamma(u - u_n)) \leq \varepsilon_n \parallel u_n - u \parallel$$

$$- (Au, u_n - u). \qquad (4.1.21)$$

Using the upper semi-continuity of $H^0(u; v)$ as a function of (u, v) on $Y \times Y$, we obtain

$$\liminf -H^0(\gamma(u_n); \gamma(u - u_n)) \geq 0.$$

Passing to the limit inferior in (4.1.21) we obtain

$$\liminf \parallel u_n - u \parallel^2 \leq 0$$

so that $u_n \to u$ in X along a subsequence. ∎

The (HPS)-condition results here from the coercivity of I. Note also that the assumptions required by Proposition 4.1.9 ensure that I is bounded from below. Therefore we may use Theorem 4.1.6 to conclude to the existence of at least one $u \in X$ satisfying the hemivariational inequality:

$$(Au, v) + H \mid_X^0 (u; v) \geq 0, \ \forall v \in X.$$

Proposition 4.1.10   Suppose that A is coercive and $J : X \to \mathbb{R}$ satisfies

$$J(u) \geq c_0 \max_{w \in \partial J(u)} \langle w, u \rangle + c_1, \ \forall u \in X$$

where $c_0 \in (0, \frac{1}{2})$, $c_1 \in \mathbb{R}$. Then the function I defined in (4.1.15) satisfies the (HPS) condition.

Proof. Let $\{u_n\}$ be a sequence satisfying (4.1.18) and (4.1.19) with $\varepsilon_n \to 0$. We set $v = u_n$ in (4.1.19) to get

$$(Au_n, u_n) + J^0(u_n; u_n) \geq -\varepsilon_n \parallel u_n \parallel .$$

Thus

$$(Au_n, u_n) + \max_{w \in \partial J(u_n)} \langle w, u_n \rangle \geq -\varepsilon_n \parallel u_n \parallel .$$

Therefore

$$\begin{aligned} I(u_n) &\geq \frac{1}{2}(Au_n, u_n) + c_0 \max_{w \in \partial J(u_n)} \langle w, u_n \rangle + c_1 \\ &= (\frac{1}{2} - c_0)(Au_n, u_n) + c_0((Au_n, u_n) \\ &\quad + \max_{w \in \partial J(u_n)} \langle w, u_n \rangle) + c_1 \\ &\geq \alpha(\frac{1}{2} - c_0) \parallel u_n \parallel^2 - c_0 \varepsilon_n \parallel u_n \parallel + c_1 \end{aligned}$$

and the boundedness of $\{u_n\}$ follows from (4.1.19). We conclude as in the proof of Proposition 4.1.9. ∎

Let X^0, X^+, X^- be the closed subspaces of X on which A is null, positive definite, and negative definite, i.e.

$$\begin{aligned} (Az^0, z^0) &= 0, \ \forall \, z^0 \in X^0, \\ (Az^+, z^+) &> 0, \ \forall \, z^+ \in X^+ \backslash \{0\}, \\ (Az^-, z^-) &< 0, \ \forall \, z^- \in X^- \backslash \{0\}. \end{aligned}$$

Then X has a splitting into orthogonal subspaces

$$X = X^0 \oplus X^+ \oplus X^-,$$

with (if $dim \, X^+ \geq 1$ and $dim \, X^- \geq 1$)

$$(Az^+, z^+) \geq \lambda^+ \parallel z^+ \parallel^2, \ \forall \, z^+ \in X^+$$

and

$$(Az^-, z^-) \leq \lambda^- \parallel z^- \parallel^2, \ \forall \, z^- \in X^-,$$

where λ^+, λ^- are the smallest positive and the largest negative eigenvalue of A respectively.

Proposition 4.1.11   Suppose that X^0 is finite-dimensional and $\lambda^- < 0 < \lambda^+$. Assume that

$$\mid H(x) - H(y) \mid \leq K \parallel x - y \parallel_Y, \ (K > 0) \ \forall \, x, \, y \in Y,$$

and that

$$J(z^0) \to +\infty \text{ as } \parallel z^0 \parallel \to +\infty, \ z^0 \in X^0. \tag{4.1.22}$$

Then the function I defined in (4.1.15) satisfies the (HPS) condition.

Proof. Let $\{u_n\}$ be a sequence satisfying (4.1.18) and (4.1.19). We denote by P^0, P^- and P^+ the projections onto X^0, X^- and X^+, respectively. Moreover, for $u \in X$ we set $u^0 = P^0 u$, $u^- = P^- u$ and $u^+ = P^+ u$. We put $v = -u_n^+$ in (4.1.19) to get

$$(Au_n, u_n^+) - J^0(u_n, -u_n^+) \leq \varepsilon_n \parallel u_n^+ \parallel .$$

Therefore

$$(Au_n^+, u_n^+) - H^0(\gamma(u_n); -\gamma(u_n^+)) \leq \varepsilon_n \parallel u_n^+ \parallel$$

and thus

$$\lambda^+ \parallel u_n^+ \parallel^2 \leq (K' + \varepsilon_n) \parallel u_n^+ \parallel$$

which means that u_n^+ is bounded. Here $K' = cK$ where c denotes the constant of continuity of the embedding $\gamma : X \to Y$. We set now $v = u_n^-$ in (4.1.19) to get

$$-(Au_n, u_n^-) - J^0(u_n, u_n^-) \leq \varepsilon_n \parallel u_n^- \parallel .$$

Thus

$$-(Au_n^-, u_n^-) \leq \varepsilon_n \parallel u_n^- \parallel + H^0(\gamma(u_n); \gamma(u_n^-))$$

so that

$$-\lambda^- \parallel u_n^- \parallel^2 \leq (K' + \varepsilon_n) \parallel u_n^- \parallel$$

and recalling that $\lambda^- < 0$, we conclude that the sequence u_n^- is bounded too. Relation (4.1.18) implies that for n great enough

$$I(u_n) = \frac{1}{2}(Au_n^+, u_n^+) + \frac{1}{2}(Au_n^-, u_n^-) + J(u_n) \leq c + \frac{1}{n}. \qquad (4.1.23)$$

We have

$$J(u_n) = J(u_n) - J(u_n^0) + J(u_n^0)$$

and thus

$$J(u_n) \geq -K \parallel \gamma(u_n^+) + \gamma(u_n^-) \parallel_Y + J(u_n^0).$$

Thus (4.1.23) implies

$$\frac{1}{2}(Au_n^+, u_n^+) + \frac{1}{2}(Au_n^-, u_n^-) - K' \parallel u_n^+ \parallel - K' \parallel u_n^- \parallel + J(u_n^0) \leq c + \frac{1}{n}.$$

Thus u_n^0 has to be bounded due to the boundedness of u_n^-, u_n^+ and assumption (4.1.22).

By considering eventually a subsequence, we may suppose that $u_n \rightharpoonup u$ in X and $\gamma(u_n) \to \gamma(u)$ in Y. Moreover, we have

$$(Au_n, v) + H^0(\gamma(u_n); \gamma(v)) \geq -\varepsilon_n \parallel v \parallel, \ \forall \, v \, \in \, X \qquad (4.1.24)$$

since

$$H^0(\gamma(u_n); \gamma(v)) \geq J^0(u_n; v), \ \forall \, v \, \in \, X.$$

We set $v = u^+ - u_n^+$ in (4.1.24) to get

$$(Au_n^+, u_n^+ - u^+) \leq H^0(\gamma(u_n); \gamma(u^+ - u_n^+)) + \varepsilon_n \parallel u_n^+ - u^+ \parallel$$

and thus

$$\lambda^+ \parallel u_n^+ - u^+ \parallel^2 \leq H^0(\gamma(u_n); \gamma(u^+ - u_n^+))$$
$$+\varepsilon_n \parallel u_n^+ - u^+ \parallel -(Au^+, u_n^+ - u^+).$$

Taking the limit superior and using the upper semi-continuity of $H^0(z; w)$ as a function of (z, w) on $Y \times Y$, we obtain that $u_n^+ \to u^+$ in X. Setting now $v = u_n^- - u^-$ in (4.1.24), we obtain

$$(Au_n^-, u_n^- - u^-) \geq -H^0(\gamma(u_n); \gamma(u_n^- - u^-))$$

$$-\varepsilon_n \parallel u_n^- - u^- \parallel$$

and thus

$$-\lambda^- \parallel u_n^- - u^- \parallel^2 \leq \varepsilon_n \parallel u_n^- - u^- \parallel$$
$$+H^0(\gamma(u_n); \gamma(u_n^- - u^-)) + (Au^-, u_n^- - u^-).$$

Taking the limit superior, we obtain that $u_n^- \to u^-$ in X. The space X^0 is finite dimensional and thus $u_n^0 \to u^0$ in X. Thus $u_n = u_n^+ + u_n^0 + u_n^- \to u^+ + u^0 + u^- = u$. $\blacksquare$

Remark 4.1.12 i) It is easy to see that if we replace assumption (4.1.22) by

$$J(z^0) \to -\infty \quad \text{as} \quad \parallel z^0 \parallel \to +\infty, \ z^0 \in X^0$$

then Proposition 4.1.11 remains true.

ii) The function $f(x) = \min\{1, e^x\}$ is locally Lipschitz on $\mathbb{R}$. This function does not satisfy the (PS)-condition. Particularly, if $\{u_k\}$ is a sequence such that $u_k \to -\infty$, $f(u_k) \to 0$ then for k great enough $\partial f(u_k) = e^{u_k} \to 0$ but it is impossible to find a convergent subsequence.

Let Ω be a nonempty open subset of $\mathbb{R}^n(n \geq 1)$ and let X be a real Hilbert space such that $X \hookrightarrow L^2(\Omega)$ with dense and compact embedding.

Let $a : X \times X \to \mathbb{R}$ be a bilinear, symmetric and continuous form satisfying the coercivity condition

$$a(u, u) \geq \alpha \parallel u \parallel^2, (\alpha > 0), \forall u \in X.$$

Denote by $A : X \to X$ the linear operator defined by

$$(Au, v) = a(u, v), \ \forall u, v \in X.$$

Let F and G be two functions satisfying the following assumptions; $F : \mathbb{R} \to \mathbb{R} \cup \{+\infty\}$ is convex, l.s.c. and $F(0) = 0$; $G : \mathbb{R} \to \mathbb{R}$ is of class $C^1, G(0) = 0, G'(t) = g(t)$ and $\mid g(t) \mid \leq c_1 + c_2 \mid t \mid$, $\forall t \in \mathbb{R}$, where c_1, c_2 are positive constants. We consider the function $I : X \to \mathbb{R} \cup \{+\infty\}$ satisfying (H_2), i.e.

$$I = \Phi + \Psi, \tag{4.1.25}$$

with

$$\Phi(u) = -\int_\Omega G(u)dx, \ u \in X \tag{4.1.26}$$

and

$$\Psi(u) = \begin{cases} \frac{1}{2}a(u, u) + \int_\Omega F(u)dx & \text{if } F(u) \in L^1(\Omega) \\ +\infty & \text{otherwise.} \end{cases} \tag{4.1.27}$$

We assume that $D(\Psi) = \{u \in X : F(u) \in L^1(\Omega)\}$ is weakly closed in X. The functional Ψ is proper convex and l.s.c. (see Section 1.3.4). The assumptions required on G ensure that $\Phi \in C^1(X; \mathbb{R})$. Let $\lambda_1 > 0$ be the first eigenvalue of the problem: Find $(\lambda, u) \in \mathbb{R} \times X$ such that

$$(Au, v) = \lambda \int_\Omega uv dx, \ \forall v \in X.$$

Proposition 4.1.13   Suppose that

$$\liminf_{|t| \to +\infty} \frac{F(t) - G(t)}{t^2} > -\frac{1}{2}\lambda_1. \tag{4.1.28}$$

Then I satisfies the (VPS) condition.

Proof. Let $\{u_n\}$ be a sequence such that

$$I(u_n) \to c, \tag{4.1.29}$$

and

$$(\Phi'(u_n), v - u_n) + \Psi(v) - \Psi(u_n) \geq -\varepsilon_n \parallel v - u_n \parallel, \forall\, v \in X, \quad (4.1.30)$$

with $\varepsilon_n \to 0$.

Choose $R > 0$ and $\lambda < \lambda_1$ such that $(F(t) - G(t))t^{-2} \geq -\frac{1}{2}\lambda$ for $|\,t\,| > R$. Then

$$
\begin{aligned}
I(u) &= \frac{1}{2}(Au, u) + \int_{|u|>R}(F(u) - G(u))dx + \int_{|u|\leq R}(F(u) - G(u))dx \\
&\geq \frac{1}{2}(Au, u) - \frac{1}{2}\lambda \int_{\Omega} u^2 dx - C
\end{aligned}
$$

for some positive constant $C > 0$. Here we have used Theorem 1.1.11 to assert that

$$\int_{|u|\leq R} F(u)dx \geq -\theta \int_{|u|\leq R} |\,u\,|\,dx - \beta \int_{|u|\leq R} dx$$

for some positive constants θ, β. Moreover, for all $t > 0$, there exists $\zeta(t) \in (0, t)$ such that

$$|\,G(t)\,| \leq |\,g(\zeta(t))\,|\,|\,t\,|.$$

Thus

$$-\int_{|u|\leq R} G(u)dx \geq -\int_{|u|\leq R} c_1 |\,u\,|\,dx - \int_{|u|\leq R} c_2 |\,u\,|^2\,dx.$$

It results that

$$\int_{|u|\leq R}(F(u) - G(u))dx \geq -C$$

for some constant $C > 0$.

Thus

$$I(u) \geq \frac{\alpha}{2}(1 - \frac{\lambda}{\lambda_1}) \parallel u \parallel^2 -C.$$

Therefore $I(u) \to +\infty$ as $\parallel u \parallel \to +\infty$. It results that the sequence $\{u_n\}$ is bounded. We may therefore assume that for a subsequence $u_n \rightharpoonup u$ in X, $u_n \to u$ in $L^2(\Omega)$, $u_n(x) \to u(x)$ a.e. x in Ω and $\Psi(u) \leq \liminf_{n\to\infty} \Psi(u_n)$. Our conditions on g ensure that the Nemyckii operator $Nu(x) = g(u)(x)$ is continuous on $L^2(\Omega)$. It results that Φ' is completely continuous on X. Thus $\Phi'(u_n) \to \Phi(u)$. Setting $v = u$ in (4.1.30), we use the fact that $D(\Psi)$ is weakly closed to write

$$(\Phi'(u_n), u_n - u) + \frac{1}{2}a(u_n, u_n - u) - \int_{\Omega} F(u)dx + \int_{\Omega} F(u_n)dx \leq \varepsilon_n \parallel u - u_n \parallel.$$

Thus

$$\frac{1}{2}\alpha \parallel u_n - u \parallel^2 + \int_\Omega F(u_n)dx - \int_\Omega F(u)dx$$

$$\leq \ \varepsilon_n \parallel u - u_n \parallel -\frac{1}{2}a(u, u_n - u) - (\Phi'(u_n), u_n - u).$$

Taking the limit superior as $n \to +\infty$, we obtain that $u_n \to u$ in X. ∎

4.2   A DEFORMATION RESULT

The aim of this Section is to prove a deformation result for functionals of type (H). To prove our deformation result two preliminary properties are necessary.

Lemma 4.2.1   Let the functional $I : X \to (-\infty, +\infty]$ satisfy (H) and $(PS)_c$ for some $c \in \mathbb{R}$ and let U be a neighborhood of $K_c(I)$. Then, for each $\bar{\varepsilon} > 0$, there is an $\varepsilon \in (0, \bar{\varepsilon})$ such that, corresponding to every $u_0 \in I^{-1}([c - \varepsilon, c + \varepsilon])\backslash U$, there exists an $v_0 \in X$ satisfying

$$\Phi^0(u_0; v_0 - u_0) + \Psi(v_0) - \Psi(u_0) < -3\varepsilon \parallel v_0 - u_0 \parallel \ . \qquad (4.2.1)$$

Proof. If (4.2.1) were not true, then we could find sequences $\{\varepsilon_n\} \subset \mathbb{R}$ and $\{u_n\} \subset X\backslash U$ such that $\varepsilon_n \downarrow 0$, $I(u_n) \to c$ and

$$\Phi^0(u_n; v - u_n) + \Psi(v) - \Psi(u_n) \geq -3\varepsilon_n \parallel v - u_n \parallel, \ \forall \, v \in X.$$

Then properties $(PS)_c$ and Proposition 4.1.3 imply that up to a subsequence $u_n \to u \in K_c(I)$. Thus $u_n \in U$ if n is sufficiently large, which is a contradiction. ∎

Lemma 4.2.2   Under the assumptions of Lemma 4.2.1 let an $\varepsilon \in (0, \bar{\varepsilon})$ be fixed as there stated. Then for each $u_0 \in I_{c+\varepsilon}\backslash U$, there exist $v_0 \in X$ and a neighborhood U_0 of v_0 in X such that Φ is Lipschitz continuous on U_0 of Lipschitz constant $K > 0$ and the relations below hold:

$$\Phi^0(u; v_0 - u) + \Psi(v_0) - \Psi(w) \leq K(\parallel u - v_0 \parallel + \parallel w - v_0 \parallel),$$

$$\forall \, u, \ w \in U_0, \qquad (4.2.2)$$

and

$$\Phi^0(u; v_0 - u) + \Psi(v_0) - \Psi(w) \leq -3\varepsilon \parallel v_0 - w \parallel,$$

$$\forall\, u,\, w \,\in\, U_0 \ \text{ with } \ I(w) \geq c - \varepsilon. \tag{4.2.3}$$

If $u_o \in K(I)$ one can take $v_0 = u_0$. Otherwise the data v_0, U_0 and a number $\delta_0 > 0$ can be chosen so that $v_0 \notin \overline{U}_0$ and

$$\Phi^0(u; v_0 - u) + \Psi(v_0) - \Psi(w) \leq -\delta_0 \parallel v_0 - w \parallel, \ \forall\, u,\, w \,\in\, U_0. \tag{4.2.4}$$

Proof. Case n° 1: $u_0 \in K(I)$. By (4.1.1) if follows that

$$\Phi^0(u; u_0 - u) + \Psi(u_0) - \Psi(w) \leq \Phi^0(u; u_0 - u) + \Phi^0(u_0; w - u_0),$$

$$\forall\, u,\, w \,\in\, X.$$

Thus for all u, w close enough to u_0,

$$\Phi^0(u; u_0 - u) + \Psi(u_0) - \Psi(w) \leq K(\parallel u - u_0 \parallel + \parallel w - u_0 \parallel).$$

Therefore (4.2.2) holds with $v_0 = u_0$.

We claim that if $u_0 \in K(I)$ with $u_0 \in I_{c+\varepsilon} \backslash U$, then one has

$$I(u_0) < c - \varepsilon. \tag{4.2.5}$$

Suppose that this is not true. Then we would have $u_0 \in I^{-1}([c - \varepsilon, c + \varepsilon]) \backslash U$. Then Lemma 4.2.1 yields a point $\overline{v}_0 \in X$ satisfying (4.2.1). This contradicts the assumption that $u_0 \in K(I)$ and the claim (4.2.5) is proved. In order to justify (4.2.3) with $v_0 = u_0$, when $u_0 \in K(I)$, we suppose that every neighborhood of u_0 contains points u with $I(u) \geq c - \varepsilon$. Let us show that there exist a constant $d > 0$ and a neighborhood U_0 of u_0 such that

$$\Psi(u) - \Psi(u_0) \geq d, \ \forall\, u \,\in\, U_0 \ \text{ with } \ I(u) \geq c - \varepsilon. \tag{4.2.6}$$

Assuming the contrary there would exist a sequence $u_n \to u_0$ in X that fulfills $\Psi(u_n) \leq \Psi(u_0) + d_n$ where $d_n > 0, d_n \to 0$ and $I(u_n) \geq c - \varepsilon$. Then, due to the continuity of Φ, we obtain

$$c - \varepsilon \leq \Phi(u_n) + \Psi(u_n) \leq \Phi(u_n) + \Psi(u_0) + d_n < c - \varepsilon$$

for n large enough, where the last inequality above follows from (4.2.5). The contradiction ensures that (4.2.6) is true.

Taking into account (4.2.6), we obtain

$$\Phi^0(u; u_0 - u) + \Psi(u_0) - \Psi(w) \leq \Phi^0(u; u_0 - u) - d,$$

for all $u, w \in U_0$, $I(w) \geq c - \varepsilon$. Moreover, for a possibly smaller neighborhood U_0 of u_o we have also

$$\Phi^0(u; u_0 - u) - d \leq -3\varepsilon \parallel u_0 - w \parallel, \ \forall \, u, \, w \, \in \, U_0, \ I(w) \geq c - \varepsilon.$$

Indeed, if we suppose the contrary then we can find sequences $u_n \to u_0$ and $w_n \to u_0$ such that $I(w_n) \geq c - \varepsilon$ and

$$\Phi^0(u_n; u_0 - u_n) > -3\varepsilon \parallel u_0 - w_n \parallel + d.$$

Taking the lim sup above and using the upper semicontinuity of $\Phi^0(\cdot; \cdot)$ we obtain $d \leq 0$, which is a contradiction. Thus (4.2.3) is checked .

Case n°2: $u_0 \notin K(I)$. Let us treat firstly the case $I(u_0) < c - \varepsilon$. According to (4.1.1) the assumption $u_0 \notin K(I)$ implies the existence of $v_0 \in X$ such that

$$\Phi^0(u_0; v_0 - u_0) + \Psi(v_0) - \Psi(u_0) < 0. \tag{4.2.7}$$

Notice that the point $v_0 \in X$ in (4.2.7) can be chosen arbitrarily close to u_0. To see this we use the convexity of Ψ and the positive homogeneity of $\Phi^0(u_0; \cdot)$, together with (4.2.7). Indeed, we can write

$$\Phi^0(u_0; tv_0 + (1 - t)u_0 - u_0) + \Psi(tv_0 + (1 - t)u_0) - \Psi(u_0)$$

$$\leq t(\Phi^0(u_0; v_0 - u_0) + \Psi(v_0) - \Psi(u_0)) < 0, \ \forall \, t \, \in \, (0, 1).$$

This shows that v_0 in (4.2.7) can be taken arbitrarily close to u_0. We choose it so that $\parallel v_0 - u_0 \parallel \leq \overline{\delta} := \min\{d/3K, d/18\varepsilon\}$. From (4.2.7), we get

$$\Psi(v_0) - \Psi(u_0) < -\Phi^0(u_0; v_0 - u_0) \leq K \parallel v_0 - u_0 \parallel \leq \frac{d}{3}. \tag{4.2.8}$$

Then (4.2.6) (which holds true since we have supposed that $I(u_0) < c - \varepsilon$) and (4.2.8) allow to write

$$\Phi^0(u; v_0 - u) + \Psi(v_0) - \Psi(w)$$

$$\leq \ K \parallel v_0 - u \parallel + \Psi(v_0) - \Psi(u_0) + \Psi(u_0) - \Psi(w)$$

$$\leq \ K \parallel v_0 - u \parallel + \frac{d}{3} - d,$$

for all $u, w \in U_0$ with $I(w) \geq c - \varepsilon$ and $\parallel v_0 - u_0 \parallel$ sufficiently small ($\leq \overline{\delta}$). Thus with U_0 possibly smaller, we can say that

$$K \parallel v_0 - u \parallel - \frac{2d}{3} \leq -3\varepsilon \parallel v_0 - w \parallel .$$

If not we obtain sequences $u_n \to u_0$ and $w_n \to u_0$ such that

$$K \parallel v_0 - u_n \parallel > \frac{2d}{3} - 3\varepsilon \parallel v_0 - w_n \parallel .$$

and thus

$$\frac{d}{3} \geq K \parallel v_0 - u_0 \parallel \geq \frac{2d}{3} - 3\varepsilon \parallel v_0 - u_0 \parallel \geq \frac{2d}{3} - 3\varepsilon \left(\frac{d}{18\varepsilon} \right) = \frac{d}{2},$$

which is a contradiction. Thus

$$\Phi^0(u; v_0 - u) + \Psi(v_0) - \Psi(w) \leq -3\varepsilon \parallel v_0 - w \parallel,$$

for all $u, w \in U_0$ with $I(w) \geq c - \varepsilon$. This is just (4.2.3). We pass to justify (4.2.4) in the case where $u_0 \notin K(I)$ and $I(u_0) < c - \varepsilon$. In this respect, from (4.2.7) it follows that we can admit $v_0 \notin \overline{U}_0$ (U_0 possibly smaller), and a constant $\delta_0 > 0$ can be found to have

$$\Phi^0(u_0; v_0 - u_0) + \Psi(v_0) - \Psi(u_0) < -\delta_0 \parallel v_0 - u_0 \parallel . \qquad (4.2.9)$$

We establish (4.2.4) with this $\delta_0 > 0$ arguing by contradiction. To this end let $u_n \to u_0$ and $w_n \to u_0$ fulfill

$$\Phi^0(u_n; v_0 - u_n) + \Psi(v_0) - \Psi(w_n) \geq -\delta_0 \parallel v_0 - w_n \parallel .$$

Then the lower semicontinuity of Ψ and the upper semicontinuity of $\Phi^0(\cdot; \cdot)$ enable us to conclude

$$\Phi^0(u_0; v_0 - u_0) + \Psi(v_0) - \Psi(u_0) \geq -\delta_0 \parallel v_0 - u_0 \parallel .$$

Since we arrived at a contradiction with (4.2.9), the proof of (4.2.4) is complete in the case specified above. We proceed now to the situation where $u_0 \notin K(I)$ and $I(u_0) \geq c - \varepsilon$. Notice that we are in position to apply Lemma 4.2.1 since $u_0 \in I_{c+\varepsilon}$ by assumption. Then arguing as above on the basis of (4.2.1) in place of (4.2.9) one obtains (4.2.3) and (4.2.4) with $\delta_0 = -3\varepsilon$. This completes the proof of Lemma 4.2.2.   ∎

Our deformation result is now stated.

Theorem 4.2.3   Let the functional $I : X \to (-\infty, +\infty]$ and the number $c \in \mathbb{R}$ be such that conditions (H) and (PS)$_c$ are satisfied. Let U be a neighborhood of $K_c(I)$ and let $\bar\varepsilon > 0$ be a fixed number. Then there exists $\varepsilon \in (0, \bar\varepsilon)$ such that for every compact subset $A \subset X \backslash U$ with

$$c \leq \sup_A I \leq c + \varepsilon \qquad (4.2.10)$$

there exist a closed neighborhood W of A in X and a deformation $\alpha :$ $W \times [0, \bar{s}] \to X$, $\bar{s} > 0$, i.e. a continuous map with $\alpha(\cdot, 0) = id_W$, satisfying

$$\| u - \alpha(u, s) \| \le s, \ \forall u \in W, \ \forall s \in [0, \bar{s}], \tag{4.2.11}$$

$$I(\alpha(u, s)) - I(u) \le Ms, \ \forall u \in W, \ \forall s \in [0, \bar{s}], \tag{4.2.12}$$

with a constant $M > 0$ independent of u and s;

$$I(\alpha(u, s)) - I(u) \le -2\varepsilon s, \ \forall u \in W, \ \text{with} \tag{4.2.13}$$

$$I(u) \ge c - \varepsilon, \ \forall s \in [0, \bar{s}];$$

$$\sup_{u \in A} I(\alpha(u, s)) - \sup_{u \in A} I(u) \le -2\varepsilon s, \ \forall s \in [0, \bar{s}]. \tag{4.2.14}$$

Furthermore, if W_0 is a closed subset of X with $W_0 \cap K(I) = \emptyset$, W and α can be chosen so that

$$I(\alpha(u, s)) - I(u) \le 0, \ \forall u \in W \cap W_0, \ \forall s \in [0, \bar{s}]. \tag{4.2.15}$$

Proof. Consider an $\varepsilon \in (0, \bar{\varepsilon})$ determined by Lemma 4.2.2. By Lemma 4.2.2 and relation (4.2.10), for each point $u_0 \in A$ there exists an open convex neighborhood U_0 of u_0 in X and some $v_0 \in X$ verifying the properties stated in Lemma 4.2.2. In addition, we can suppose

$$u_0 \in K(I) \implies U_0 \cap W_0 = \emptyset. \tag{4.2.16}$$

The compactness of A insures the existence of a finite open covering $\{U_i\}_{1 \le i \le m}$ of A in X, where every set U_i corresponds to a point $u_i \in A$ in the same way as does U_0 for u_0 above. Let v_i denote the point corresponding to u_i as v_0 was obtained by means of u_0. Taking, if necessary, a finer covering of A in X we may assume

$$u_i \in K(I) \implies d(u_i, U_j) > 0, \ \forall j \ne i. \tag{4.2.17}$$

We introduce

$$V = \bigcup_{i=1}^{m} U_i \tag{4.2.18}$$

and, for each $i = 1, \cdots, m$, $\sigma_i : V \to \mathbb{R}$ by

$$\sigma_i = \frac{\rho_i}{\sum_{j=1}^{m} \rho_j}, \tag{4.2.19}$$

where $\rho_i : X \to \mathbb{R}$ is given by

$$\rho_i(x) = d(x, X \backslash U_i), \ \forall \, x \, \in \, X.$$

Now, we define $\alpha : V \times [0, +\infty) \to X$. If $u \in U_{i_0} \backslash \cup_{j \neq i_0} U_j$ with $u_{i_0} \in A \cap K(I)$ we put

$$\alpha(u, s) := \begin{cases} u + \dfrac{s(u_{i_0} - u)}{\|u_{i_0} - u\|}, & 0 \, \leq \, s <\| \, u_{i_0} - u \, \| \\[2mm] u_{i_0}, & s \geq \| \, u_{i_0} - u \, \| \end{cases} \qquad (4.2.20)$$

and for all other $u \in V$

$$\alpha(u, s) = u + s \sum_i \sigma_i(u) \frac{(v_i - u)}{\| \, v_i - u \, \|} \, . \qquad (4.2.21)$$

In the sum of (4.2.21) we count only the terms with $\sigma_i(u) \neq 0$, i.e. $u \in U_i$. If $u_i \not\subset K(I)$ then from Lemma 4.2.2, $v_i \notin \overline{U}_i$. If $u_i \in K(I)$ then $v_i = u_i$ and from (4.2.17), we obtain $v_i \notin \overline{U}_j$, $j \neq i$. Thus $v_i \neq u$ and (4.2.21) is well-defined. Moreover, we note that for

$$u \in U_{i_0} \backslash \underset{j \neq i_o}{\cup} U_j$$

and

$$u \neq u_{i_0} \in A \cap K(I) \, ,$$

by (4.2.18) and (4.2.19), the expression of $\alpha(u, s)$ in (4.2.21) makes sense and, if $s <\| \, u_{i_0} - u \, \|$, coincides with (4.2.20). Simple computations shows the existence of $\bar{s} > 0$ such that $\alpha : V \times [0, \bar{s}] \to X$ is continuous. Clearly, the relations $\alpha(\cdot, 0) = id_V$ and (4.2.11) are satisfied. We check now (4.2.12), (4.2.13) and (4.2.15) in the case where $\alpha(u, s)$ is expressed by formula (4.2.21). Let us write (4.2.21) in the form

$$\alpha(u, s) = u + s\overline{w}, \qquad (4.2.22)$$

where

$$\overline{w} = \sum_i \frac{\sigma_i(u)(v_i - u)}{\| \, v_i - u \, \|}.$$

The Lebourg's mean value theorem (see Theorem 1.2.12), Proposition 1.2.7 and relation (4.2.22) imply

$$I(\alpha(u, s)) = \Phi(u + s\overline{w}) + \Psi(u + s\overline{w})$$

$$\leq \Phi(u) + s\Phi^0(u + r\overline{w}; \overline{w}) + \Psi(u + s\overline{w}), \qquad (4.2.23)$$

for some $r = r(s) \in (0, s)$. From the subadditivity and the positive homogeneity of $\Phi^0(u + r\overline{w}; \cdot)$ we get

$$\Phi^0(u + r\overline{w}; \overline{w}) \leq \sum_i \frac{\sigma_i(u)}{\| v_i - u \|} \Phi^0(u + r\overline{w}; v_i - u)$$

$$\leq \sum_i \frac{\sigma_i(u)}{\| v_i - u \|} (\Phi^0(u + r\overline{w}; v_i - u - r\overline{w})$$

$$+ r\Phi^0(u + r\overline{w}; \overline{w})).$$

Since $v_i \notin \overline{U}_i$ and Φ is Lipschitz continuous on the compact set A one finds a closed neighborhood W of A with $W \subset V$ and a constant $C > 0$ such that

$$\Phi^0(u + r\overline{w}; \overline{w}) \leq \sum_i \frac{\sigma_i(u)}{\| v_i - u \|} \Phi^0(u + r\overline{w}, v_i - u - r\overline{w}) + Cs, \quad (4.2.24)$$

$$\forall u \in W, \forall 0 \leq s \leq \bar{s},$$

provided $\bar{s} > 0$ is small enough and u is admissible for (4.2.21). The convexity of Ψ allows us to write

$$\Psi(u + s\overline{w}) = \Psi((1 - s \sum_i \frac{\sigma_i(u)}{\| v_i - u \|})u + s \sum_i \frac{\sigma_i(u)}{\| v_i - u \|} v_i)$$

$$\leq \Psi(u) + s \sum_i \frac{\sigma_i(u)}{\| v_i - u \|} (\Psi(v_i) - \Psi(u)), \quad (4.2.25)$$

where $0 \leq s \leq \bar{s}$, with $\bar{s}$ sufficiently small to have actually in (4.2.25) a convex combination. Then (4.2.23)-(4.2.25) lead to

$$I(\alpha(u, s)) \leq I(u) + s \sum_{i=1}^{m} \frac{\sigma_i(u)}{\| v_i - u \|} (\Phi^0(u + r\overline{w}; v_i - u - r\overline{w}) +$$

$$+ \Psi(v_i) - \Psi(u)) + Cs^2. \quad (4.2.26)$$

Let $\{U_i'\}_{1 \leq i \leq m}$ be an open covering of A with $\overline{U_i'} \subset U_i$ and U_i' bounded; $1 \leq i \leq m$. We can choose W and $\bar{s}$ so small that $W \subset \cup_{1 \leq i \leq m} U_i'$ and if $x \in U_i' \cap W$, $0 \leq s \leq \bar{s}$ and $\| y \| \leq 1$ then $x + sy \in U_i$ for all $i = 1, \cdots, m$. Thus using (4.2.26) together with (4.2.2), we obtain:

$$I(\alpha(u, s)) - I(u) \leq s \sum_{i=1}^{m} \frac{\sigma_i(u)}{\| v_i - u \|} K(\| u + r\overline{w} - v_i \| + \| u - v_i \|) + Cs^2.$$

Thus (4.2.12) is satisfied. Analogously, from (4.2.3) and (4.2.26) we infer that (4.2.13) holds provided that we choose $\bar{s} \leq \frac{\varepsilon}{C}$. It is clear that combining (4.2.4), (4.2.16) and (4.2.26) one obtains (4.2.15) provided that we choose $\bar{s} \leq \frac{\delta_0}{C}$.

Let us check now (4.2.12), (4.2.13), (4.2.15) in the case where $\alpha(u, s)$ is given by (4.2.20). In view of (4.2.16) it is clear that (4.2.15) is vacuous in this situation. Notice that if $s <\| u_{i_0} - u \|$ in (4.2.20), $\alpha(u, s)$ is expressed by (4.2.22) with

$$\overline{w} = \| u_{i_0} - u \|^{-1} (u_{i_0} - u) \ .$$

A careful examination of arguments centered around (4.2.23) - (4.2.26) shows readily that using the same procedure we deduce (4.2.12), (4.2.13). We suppose that $s \geq \| u_{i_0} - u \|$ holds in (4.2.20). Then we get

$$\begin{aligned}
I(\alpha(u, s)) - I(u) &= I(u_{i_0}) - I(u) \\
&= \Phi(u_{i_0}) - \Phi(u) + \Psi(u_{i_0}) - \Psi(u) \\
&\leq \Phi^0(u_{i_0} + \lambda(u - u_{i_0}); u_{i_0} - u) + \Psi(u_{i_0}) - \Psi(u),
\end{aligned}$$

where $\lambda \in (0, 1)$ is given by the Lebourg's mean value theorem. Thus

$$I(\alpha(u, s)) - I(u) \leq \Phi^0(u_{i_0} + \lambda(u - u_{i_0}); u_{i_0} - (u_{i_0} + \lambda(u - u_{i_0})))$$

$$+ \Phi^0(u_{i_0} + \lambda(u - u_{i_0}); (1 - \lambda)(u_{i_0} - u)) + \Psi(u_{i_0}) - \Psi(u)$$

$$\leq K \| u - u_{i_0} \| + \Phi^0(u_{i_0} + \lambda(u - u_{i_0}); u_{i_0} - (u_{i_0} + \lambda(u - u_{i_0})))$$

$$+ \Psi(u_{i_0}) - \Psi(u),$$

where $K > 0$ denotes the Lipschitz constant of Φ on U_{i_0}. Then, by using relation (4.2.2) appropriately, we obtain:

$$I(\alpha(u, s)) - I(u) \leq K(\lambda \| u - u_{i_0} \| + 2 \| u - u_{i_0} \|)$$

which shows that (4.2.12) holds true because $\| u - u_{i_0} \| \leq s$. Since $u_{i_0} \in K(I)$, by (4.2.5) it is known that

$$I(\alpha(u, s)) = I(u_{i_0}) < c - \varepsilon,$$

so for a sufficiently small $s > 0$,

$$I(u_{i_0}) + 2\varepsilon s < c - \varepsilon.$$

Now, with the additional assumption that $I(u) \geq c - \varepsilon$ we arrive at (4.2.13).

It remains to prove (4.2.14). If we have

$$\sup_{u \in A} I(\alpha(u, s)) \leq c - \frac{\varepsilon}{2},$$

then taking $s \leq \frac{1}{4}$, by the first inequality of (4.2.10), it results that

$$\sup_{u \in A} I(\alpha(u, s)) - \sup_{u \in A} I(u) \leq c - \frac{\varepsilon}{2} - c \leq -\frac{\varepsilon}{2} 4s = -2\varepsilon s.$$

If the inequality below occurs

$$\sup_{u \in A} I(\alpha(u, s)) > c - \frac{\varepsilon}{2},$$

then by virtue of (4.2.12), (4.2.13) and with s small enough it is seen that

$$c - \frac{1}{2}\varepsilon \ < \ \sup_{u \in A} I(\alpha(u, s)) = \sup_{\substack{u \in A \\ I(u) \geq c - \varepsilon}} I(\alpha(u, s))$$

$$\leq \ \sup_{\substack{u \in A \\ I(u) \geq c - \varepsilon}} I(u) - 2\varepsilon s \leq \sup_{u \in A} I(u) - 2\varepsilon s \ .$$

Consequently, (4.2.14) follows. This completes the proof of Theorem 4.2.1. ∎

Corollary 4.2.4   Assume in addition to the requirements of Theorem 4.1.6 that the functionals Φ and Ψ are even, i.e., $\Phi(u) = \Phi(-u)$ and $\Psi(u) = \Psi(-u)$ for all $u \in X$. If the set A is symmetric with respect to the origin, i.e., $A = -A$ then W can be chosen to be symmetric with respect to the origin and the deformation $\alpha : W \times [0, \bar{s}] \to X$ can be constructed so that $\alpha(\cdot, s) : W \to X$ is odd for each $s \in [0, \bar{s}]$, i.e., $\alpha(-u, s) = -\alpha(u, s), \ \forall u \in W$.

Proof.   Since I is even, its critical points arise in symmetric pairs $(u, -u)$. Thus, in view of the symmetry of A, the elements U_i of the open covering considered in the proof of Theorem 4.2.3 can be taken pairwise symmetric with respect to the origin. Then V in (4.2.18) satisfies $V = -V$, hence W can be chosen with the same property.

Now, starting with a deformation α as given by Theorem 4.2.3, we define $\beta : W \times [0, \bar{s}] \to X$ as

$$\beta(u, s) = \frac{1}{2}(\alpha(u, s) - \alpha(-u, s)), \ \forall \, (u, s) \in W \times [0, \bar{s}]. \tag{4.2.27}$$

Then one can follow an analogous reasoning to the one developed in the proof of Theorem 4.2.3 based in fact on Lebourg's mean value theorem applied to Φ and the convexity of Ψ (that can be used since β above appears as a convex combination). We omit the details of the effective verification of assertions (4.2.11)-(4.2.15) with $\alpha(u,s)$ replaced by $\beta(u,s)$. ∎

Remark 4.2.5   Following the pattern of Corollary 4.2.4, under appropriate conditions, we can construct deformations $\alpha(u,s)$ with properties (4.2.11)-(4.2.15) such that each mapping $\alpha(\cdot,s)$ is equivariant with respect to the action of a topological group G on X, i.e.

$$\alpha(gu,s) = g\alpha(u,s), \ \forall \, u \, \in \, X, \ \forall \, g \, \in \, G.$$

The situation treated in Corollary 4.2.4 corresponds to the group $G = \mathbb{Z}_2 = \{id_X, -id_X\}$. Related results for the pure locally Lipschitz case, i.e. $\Psi = 0$ are given in Chapter 9.

4.3   MINIMAX PRINCIPLES FOR FUNCTIONALS OF TYPE (H)

Based on the deformation result in Theorem 4.2.3 we are now able to provide a general minimax principle for the nonsmooth functionals verifying hypothesis (H). We recall the definition of linking.

Definition 4.3.1   Let S be a nonempty subset of the Banach space X and let Q be a compact topological submanifold of X with nonempty boundary ∂Q (in the sense of manifolds with boundary). We say that S and Q link if the next properties hold

$$S \cap \partial Q = \emptyset \tag{4.3.1}$$

and

$$f(Q) \cap S \neq \emptyset \tag{4.3.2}$$

whenever $f \in \Gamma$, where

$$\Gamma := \{f \in C^0(Q;X) : \ f \mid_{\partial Q} = id_{\partial Q}\} \,. \tag{4.3.3}$$

Example 4.3.2   Suppose that $X = W \oplus S$ where W, S are closed subspaces of X with $dim\, W < +\infty$. One sets

$$Q = \{w \in W :\| w \|\leq \rho\}\ (\rho > 0).$$

Then S and Q link. Indeed, it is clear that $S \cap \partial Q = \emptyset$. Suppose now by contradiction that there exists $f \in \Gamma$ such that $f(Q) \cap S = \emptyset$. Then the mapping $r : Q \to \partial Q$ defined by

$$u \to r(u) := \rho \frac{P_W f(u)}{\| P_W f(u) \|},$$

where P_W denotes the orthogonal projection on W, is a retraction from Q onto ∂Q, i.e. a continuous mapping such that $r(u) = u$ on ∂Q. This is a contradiction to a standard topological result (Brouwer fixed point theorem) since $dim\, W < +\infty$.

Example 4.3.3   Suppose that $X = W \oplus Z$ where W and Z are closed subspaces of X with $dim\, W < +\infty$. One sets

$$S = \{z \in Z :\| z \|= r\}$$

and

$$Q = \{w + \lambda e : w \in W, \| w \|\leq \rho, 0 \leq \lambda \leq R\},$$

with $e \in Z, \| e \|= 1$, $R > 0$, $r \in (0, R)$, $\rho > 0$. Then S and Q link. Here also $S \cap \partial Q = \emptyset$. If we suppose by contradiction that there exists $f \in \Gamma$ such that $f(Q) \cap S = \emptyset$ then we may define the mapping $\bar{r} : Q \to \partial Q$,

$$u \to \bar{r}(u) = \theta(P_W f(u)+ \| (I - P_W)f(u) \| e),$$

where θ denotes a retraction from $W \oplus \mathbb{R}\, e \backslash \{re\}$ onto ∂Q. (The retraction θ can be constructed as the radial projection of center re from $W \oplus \mathbb{R}\, e \backslash \{re\}$ onto ∂Q). The mapping $\bar{r}$ is a retraction from Q onto ∂Q and we obtain a contradiction as in Example 4.3.2.

The theorem below is our main result in this Section.

Theorem 4.3.4   Let the functional $I : X \to (-\infty, +\infty]$ on the Banach space X satisfy assumptions (H) and (PS). Let S and Q link in the sense of Definition 4.3.1. Assume further that

$$\sup_Q I \in \mathbb{R}, \quad b := \inf_S I \in \mathbb{R}, \quad a := \sup_{\partial Q} I < b \,. \tag{4.3.4}$$

Then the number

$$c := \inf_{f \in \Gamma} \sup_{x \in Q} I(f(x)), \tag{4.3.5}$$

where

$$\Gamma = \{f \in C^0(Q; X) : \ f\,|_{\partial Q} = id_{\partial Q}\}, \tag{4.3.6}$$

is a critical value of I with

$$c \geq b. \tag{4.3.7}$$

This means in particular that $K_c(I)$ is nonempty.

Proof. According to the linking property of Definition 4.3.1 we know that $f(Q) \cap S \neq \emptyset$ for all $f \in \Gamma$. By (4.3.4) and (4.3.5) this ensures that (4.3.7) is true. Indeed, for $f \in \Gamma$ be given, there exists $x \in S$ and $y \in Q$ such that $f(y) = x$. We have

$$\inf_S I \leq I(x) = I(f(y)) \leq \sup_{y \in Q} I(f(y))$$

and thus

$$b \leq \inf_{f \in \Gamma} \sup_{y \in Q} I(f(y)).$$

In addition, the first relation in (4.3.4) yields $c < \infty$ by putting $f = id_Q$ in (4.3.5) and (4.3.6).

We have to show that c is a critical value of I. Arguing by contradiction we assume that $K_c(I) = \emptyset$. Then we can take $U = \emptyset$ in Theorem 4.1.6. Applying Theorem 4.1.6 with

$$\bar{\varepsilon} = c - a, \tag{4.3.8}$$

which is a positive number by (4.3.4), (4.3.7), we obtain an $\varepsilon \in (0, \bar{\varepsilon})$. Notice that

$$\partial Q \subset I_{c - \frac{\varepsilon}{2}}. \tag{4.3.9}$$

Since, (4.3.9) follows easily from (4.3.4), (4.3.7) and (4.3.8). Indeed, if $x \in \partial Q$ then

$$I(x) \leq a = \frac{a}{2} + \frac{a}{2} < \frac{a}{2} + \frac{b}{2} \leq \frac{c}{2} + \frac{a}{2} = c - \frac{c}{2} + \frac{a}{2} = c - \frac{\bar{\varepsilon}}{2} \leq c - \frac{\varepsilon}{2}.$$

The class Γ in (4.3.6) does not generally satisfy the property $\alpha(f(\cdot), s) \in \Gamma$ whenever $f \in \Gamma$, where α is the deformation in Theorem 4.2.3. Therefore it is necessary to enlarge Γ to fulfill the foregoing requirement. Let us set

$$\Gamma_1 := \{f \in C^0(Q; X) : \ f\,|_{\partial Q} \ \text{and} \ id_{\partial Q} \ \text{are homotopic as}$$

$$\text{maps from } \partial Q \text{ into } I_{c-\frac{\varepsilon}{4}} \text{ and } f(\partial Q) \subset I_{c-\frac{\varepsilon}{2}}\} . \qquad (4.3.10)$$

In other words $f \in \Gamma_1$ if and only if $f \in C^0(Q; X)$ and there exists $h \in C^0([0,1] \times Q; \mathbb{R})$ such that

$$
\begin{aligned}
I(h(t,x)) &\leq c - \frac{\varepsilon}{4}, \ \forall \, x \in \partial Q, \ t \in [0,1],\\
h(1,x) &= f(x), \ \forall \, x \in \partial Q,\\
h(0,x) &= x, \ \forall \, x \in \partial Q,
\end{aligned}
$$

and

$$I(f(x)) \leq c - \tfrac{\varepsilon}{2}, \ \forall \, x \in \partial Q.$$

Taking into account (4.3.9), the family (4.3.10) is well defined and non-empty containing at least id_Q. Corresponding to the family Γ_1 in (4.3.10) we introduce the minimax value

$$c_1 = \inf_{f \in \Gamma_1} \sup_{x \in Q} I(f(x)) . \qquad (4.3.11)$$

We claim that

$$c = c_1 . \qquad (4.3.12)$$

Since $\Gamma \subset \Gamma_1$, from (4.3.5), (4.3.11) it is clear that $c_1 \leq c$. Assuming by contradiction that $c_1 < c$, there would exist $f \in \Gamma_1$ provided

$$\sup_{x \in Q} I(f(x)) < c .$$

Recalling that $f \mid_{\partial Q}$ and $id_{\partial Q}$ are homotopic in $I_{c-\frac{\varepsilon}{4}}$, the (absolute) Homotopy Extension Property (see Proposition I.9.2 of Hu [193] or Spanier [390]) guarantees that $id_{\partial Q}$ can be extended to some $g \in C^0(Q; X)$, so $g \in \Gamma$, such that

$$\sup_{x \in Q} I(g(x)) < c .$$

This fact is in contradiction with the definition of c in (4.3.5). Hence the claim (4.3.12) is checked.

The next objective is to show that Γ_1 in (4.3.10) is a closed subset of the Banach space $C^0(Q; X)$ with respect to the usual uniform norm (recall that Q is compact in X), i.e.

$$\parallel f \parallel = \sup_{x \in Q} \parallel f(x) \parallel, \ \forall \, f \in C^0(Q; X).$$

To verify this assertion let us take a sequence $\{f_n\} \subset \Gamma_1$ which converges in $C^0(Q; X)$, say $f_n \to f$. By (4.3.10) one has

$$I(f_n(x)) \leq c - \frac{\varepsilon}{2}, \ \forall \, x \in \partial Q, \ \forall \, n \geq 1.$$

Since I is l.s.c., we derive

$$I(f(x)) \leq \liminf_{n \to \infty} I(f_n(x)) \leq c - \frac{\varepsilon}{2}, \ \forall \, x \, \in \, \partial \, Q. \qquad (4.3.13)$$

Using Lebourg's mean value theorem for Φ and the convexity of Ψ it is seen that

$$I(tf_n(x) + (1-t)f(x)) = \Phi(tf_n(x) + (1-t)f(x))$$

$$+ \Psi(tf_n(x) + (1-t)f(x))$$

$$\leq \Phi(f(x)) + \bar{t}\Phi^0(f(x); f_n(x) - f(x)) + t\Psi(f_n(x)) + (1-t)\Psi(f(x)),$$

$$\forall \, x \, \in \, \partial \, Q, \, \forall \, t \, \in \, [0,1],$$

for some $\bar{t} \in (0,1)$. Since $f(\partial Q)$ is compact, the sequence $\{f_n\}$ converges uniformly to f and Φ being locally Lipschitz, it follows that Φ is Lipschitz continuous near $f(\partial Q)$ of Lipschitz constant $K > 0$. Consequently, it turns out that

$$I(tf_n(x) + (1-t)f(x)) \leq \Phi(f(x)) + K \parallel f_n(x) - f(x) \parallel$$

$$+ t\Psi(f_n(x)) + (1-t)\Psi(f(x)), \forall \, x \, \in \, \partial \, Q, \, \forall \, t \, \in \, [0,1].$$

Taking into account the Lipschitz continuity of Φ around $f(\partial Q)$ we obtain that

$$\Phi(f(x)) \leq \Phi(f_n(x)) + K \parallel f_n(x) - f(x) \parallel, \forall \, x \, \in \, \partial \, Q.$$

It results that

$$I(tf_n(x) + (1-t)f(x)) \leq t\Phi(f_n(x)) + (1-t)\Phi(f(x))$$

$$+ 3K \parallel f_n(x) - f(x) \parallel + t\Psi(f_n(x)) + (1-t)\Psi(f(x))$$

$$= tI(f_n(x)) + (1-t)I(f(x)) + 3K \parallel f_n(x) - f(x) \parallel,$$

for all $x \in \partial Q$ and $t \in [0,1]$. Then, using that $f_n \in \Gamma_1$ and (4.3.13), we obtain

$$I(tf_n(x) + (1-t)f(x)) \leq c - \frac{\varepsilon}{2} + 3K \parallel f_n(x) - f(x) \parallel, \forall \, x \, \in \, \partial \, Q.$$

Since $\{f_n\}$ converges uniformly to f it follows

$$I(tf_n(x) + (1-t)f(x)) \leq c - \frac{\varepsilon}{4}, \ \forall \, x \, \in \, \partial \, Q, \qquad (4.3.14)$$

provided that n is sufficiently large. From (4.3.14) we deduce that $f_n \mid_{\partial Q}$ and $f \mid_{\partial Q}$ are homotopic in $I_{c-\frac{\varepsilon}{4}}$. Hence, by (4.3.10) and $f_n \in \Gamma_1$, we conclude that $f \mid_{\partial Q}$ and $id_{\partial Q}$ are homotopic in $I_{c-\frac{\varepsilon}{4}}$. This property combined with (4.3.13) entails that $f \in \Gamma_1$. Consequently, we proved that Γ_1 is a closed subset of $C^0(Q; X)$, so a complete metric space.

Let us introduce the functional $\Pi : C^0(Q; X) \to (-\infty, +\infty]$ by

$$\Pi(f) = \sup_{x \in Q} I(f(x)), \ \forall f \in C^0(Q; X). \tag{4.3.15}$$

It is straightforward to see that Π is l.s.c. This is the consequence of the lower semicontinuity of I. Therefore we know that $\Pi : \Gamma_1 \to (-\infty, +\infty]$ is l.s.c.. As established previously Γ_1 is a complete metric space being a closed subset of $C^0(Q; X)$. It is allowed to apply to $\Pi : \Gamma_1 \to (-\infty, +\infty]$ Ekeland's variational principle (see Theorem 1.1.13). Notice that Π on Γ_1 is bounded from below by $c \in \mathbb{R}$ because (4.3.11), (4.3.12) hold. Ekeland's variational principle gives rise to some $f \in \Gamma_1$ satisfying

$$c \leq \Pi(f) \leq c + \varepsilon \tag{4.3.16}$$

and

$$\Pi(g) - \Pi(f) \geq -\varepsilon \parallel g - f \parallel, \ \forall g \in \Gamma_1. \tag{4.3.17}$$

Now we invoke Theorem 4.2.3 that provides the deformation $\alpha : W \times [0, \bar{s}] \to X$ corresponding to the compact set $A = f(Q)$. We point out that the specified set A satisfies the required condition (4.2.10) due to (4.3.16). Let us show that for $\bar{s} > 0$ small enough we have

$$\alpha(f(\cdot), s) \in \Gamma_1, \ \forall s \in [0, \bar{s}]. \tag{4.3.18}$$

Looking at (4.3.10), we see that in order to prove (4.3.18) it suffices to establish that $\alpha(f(\cdot), s) \mid_{\partial Q}$ and $f \mid_{\partial Q}$ are homotopic in $I_{c-\frac{\varepsilon}{2}}$. Clearly, a homotopy in X between the involved mappings is $(x, t) \in \partial Q \times [0, 1] \to \alpha(f(x), ts) \in X$. Therefore, it is sufficient to show that

$$I(\alpha(f(x), s)) \leq c - \frac{\varepsilon}{2}, \ \forall x \in \partial Q, \ \forall 0 \leq s \leq \bar{s}. \tag{4.3.19}$$

If one has for $x \in \partial Q$ that

$$I(f(x)) \in [c - \varepsilon, c - \frac{\varepsilon}{2}] \, ,$$

then by (4.2.13) we get

$$I(\alpha(f(x), s)) \leq I(f(x)) - 2\varepsilon s \leq c - \frac{\varepsilon}{2} - 2\varepsilon s < c - \frac{\varepsilon}{2} \, .$$

If for $x \in \partial Q$ holds

$$I(f(x)) < c - \varepsilon,$$

then (4.2.12) implies

$$I(\alpha(f(x), s)) \leq I(f(x)) + Ms < c - \varepsilon + Ms \leq c - \frac{\varepsilon}{2}$$

provided that $\bar{s} > 0$ is sufficiently small. We thus justified that (4.3.19) is valid, so (4.3.18) holds true.

From (4.2.11), (4.2.14), (4.3.15), (4.3.17) and (4.3.18) we then deduce

$$
\begin{aligned}
-2\varepsilon s \;\geq\; & \Pi(\alpha(f(\cdot), s)) - \Pi(f) \\
\geq\; & -\varepsilon \parallel \alpha(f(\cdot), s) - f \parallel \geq -\varepsilon s, \; \forall\, 0 \leq s \leq \bar{s}.
\end{aligned}
$$

We arrived at a contradiction which proves that our initial assumption that c in (4.3.5) is not a critical value of I is false. This achieves the proof of Theorem 4.3.4. ∎

Corresponding to different situations of linking in Definition 4.3.1, Theorem 4.3.4 implies useful minimax results for functionals satisfying hypothesis (H).

Corollary 4.3.5   (Mountain Pass Theorem). Assume that the functional $I : X \to (-\infty, +\infty]$ on the Banach space X satisfies (H), (PS), $I(0) = 0$,

(i) there exist constants $\alpha > 0$ and $\rho > 0$ such that $I(u) \geq \alpha$ for all $\parallel u \parallel = \rho$;

(ii) there exists $e \in X$ with $\parallel e \parallel > \rho$ and $I(e) \leq 0$.

Then the number

$$c = \inf_{f \in \Gamma} \; \sup_{t \in [0,1]} I(f(t)),$$

where

$$\Gamma = \{f \in C^0([0, 1]; X) : f(0) = 0, f(1) = e\},$$

is a critical value of I with $c \geq \alpha$.

Proof. It is sufficient to apply Theorem 4.3.4 and Example 4.3.3 (with $Z = X, W = \{0\}, r = \rho, R = 1$) for

$$S = \{u \in X : \parallel u \parallel = \rho\}$$

and

$$Q = [0, e] = \{te : \; t \in [0, 1]\}.$$

The first condition in (4.3.4) holds by the convexity of Ψ. ∎

Corollary 4.3.6   (Saddle-Point Theorem). Let a Banach space X with $X = X_1 \oplus X_2$, $dim\, X_1 < \infty$, and let a functional $I : X \to (-\infty, +\infty]$ satisfy (H), (PS),

(i) there exist constants $\rho > 0$ and $\alpha \in \mathbb{R}$ such that $I(u) \leq \alpha$, $\forall\, u \in X_1$, with $\| u \| = \rho$;

(ii) there exists a constant $\beta \in \mathbb{R}$ such that $\beta > \alpha$ and $I(u) \geq \beta$, $\forall\, u \in X_2$.

Then the number

$$c = \inf_{f \in \Gamma} \sup_{x \in Q} I(f(x)), \tag{4.3.20}$$

where

$$\Gamma = \{f \in C^0(Q; X) : \ f \mid_{\partial Q} = id_{\partial Q}\} \tag{4.3.21}$$

with

$$Q = \{x \in X_1 : \| x \| \leq \rho\} \ \text{ and } \ \partial Q = \{x \in X_1 : \| x \| = \rho\},$$

is a critical value of I satisfying $c \geq \beta$.

Proof. One applies Theorem 4.3.4 by taking $S = X_2$ and Q as specified in the statement. The linking property in Definition 4.3.1 follows by using Example 4.3.2 (with $W = X_1$ and $S = X_2$). ∎

Corollary 4.3.7   (Generalized Mountain Pass Theorem). Assume that $X = X_1 \oplus X_2$ with $dim\, X_1 < \infty$ and the functional $I : X \to (-\infty, +\infty]$ satisfies (H), (PS),

(i) there exist constants $\rho > 0$ and $\alpha > 0$ such that $I(u) \geq \alpha$ for all $u \in X_2$ with $\| u \| = \rho$;

(ii) there exist a constant $r > \rho$ and a point $e \in X_2$, $\| e \| = 1$ such that $I(u) \leq 0$ for all $u \in \partial Q$, where $Q = \{u \in X_1 : \| u \| \leq r\} \oplus \{te : t \in [0, r]\}$. Then the number

$$c = \inf_{f \in \Gamma} \sup_{x \in Q} I(f(x)),$$

where

$$\Gamma = \{f \in C^0(Q; X) : \ f \mid_{\partial Q} = id_{\partial Q}\},$$

is a critical value of I with $c \geq \alpha$.

Proof. Theorem 4.3.4 is applied for

$$S = \{u \in X_2 : \| u \| = \rho\}$$

and Q as described in the statement. The fact that S and Q link in the sense of Definition 4.3.1 follows from Example 4.3.3 (with the notations $W = X_1, Z = X_2, r = \rho, R = r, \rho = r$). $\blacksquare$

4.4   MULTIPLICITY THEOREMS FOR EVEN FUNCTIONALS OF TYPE (H)

In the same spirit one can derive from Corollary 4.2.4 minimax principles with additional multiplicity information for even functionals of type (H). The approach uses the concept of genus of Krasonel'skii [231] that we recall now.

Let X be a real Banach space and $\sum$ the collection of all symmetric (with respect to the origin) subsets of $X \backslash \{0\}$ which are closed in X. A nonempty set $A \in \sum$ is said to have genus k (we denote $\gamma(A) = k$) if k is the smallest integer with the property that there exists an odd continuous mapping $\eta : A \to \mathbb{R}^k \backslash \{0\}$. If there is no such k then we say that $\gamma(A) = +\infty$ and if $A = \emptyset$ then we set $\gamma(A) = 0$. Let us denote by $d(u, A)$ the distance from a point $u \in X$ to the set A and let

$$N_\delta(A) = \{u \in X : d(u, A) \leq \delta\}.$$

The main properties of the genus are now listed.

Proposition 4.4.1   Let $A, B \in \sum$.

1. If there exists an odd continuous mapping $f : A \to B$, then $\gamma(A) \leq \gamma(B)$.

2. If $A \subset B$, then $\gamma(A) \leq \gamma(B)$.

3. $\gamma(A \cup B) \leq \gamma(A) + \gamma(B)$.

4. If $\gamma(B) < +\infty$ then $\gamma(\overline{A \backslash B}) \geq \gamma(A) - \gamma(B)$.

5. If $U \subset \mathbb{R}^k$ is an open, bounded and symmetric neighborhood of the origin, then $\gamma(\partial U) = k$.

6. If N is an open, bounded and symmetric neighborhood of the origin in $\mathbb{R}^k$ and if A is homeomorphic to ∂N by an odd homeomorphism then $\gamma(A) = k$.

7. If A is compact, then $\gamma(A) < +\infty$ and $\gamma(N_\delta(A)) = \gamma(A)$ for all sufficiently small $\delta > 0$.

8. If $\gamma(A) > 1$, then A contains infinitely many distinct points.

9. Let Y be a closed subspace of finite codimension of X. If $A \cap Y = \emptyset$, then $\gamma(A) \le codim\ Y$.

Proof. 1) and 2) follow directly from the definition of genus. 3) Suppose $\gamma(A) = a$, $\gamma(B) = b$ and $a, b < \infty$. If either $\gamma(A) = +\infty$ or $\gamma(B) = +\infty$ then the result is trivial. Let $f : A \to \mathbb{R}^a \setminus \{0\}$ and $g : B \to \mathbb{R}^b \setminus \{0\}$ be odd and continuous. By Tietze's Extension Theorem (see e.g. [106]) there are two mappings, $\tilde{f} : X \to \mathbb{R}^a$ and $\tilde{g} : X \to \mathbb{R}^b$, such that $f = \tilde{f}_{|A}$, $g = \tilde{g}_{|B}$. We may suppose that $\tilde{f}$ and $\tilde{g}$ are odd (otherwise we could take $\tilde{f}_1(u) = \frac{1}{2}(\tilde{f}(u) - \tilde{f}(-u))$ and $\tilde{g}_1(u) = \frac{1}{2}(\tilde{g}(u) - \tilde{g}(-u))$). We consider the mapping $h : A \cup B \to \mathbb{R}^a \times \mathbb{R}^b$; $u \to h(u) = (\tilde{f}(u), \tilde{g}(u))$. It is clear that h is odd. Moreover, if $u \in A \cup B$ then

$$h(u) \ne (0, 0).$$

Therefore $\gamma(A \cup B) \le a + b$. 4) Using 2), 3) and the inclusion $A \subset \overline{A \setminus B} \cup B$, we obtain $\gamma(A) \le \gamma(\overline{A \setminus B}) + \gamma(B)$. 5) It is clear that $\gamma(\partial U) \le k$. Using Borsuk-Ulam Theorem (see Theorem 5.1.1) there is no odd continuous mapping from ∂U to $\mathbb{R}^{k-1} \setminus \{0\}$. Therefore $\gamma(\partial U) = k$. 6) There exists an odd homeomorphism $\eta : A \to \partial N$. By 5) there exists an odd and continuous mapping $\theta : \partial N \to \mathbb{R}^k \setminus \{0\}$. Thus $\theta \circ \eta : A \to \mathbb{R}^k \setminus \{0\}$ is an odd and continuous mapping and $\gamma(A) \le k$. Suppose that $\gamma(A) = k' < k$. Then there exists an odd and continuous mapping $\sigma : A \to \mathbb{R}^{k'} \setminus \{0\}$. Then the mapping $\sigma \circ \eta^{-1} : \partial N \to \mathbb{R}^{k'} \setminus \{0\}$ is odd and continuous and $\gamma(\partial N) \le k'$ which is a contradiction. Therefore $\gamma(A) = k$. 7) For $x \in A$, set $r(x) = \frac{1}{2}\|x\|$ and $T_x = B_{r(x)}(x) \cup B_{r(x)}(-x)$ where $B_\rho(z)$ denotes the closed ball centered at z and having radius ρ. We have $A \subset \cup_{x \in A} T_x$ and by the compactness of A,

$$A \subset \overset{k}{\underset{i=1}{\cup}} T_{x_i},$$

for some finite set of points $x_1, \cdots, x_k \in A$. It is easy to show that $\gamma(T_{x_i}) = 1$ $(i = 1, \cdots, k)$ (see Example 4.4.2, i)). Thus

$$\gamma(A) \leq \sum_{i=1}^{k} \gamma(T_{x_i}) = k < +\infty.$$

Let now $\sigma : A \to \mathbb{R}^k \setminus \{0\}$ be odd an continuous. Using Tietze's Extension Theorem, we obtain an odd and continuous mapping $\tilde{\sigma} :$ $N_\delta(A) \to \mathbb{R}^k \setminus \{0\}$. Thus $\gamma(N_\delta(A)) \leq k$. Moreover $A \subset N_\delta(A)$ so that $k \leq \gamma(N_\delta(A))$. Therefore $\gamma(N_\delta(A)) = k$. 8) Suppose on the contrary that A is finite. Then we could write $A = B \cup (-B)$ where B is closed and $B \cap (-B) = \emptyset$. Then $\gamma(A) = 1$ (see Example 4.4.2, i)) which is a contradiction. 9) We set $X = Y \oplus Z$. We denote by $P : X \to Z$ the projector of X onto Z. It is clear that P is odd and continuous. If $x \in A$ then $P(x) \neq 0$. Indeed, suppose that $P(x) = 0$. Then $x \in Y$ and thus $A \cap Y \neq \emptyset$. Since $dim\ Z = k := codim\ Y$, we can find an odd isomorphism $\sigma : Z \to \mathbb{R}^k$. Thus the mapping $\sigma \circ P : A \to \mathbb{R}^k \setminus \{0\}$ is odd and continuous and therefore $\gamma(A) \leq k = codim\ Y$. $\blacksquare$

Example 4.4.2   i) Suppose that $B \subset X$ is closed and $B \cap (-B) = \emptyset$. If $A = B \cup (-B)$ then $\gamma(A) = 1$. Indeed, the function

$$\phi : A \to \mathbb{R} \setminus \{0\}; x \to \phi(x) = \begin{cases} +1 & \text{if } x \in B \\ -1 & \text{if } x \in -B \end{cases}$$

is odd and continuous. ii) $\gamma(S^{k-1}) = k$. This follows from Proposition 4.4.1 (property 5). Let us give the proof with $k = 2$. We see that the closed sets B_1, B_2, B_3, B_4 in Fig. 4.4.1 satisfy

$$S^1 = \bigcup_{i=1}^{4} B_i,$$

$$B_1 \cap B_2 = \emptyset,$$

and

$$B_2 \cap B_3 = \emptyset.$$

Thus $\gamma(S^1) \leq \gamma(B_1 \cup B_2) + \gamma(B_2 \cup B_3) \leq 2$. We can proceed by induction to get $\gamma(S^{k-1}) \leq k$. The fact that $\gamma(S^{k-1}) \geq k$ is a consequence of the Borsuk-Ulam theorem (see Theorem 6.1.1).

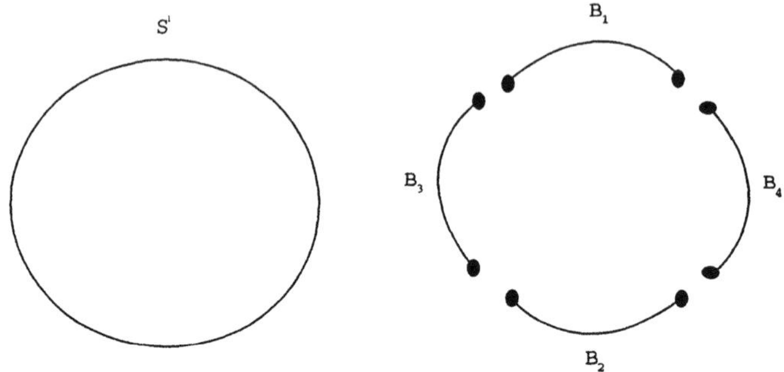

Figure 4.4.1.   Example 4.4.2 ii).

Let S be the collection of all nonempty closed and bounded subsets of X. In S we introduce the Hausdorff metric d_H given by

$$d_H(A, B) = \max\{\sup_{a \in A} d(a, B), \sup_{b \in B} d(b, A)\}.$$

Let Γ be the subcollection of S consisting of all nonempty compact symmetric subsets of X. We set

$$\Gamma_j = cl\{A \in \Gamma : 0 \notin A, \gamma(A) \geq j\}.$$

Here "cl" denotes the closure in S with respect to d_H. We note that (Γ, d_H) and (Γ_j, d_H) are complete metric spaces.

Corollary 4.4.3   Assume that $I : X \to (-\infty, +\infty]$ satisfies (H), (PS), $I(0) = 0$ and the functions Φ and Ψ entering (H) are even. Assume also that

$$c_j := \inf_{A \in \Gamma_j} \sup_{u \in A} I(u) \in (-\infty, 0), \ \forall\, j = 1, \cdots, k. \qquad (4.4.1)$$

Then I possesses at least k distinct pairs of symmetric nontrivial critical points (whose corresponding critical values are the numbers c_j).

Proof.   Let us first prove that if $A \in \Gamma_j$ and $0 \notin A$ then $\gamma(A) \geq j$. Indeed, let $\{A_n\}$ be a sequence in Γ_j such that $A_n \to A$ (for the Hausdorff distance), $0 \notin A_n$ and $\gamma(A_n) \geq j$. By (7) of Proposition 4.4.1, there exists $\delta > 0$ such that $\gamma(A) = \gamma(N_\delta(A))$. We have $A_n \subset N_\delta(A)$ for almost all n since $A_n \to A$. Thus $j \leq \gamma(A_n) \leq \gamma(N_\delta(A)) = \gamma(A)$. Given $j, 1 \leq j \leq k$, we suppose that $c_j = \cdots = c_{j+p} = c$ for some $p \geq 0, p \in \mathbb{N}$. Note that $0 \notin K_c(I)$ since $c < 0$. We claim that

$$\gamma(K_c(I)) \geq p + 1.$$

Suppose by contradiction that $\gamma(K_c(I)) \leq p$. The set $K_c(I)$ is compact and for ρ sufficiently small we have

$$\gamma(N_{2\rho}(K_c(I))) = \gamma(K_c(I)).$$

We set

$$\Pi(A) = \sup_{u \in A} I(u).$$

The mapping $\Pi : \Gamma_j \to \mathbb{R} \cup \{+\infty\}$ is l.s.c. Indeed, suppose that $A_n \to A$. For each $u \in A$ there exist a sequence $u_n \to u$ with $u_n \in A_n$ and

$$I(u) \leq \liminf_{n \to \infty} I(u_n).$$

Thus

$$I(u) \leq \liminf_{n \to \infty} \Pi(A_n).$$

This last inequality is true for all $u \in A$ and thus

$$\Pi(A) \leq \liminf_{n \to \infty} \Pi(A_n).$$

Choose $\bar{\varepsilon} = \min\{1, \rho, -c\}$ and apply Corollary 4.2.4 to obtain an $\varepsilon < \bar{\varepsilon}$. Let $A_1 \in \Gamma_{j+p}$ such that $\Pi(A_1) \leq c + \varepsilon^2$. Since $c + \varepsilon^2 \leq c + \varepsilon < 0$, it results that $0 \notin A_1$. We know that $\gamma(A_1) \geq j + p$. Let

$$A_2 = \overline{A_1 \backslash N_{2\rho}(K_c(I))}.$$

Then

$$\Pi(A_2) = \sup_{u \in A_1 \backslash N_{2\rho}(K_c(I))} I(u) \leq \Pi(A_1)$$

and from part (4) of Proposition 4.4.1, we obtain

$$\begin{aligned}
\gamma(A_2) &\geq \gamma(A_1) - \gamma(N_{2\rho}(K_c(I))) \\
&\geq j + p - p \\
&= j.
\end{aligned}$$

Thus $A_2 \in \Gamma_j$. Using Theorem 1.1.13 with $\delta = \varepsilon^2$ and $\lambda = \frac{1}{\varepsilon}$, we obtain an $A \in \Gamma_j$ such that

$$\Pi(A) \leq c + \varepsilon^2,$$

$$d_H(A, A_2) \leq \varepsilon$$

and

$$\Pi(B) - \Pi(A) \geq -\varepsilon d_H(A, B), \ \forall \, B \in \Gamma_j.$$

Since $\varepsilon < \rho$ and $A \in \Gamma_j$ we get $A \cap N_\rho(K_c(I)) = \emptyset$ and $\Pi(A) \geq c$ by definition of c. Thus A satisfies the conditions of Corollary 4.2.4 with

$U = N_\rho(K_c(I))$. We obtain an odd deformation $\alpha : W \times [0, \bar{s}] \to X$ as stated in Theorem 4.2.3. We have $0 \notin A$ since $c + \varepsilon^2 < 0$. Thus $\gamma(A) \geq j$. Let us define B by setting

$$B = \alpha(A, s),$$

with s small. Then from (1) of Proposition 4.4.1, we obtain that

$$\gamma(B) \geq \gamma(A) \geq j.$$

We have

$$I(\alpha(u, s)) \leq Ms + I(u), \ \forall \, u \in A$$

and thus

$$I(\alpha(u, s)) \leq Ms + c + \varepsilon^2, \ \forall \, u \in A.$$

It results that for s small enough

$$I(\alpha(u, s)) < 0, \ \forall \, u \in A$$

so that $0 \notin B$. Thus $B \in \Gamma_j$. Using (4.2.11) and (4.2.14) with the notations here used, we obtain

$$-2\varepsilon s \geq \Pi(B) - \Pi(A) \geq -\varepsilon d_H(A, B) \geq -\varepsilon s.$$

This is a contradiction. We have thus shown that $\gamma(K_c(I)) \geq p + 1$. In particular $\gamma(K_{c_j}(I)) \geq 1$. So each $K_{c_j}(I)$ contains at least two points, u_j and $-u_j$. This gives the required number of critical points if all c_j are distinct. If they are not, $p > 0$ for some j. Hence $\gamma(K_{c_j}(I)) \geq 2$ and $K_{c_j}(I)$ is an infinite set according to (8) of Proposition 4.4.1.   ∎

Corollary 4.4.4   (Symmetric version of Mountain Pass Theorem). Assume that $I : X \to (-\infty, +\infty]$ satisfies (H) with Φ and Ψ even, (PS), $I(0) = 0$ and the conditions:

(i) There exists a subspace X_1 of X of finite codimension and numbers $\beta > 0$, $\rho > 0$ such that

$$I(u) \geq \beta \quad \text{whenever} \quad u \in X_1, \ \| u \| = \rho. \tag{4.4.2}$$

(ii) There exists a finite dimensional subspace X_2 of X, $dim \, X_2 > codim \, X_1$, such that $I(u) \to -\infty$ as $\| u \| \to \infty$, $u \in X_2$.

Then I has at least $dim \, X_2 - codim \, X_1$ distinct pairs of symmetric nontrivial critical points.

In particular, if (ii) is replaced by

(ii$'$) There exists a k-dimensional subspace X_2 as in (ii) for each positive integer k,

then I admits infinitely many distinct pairs $(u, -u)$ of nontrivial critical points.

Proof. We can assume that I has no critical points in I_{-d} for some $d > 0$, otherwise there are infinitely many critical points and there is nothing to prove. Set $m = codim\ X_1, k = dim\ X_2, Q = \{x \in X_2 :\parallel x \parallel \leq R\}$, where $R > \rho$ is chosen so that $I \leq -d$ on ∂Q. Define for $1 \leq j \leq k$,

$$
\begin{aligned}
F \quad &:= \quad \{\eta \in C^0(Q; X) : \eta \text{ is odd and } \eta_{|\partial Q} \text{ is homotopic to} \\
&\qquad id_{|\partial Q} \text{ in } I_{-d} \text{ by an odd homotopy } \}, \\
\Gamma_j \quad &:= \quad \{\eta(Q\backslash V) : \eta \in F, V \text{ is open in } Q \text{ and symmetric,}
\end{aligned}
$$

$V \cap \partial Q = \emptyset$ and for each $Y \subset V$ such that $Y \in \Sigma, \gamma(Y) \leq k - j\}$,

$$
\Delta_j := \{A \subset X : A \text{ is compact}, \ A = -A \text{ and for each open set } U \supset A
$$

$$
\text{there is } A_0 \in \Gamma_j \text{ such that } A_0 \subset U\}.
$$

It is clear that $Q \in \Delta_j$ with $A_0 = Q, V = \emptyset$ and $\eta = id_{|Q}$ so that $\Delta_j \neq \emptyset$. We set

$$
c_j := \inf_{A \in \Delta_j} \sup_{u \in A} I(u).
$$

The following properties are proved by using standard topological arguments.

$\mathbf{P_1}$. For $m + 1 \leq j \leq k, c_j \geq \beta$.

$\mathbf{P_2}$. $\Delta_{j+1} \subset \Delta_j$.

$\mathbf{P_3}$. If $A \in \Delta_j, W$ is a closed and symmetric set containing A in its interior and $\alpha : W \to X$ an odd mapping such that $\alpha_{|W \cap I_{-d}}$ is homotopic to $id_{|W \cap I_{-d}}$ in I_{-d} by an odd homotopy, then $\alpha(A) \in \Delta_j$.

$\mathbf{P_4}$. If $Z \in \Sigma$ is compact, $\gamma(Z) \leq p$ and $I_{|Z} > -d$, then there exists a $\delta > 0$ such that for each $A \in \Delta_{j+p}, A\backslash int\{N_\delta(Z)\} \in \Delta_j$.

Let us now prove these results.

Proof of $\mathbf{P_1}$. Suppose by contradiction that $c_j < \beta$. Then

$$
A \cap X_1 \cap \partial B_\rho = \emptyset
$$

for some $A \in \Delta_j$. Since $X \backslash (X_1 \cap \partial B_\rho)$ is an open set containing A, we can first find an $A_0 = \eta(Q \backslash V) \in \Gamma_j$ which does not intersect $X_1 \cap \partial B_\rho$ and then a $Y \subset V$ such that

$$Y \in \Sigma, \gamma(Y) \leq k - j \text{ and } \eta(\overline{Q \backslash Y}) \cap X_1 \cap \partial B_\rho = \emptyset.$$

Let $F(y,t), y \in \partial Q, t \in [0,1]$, be an odd homotopy joining $\eta_{|\partial Q}$ to $\mathrm{id}_{\partial Q}$ in I_{-d} and define $Y_1 = \frac{1}{2} Y$ and

$$\eta_1(y,s) = \begin{cases} \eta(y, 2s) & 0 \leq s \leq \frac{1}{2} R \\ \\ F(y, 2s/R - 1) & \frac{1}{2} R \leq s \leq R, \end{cases}$$

where $(y,s) \in \partial Q \times [0, R]$ are the polar coordinates of $x \in Q$. We have

$$\eta_1(y,s) \in I_{-d}, \ \forall \, s \geq \frac{1}{2} R$$

and

$$I_{|\partial B_\rho \cap X_1} \geq \beta > 0,$$

we deduce that

$$\eta_1(\overline{Q \backslash Y_1}) \cap X_1 \cap \partial B_\rho = \emptyset.$$

Set $\tilde{W} = \{x \in X_2 : \eta_1(x) \in B_\rho\}$ and let W be the component of $\tilde{W}$ containing the origin. We have $\eta_1(x) = x \notin B_\rho, \ \forall \, x \in \partial Q$ and thus

$$W \cap \partial Q = \emptyset.$$

It results that W is a symmetric open bounded neighborhood of 0 in X_2 and thus $\gamma(\partial W) = k$. Set $C = \{x \in X_2 : \eta_1(x) \in \partial B_\rho\}$. We have $\partial W \subset C$ and thus $\gamma(C) = k$. We obtain

$$\gamma(\overline{C \backslash Y_1}) \geq \gamma(C) - \gamma(Y_1) \geq k - (k - j) = j.$$

Let $X = X_1 \oplus \tilde{X}_1$ and denote by P the projection from X to $\tilde{X}_1$ along X_1. Since $\eta_1(\overline{Q \backslash Y_1}) \cap X_1 \cap \partial B_\rho = \emptyset$ and $\eta_1(\overline{C \backslash Y_1}) \subset \partial B_\rho$, we see that

$$Z \equiv P\eta_1(\overline{C \backslash Y_1}) \subset \tilde{X}_1 \backslash \{0\}.$$

We obtain

$$\gamma(Z) \geq \gamma(\overline{C \backslash Y_1}) \geq j.$$

On the other hand

$$\gamma(Z) \leq \dim \tilde{X}_1 = m < j.$$

That is a contradiction.

Proof of P_2. Let $A \in \Delta_{j+1}$ and choose an open set $U \supset A$. There exists $A_0 = \eta(Q\backslash V) \in \Gamma_{j+1}$ such that $A_0 \subset U$. Moreover, for each $Y \subset V$ such that $Y \in \Sigma$, $\gamma(Y) \leq k - (j+1) < k - j$. Thus $A_0 \in \Gamma_j$ and $A \in \Delta_j$.

Proof of P_3. Let $U \supset \alpha(A)$ be open. Let W_1 be an open set such that $A \subset W_1 \subset W$. Since $A \in \Delta_j$, there exists $A_0 = \eta(Q\backslash V) \in \Gamma_j$ such that $A_0 \subset W_1$. Let $\tilde{\alpha} : \eta(Q) \cup W \to X$ be an odd mapping extending α. We have

$$\eta(\partial Q) \subset I_{-d} \cap W$$

and thus

$$\tilde{\alpha} \circ \eta_{|\partial Q} \approx \eta_{|\partial Q} \approx \mathrm{id}_{\partial Q} \text{ in } I_{-d}$$

(here " $\approx$ " means "homotopic to"). Thus $\tilde{\alpha} \circ \eta \in F$. Moreover

$$\tilde{\alpha} \circ \eta(Q\backslash V) = \alpha \circ \eta(Q\backslash V) = \alpha(A_0) \subset \alpha(W_1) \subset U,$$

and thus $\alpha(A_0) \in \Gamma_j$ and $\alpha(A) \in \Delta_j$.

Proof of P_4. Let $\delta > 0$ be such that

$$\gamma(N_\delta(Z)) = \gamma(Z).$$

Denote $Z_0 = N_\delta(Z)$ and $\mathring{Z}_0 = int\{N_\delta(Z)\}$. Let $U \supset A\backslash \mathring{Z}_0$ be open and set $U_0 = U \cup \mathring{Z}_0$. Then

$$A \subset U_0.$$

Since $A \in \Delta_{j+p}$, there is an $A_0 = \eta(Q\backslash V) \in \Gamma_{j+p}$ such that $A_0 \subset U_0$. Since $I_{|z} > -d$ and $X\backslash I_{-d}$ is an open set (recall that I is l.s.c.), $I_{|z_0} > -d$ provided that δ is sufficiently small. It follows that $\eta^{-1}(Z_0) \cap \partial Q = \emptyset$ since $I_{|\partial Q} \leq -d$. It results that the set $V \cup \eta^{-1}(\mathring{Z}_0)$ is open in Q, symmetric and does not intersect ∂Q. If $Y \subset V \cup \eta^{-1}(\mathring{Z}_0)$, $Y \in \Sigma$, then there exist $Y_1, Y_2 \in \Sigma$ such that

$$Y = Y_1 \cup Y_2, Y_1 \subset V, Y_2 \subset \eta^{-1}(\mathring{Z}_0).$$

For example, we may choose $Y_1 = \{x \in Y : d(x, Q\backslash V) \geq d(x, Q\backslash \eta^{-1}(Z_0))\}$ and $Y_2 = \{x \in Y : d(x, Q\backslash V) \leq d(x, Q\backslash \eta^{-1}(Z_0))\}$. We have

$$\gamma(Y_1) \leq k - (j+p)$$

since $A_0 \in \Gamma_{j+p}$. We have also

$$\gamma(Y_2) \leq \gamma(\eta^{-1}(Z_0)) \leq \gamma(Z_0) = \gamma(Z) \leq p$$

and thus

$$\gamma(Y) \leq \gamma(Y_1) + \gamma(Y_2) \leq k - j.$$

We have also

$$A_0 \backslash \mathring{Z}_0 = \eta(Q \backslash V) \backslash \mathring{Z}_0$$
$$= \eta(Q \backslash V \cup \eta^{-1}(\mathring{Z}_0)))$$

and thus

$$A_0 \backslash \mathring{Z}_0 \in \Gamma_j.$$

Since $A \backslash \mathring{Z}_0 \subset U$ and U was chosen arbitrary, we obtain $A \backslash \mathring{Z}_0 \in \Delta_j$.

We now prove Corollary 4.4.4. By P_1 and P_2, we have

$$\beta \leq c_{m+1} \leq \cdots \leq c_k.$$

Suppose that $c_j = \cdots = c_{j+p} = c$ for some $j \in [m+1, k] \cap \mathbb{N}$, and $k - j \geq p \geq 0$. Since I is even, $K_c(I)$ is symmetric and since I satisfies the (PS) condition, $K_c(I)$ is compact. Moreover, $c > 0$ and thus $0 \notin K_c(I)$ so that $K_c(I) \in \Sigma$. We shall prove that $\gamma(K_c(I)) \geq p + 1$. Suppose on the contrary that

$$\gamma(K_c(I)) \leq p.$$

Choose $\rho' > 0$ so that

$$\gamma(N_{2\rho'}(K_c(I))) = \gamma(K_c(I))$$

and let $U = N_{\rho'}(K_c(I))$ and $\bar{\varepsilon} = \min\{1, \rho'\}$. Let $\varepsilon > 0$ be the number given in Corollary 4.2.4 (and Theorem 4.2.3). The set Δ_j is closed in $\mathcal{S}$. Indeed, suppose that $A_n \in \Delta_j$ and $A_n \to A$. Let V be an arbitrary open set containing A. Then $A_n \subset V$ for almost all n and, since $A_n \in \Delta_j$, there exists an $A_0 \in \Gamma_j$ such that $A_0 \subset V$. Hence $A \in \Delta_j$. It results that (Δ_j, d) is a complete metric space. The mapping $\Pi : \Delta_j \to \mathbb{R} \cup \{+\infty\}$ is l.s.c. and there exists an $A_1 \in \Delta_{j+p}$ such that

$$\Pi(A_1) \leq c + \varepsilon^2.$$

Let $A_2 = A_1 \backslash int\{N_{2\rho'}(K_c(I))\}$. If ρ' is sufficiently small then from P_4, we deduce that $A_2 \in \Delta_j$. Applying Theorem 1.1.13 in our context with $\delta = \varepsilon^2$ and $\lambda = \frac{1}{\varepsilon}$, we find an $A \in \Delta_j$ such that

$$\Pi(A) \leq c + \varepsilon^2,$$
$$d_H(A, A_2) \leq \varepsilon$$

and

$$\Pi(B) - \Pi(A) \geq -\varepsilon d_H(A, B), \ \forall \ B \in \Delta_j.$$

Since $\varepsilon < \rho'$, $A \cap U = \emptyset$. Moreover

$$c \leq \Pi(A) \leq c + \varepsilon^2 \leq c + \varepsilon.$$

According to Theorem 4.2.3 and Corollary 4.2.4, there exists an odd deformation $\alpha : W \times [0, \bar{s}] \to X$ satisfying (4.2.13)-(4.2.18). Let $B = \alpha(A, s)$ for s small. Choosing $W_0 = I_{-d}$, we obtain from (4.2.15) that

$$I(\alpha(u, s)) \leq I(u), \ \forall \, u \, \in \, A \cap I_{-d}$$

and thus from P_3, we deduce that $B \in \Delta_j$. Therefore

$$-2\varepsilon s \geq \Pi(B) - \Pi(A) \geq -\varepsilon d_H(A, B) \geq -\varepsilon s$$

and a contradiction occurs.

Thus $\gamma(K_c(I)) \geq p + 1$. In particular $\gamma(K_{c_j}(I)) \geq 1$, so that each $K_{c_j}(I)$ has at least two points u_j and $-u_j$. This gives the required number of critical points if all c_j are distinct. If they are not, then $p > 0$ for some j, so that $\gamma(K_{c_j}(I)) > 1$ and we conclude by part (8) of Proposition 4.4.1. The second assertion of Corollary 4.4.4 is a direct consequence of the first assertion. $\blacksquare$

4.5   EXAMPLES AND APPLICATIONS

The minmax methods developed in this Chapter will be used later in various directions. Here we give some illustrative examples.

Example 4.5.1   (A multiplicity result for a variational inequality) Let Ω be a nonempty, open bounded subset in $\mathbb{R}^N (N \geq 1, N \in \mathbb{N})$ of class $C^{0,1}$. Let $F : \mathbb{R} \to \mathbb{R} \cup \{+\infty\}$ and $G : \mathbb{R} \to \mathbb{R}$ be two functions satisfying the following assumptions; F is even, convex, l.s.c. and $F(0) = 0$; G is even, of class $C^1, G(0) = 0, G'(t) = g(t)$ and $| \, g(t) \, | \leq c_1 + c_2 \, | \, t \, |$, $\forall t \in \mathbb{R}$, where c_1, c_2 are positive constants. Moreover, we suppose that

$$\{u \in H_0^1(\Omega) : F(u) \in L^1(\Omega)\} \ \text{ is weakly closed,} \qquad (4.5.1)$$

$$\liminf_{|t| \to +\infty} \frac{(F(t) - G(t))}{t^2} > -\frac{1}{2}\lambda_1 \qquad (4.5.2)$$

and

$$\limsup_{t \to 0} \frac{(F(t) - G(t))}{t^2} < -\frac{1}{2}\lambda_k \qquad (4.5.3)$$

where λ_j denotes the j-th (counted according to its multiplicity) eigenvalue of $-\Delta$ in $H_0^1(\Omega)$. Let us also denote by e_j a corresponding eigenfunction satisfying

$$(e_i, e_j)_{1,2} = \delta_{ij}.$$

We claim that there exist k distinct pairs of solutions of the variational inequality problem: Find $u \in H_0^1(\Omega)$ such that $F(u) \in L^1(\Omega)$ and

$$\int_\Omega \nabla u \nabla(v-u)dx + \int_\Omega F(v)dx - \int_\Omega F(u)dx$$

$$\geq \int_\Omega g(u)(v-u)dx, \; \forall\, v \; \in \; H_0^1(\Omega), F(v) \in L^1(\Omega). \tag{4.5.4}$$

Let I be defined by (4.1.25)-(4.1.27) with

$$a(u,v) = \int_\Omega \nabla u \nabla v dx, \; \forall\, u, \, v \; \in \; H_0^1(\Omega). \tag{4.5.5}$$

If u is a critical point of I then $u \in D(\Psi)$ and

$$\frac{1}{2}\int_\Omega \mid \nabla v \mid^2 dx - \frac{1}{2}\int_\Omega \mid \nabla u \mid^2 dx + \int_\Omega F(v)dx - \int_\Omega F(u)dx$$

$$\geq \int_\Omega g(u)(v-u)dx, \; \forall\, v \; \in \; D(\Psi). \tag{4.5.6}$$

Here $D(\Psi) = \{u \in H_0^1(\Omega) : F(u) \in L^1(\Omega)\}$. By convexity of the mapping $v \to \frac{1}{2}\int_\Omega \mid \nabla v \mid^2 dx + F(v)$, it results that any solution of (4.5.6) is a solution of (4.5.4) too (see Section 4.1). From Proposition 4.1.13, we know that I satisfies the (PS) condition ($\equiv$ (VPS) condition). We employ Corollary 4.4.3 to conclude. It remains to prove that

$$-\infty < c_j < 0, \; \forall\, j = 1, \cdots, k,$$

with c_j defined by (4.4.1). From the proof of Proposition 4.1.13, we know that I is bounded from below and therefore $c_j > -\infty$. Let $\rho > 0$ and

$$A = \{u = \sum_{i=1}^{j} \alpha_i e_i :\parallel u \parallel_{1,2}^2 = \sum_{i=1}^{j} \alpha_i^2 = \rho^2\}.$$

Using (6) of Proposition 4.4.1 we deduce that $\gamma(A) = j$. Choose $r > 0$ and $\lambda > \lambda_k$ so that

$$(F(t) - G(t))/t^2 \leq -\frac{1}{2}\lambda \text{ as } \mid t \mid \leq r.$$

Choose now ρ sufficiently small to have $\mid u \mid_{0,\infty} \leq r$, $\forall\, u \in A$. Note that A is a subset of $X_j := span\{e_1, \cdots, e_j\}$. The space X_j is finite dimensional and therefore we may find a constant $C > 0$ such that

$$\mid u \mid_{0,\infty} \leq C \parallel u \parallel_{1,2}, \; \forall\, u \in X_j.$$

It suffices therefore to choose $\rho \leq rC^{-1}$. Recall that here $\lambda > \lambda_k > \cdots > \lambda_1 > 0$. For $u \in A$ we obtain

$$
\begin{aligned}
I(u) &= \frac{1}{2}\int_\Omega |\nabla u|^2\, dx + \int_\Omega (F(u) - G(u))dx \\
&\leq \frac{1}{2}\int_\Omega |\nabla u|^2\, dx - \frac{1}{2}\lambda \int_\Omega u^2 dx \\
&\leq \frac{1}{2}\sum_{i=1}^{j}(1 - \frac{\lambda}{\lambda_i})\int_\Omega |\nabla u|^2\, dx \\
&= \frac{\rho^2}{2}\sum_{i=1}^{j}(1 - \frac{\lambda}{\lambda_i})
\end{aligned}
$$

and thus $c_j < 0$.

Example 4.5.2   (A variational-hemivariational inequality). Let Ω be a nonempty bounded open set of class $C^{0,1}$ in $\mathbb{R}^N$ ($N \in \mathbb{N}\setminus\{0\}$). Consider the Sobolev space $H_0^1(\Omega)$ which is a Hilbert space with respect to the scalar product

$$
(u, v)_{H_0^1(\Omega)} = \int_\Omega \nabla u \nabla v dx, \ \forall\, u,\, v\, \in\, H_0^1(\Omega).
$$

The corresponding norm is here denoted by

$$
\| \cdot \|_{H_0^1} := (.,.)_{H_0^1(\Omega)}
$$

The cone of nonnegative functions in $H_0^1(\Omega)$, i.e.,

$$
K = \{u \in H_0^1(\Omega):\ u(x) \geq 0, \quad \text{a.e.} \quad x \in \Omega\} \tag{4.5.7}
$$

forms a convex and closed subset of $H_0^1(\Omega)$. Denoting by $\Psi : H_0^1(\Omega) \to (-\infty, +\infty]$ its indicator function, that is

$$
\Psi(u) \equiv \Psi_K(u) = \begin{cases} 0, & u \in K \\ +\infty, & u \notin K, \end{cases} \tag{4.5.8}
$$

it follows that Ψ is proper, convex and l.s.c., as required in assumption (H).

There are given $g \in L^2(\Omega)$ satisfying

$$
g < 0 \quad \text{a.e. in } \Omega \tag{4.5.9}
$$

and a (Carathéodory) function $j : \Omega \times \mathbb{R} \to \mathbb{R}$ such that

(a) $j(\cdot, y) : \Omega \to \mathbb{R}$ is measurable, $\forall\, y \in \mathbb{R}$;

(b) $j(x, \cdot) : \mathbb{R} \to \mathbb{R}$ is locally Lipschitz;

(c) $j(x, 0) = 0,\ \forall\, x \in \mathbb{R}$.

The function $j : \Omega \times \mathbb{R} \to \mathbb{R}$ is required to fulfill the conditions below

(j_1) $|\,\xi\,| \leq c(1 + |\,y\,|^{p-1}),\ \forall\, x \in \Omega,\ y \in \mathbb{R},\ \xi \in \partial_y j(x, y)$, with constants $c > 0$ and $1 \leq p < \frac{2N}{N-2}$ if $N \geq 3$ and an arbitrary $p > 2$ if $N = 1$ or $N = 2$;

(j_2) $\liminf_{y \to 0} y^{-2} j(x, y) \geq 0$ uniformly with respect to $x \in \Omega$;

(j_3) $\mu^{-1} j_y^0(x, y; y) \leq j(x, y)$ for all $x \in \Omega$ and $y \in \mathbb{R}$, $y \geq 0$ where μ is a constant with $\mu > 2$;

(j_4) there is $e \in H_0^1(\Omega)$ with $e \geq 0$ a.e. in Ω such that

$$\int_\Omega j(x, e(x))dx < 0.$$

Denote by λ_1 the first eigenvalue of $-\Delta$ on $H_0^1(\Omega)$.

Theorem 4.5.3   Under assumptions (4.5.9) for $g \in L^2(\Omega)$ and (j_1)-(j_4) for $j : \Omega \times \mathbb{R} \to \mathbb{R}$, for every $\lambda < \lambda_1$ the variational-hemivariational inequality: Find $u \in K$ in (4.5.7) such that

$$\int_\Omega \nabla u(\nabla v - \nabla u)dx + \int_\Omega j_y^0(x, u(x); v(x) - u(x))dx$$

$$\geq \lambda \int_\Omega u(v - u)dx + \int_\Omega g(v - u)dx,\ \forall\, u \in K, \qquad (4.5.10)$$

has a nontrivial solution.

Proof. The idea is to apply Corollary 4.3.5 for $I = \Phi + \Psi : H_0^1(\Omega) \to (-\infty, +\infty]$, with Ψ determined by (4.5.8) and $\Phi : H_0^1(\Omega) \to \mathbb{R}$ expressed as follows

$$\Phi(u) = \frac{1}{2}\|\,u\,\|_{H_0^1}^2 + J_{|_{H_0^1(\Omega)}}(u) - \frac{\lambda}{2}|\,u\,|_{0,2}^2 - \int_\Omega gudx,$$

$$\forall\, u \in H_0^1(\Omega), \qquad\qquad (4.5.11)$$

where

$$J(u) = \int_\Omega j(x, u(x))dx, \ \forall\, u \in L^2(\Omega). \qquad (4.5.12)$$

It is clear from (j_1) that Φ in (4.5.11) is locally Lipschitz. Observe also that by (j_2) one has $I(0) = 0$. The first task is to show that our functional I satisfies condition (PS) in the sense of Definition 4.1.2. To this end let a sequence $\{u_n\} \subset K$ satisfy $I(u_n) \to c \in \mathbb{R}$ and (4.1.2) for some $\varepsilon_n \to 0$ in $\mathbb{R}$. In our case of (4.5.8), (4.5.11), inequality (4.1.2) reads $u_n \in K$ and

$$\Phi^0(u_n; v - u_n) = \int_\Omega \nabla u_n \nabla(v - u_n)dx + J^0_{\big|_{H_0^1(\Omega)}}(u_n; v - u_n)$$

$$-\lambda \int_\Omega u_n(v - u_n)dx - \int_\Omega g(v - u_n)dx$$

$$\geq -\varepsilon_n \parallel v - u_n \parallel_{H_0^1}, \ \forall\, v \in K. \qquad (4.5.13)$$

Recall that

$$J^0_{\big|_{H_0^1(\Omega)}}(u; w) \leq J^0(\gamma(u); \gamma(w)) \leq \int_\Omega j_y^0(x, u(x); w(x))dx,$$

for all $u, w \in H_0^1(\Omega)$. Here γ denotes the compact embedding from $H_0^1(\Omega)$ into $L^2(\Omega)$. Let us also precise that we use as usually the straightforward identifications

$$\int_\Omega j(x, \gamma(u(x)))dx \equiv \int_\Omega j(x, u(x))dx, \ \forall\, u \in H_0^1(\Omega)$$

and

$$\int_\Omega j_y^0(x, \gamma(u(x)); \gamma(w(x)))dx \equiv \int_\Omega j_y^0(x, u(x); w(x))dx, \ \forall\, u \in H_0^1(\Omega).$$

Setting $v = 2u_n$ in (4.5.13), we derive

$$\Phi^0(u_n; u_n) \geq -\varepsilon_n \parallel u_n \parallel_{H_0^1} \, .$$

Then, for any sufficiently large n, we can write in view of (4.5.11), (4.5.13), (j_3) and Rayleigh - Ritz variational characterization of λ_1 that

$$c + 1 + \frac{1}{\mu} \parallel u_n \parallel_{H_0^1} \ \geq \ \Phi(u_n) + \frac{1}{\mu}\varepsilon_n \parallel u_n \parallel_{H_0^1}$$

$$\geq \ \Phi(u_n) - \frac{1}{\mu}\Phi^0(u_n; u_n) \geq (\frac{1}{2} - \frac{1}{\mu}) \parallel u_n \parallel^2_{H_0^1}$$

$$+ \lambda(\frac{1}{\mu} - \frac{1}{2}) \mid u_n \mid_{0,2}^2 + (\frac{1}{\mu} - 1) \int_\Omega g u_n dx$$

$$+ \int_\Omega (j(x, u_n) - \frac{1}{\mu} j_y^0(x, u_n; u_n)) dx$$

$$\geq \quad (\frac{1}{2} - \frac{1}{\mu})(1 - \lambda_1^{-1}\lambda) \parallel u_n \parallel_{H_0^1}^2$$

$$+ (\frac{1}{\mu} - 1)\lambda_1^{-\frac{1}{2}} \mid g \mid_{0,2} \parallel u_n \parallel_{H_0^1} .$$

Since we suppose that $\mu > 2$ and $\lambda < \lambda_1$ it follows the boundedness of $\{u_n\}$ in $H_0^1(\Omega)$. Then there exists $\bar{u} \in K$ such that $u_n \rightharpoonup \bar{u}$ in $H_0^1(\Omega)$ and $\gamma(u_n) \to \gamma(\bar{u})$ in $L^2(\Omega)$. Therefore from (4.5.13) with $v = \bar{u}$ we get

$$\limsup_{n \to \infty} \parallel u_n \parallel_{H_0^1}^2 \leq \parallel \bar{u} \parallel_{H_0^1}^2 + \limsup_{n \to \infty} J_{\mid H_0^1(\Omega)}^0 (u_n; \bar{u} - u_n) . \qquad (4.5.14)$$

$$\leq \parallel \bar{u} \parallel_{H_0^1(\Omega)}^2 + \limsup_{n \to \infty} J^0(\gamma(u_n); \gamma(\bar{u}) - \gamma(u_n)).$$

Since $\gamma(u_n) \to \gamma(u)$ in $L^2(\Omega)$ and $J^0(.,.)$ is upper semi-continuous on $L^2(\Omega) \times L^2(\Omega)$, we obtain that

$$\limsup_{n \to \infty} \parallel u_n \parallel_{H_0^1}^2 \leq \parallel \bar{u} \parallel_{H_0^1}^2 .$$

This yields that $u_n \to \bar{u}$ in $H_0^1(\Omega)$ along a subsequence which enables us to conclude that $I = \Phi + \Psi$ satisfies condition (PS).

Let us check that condition (i) in Corollary 4.3.5 holds. We carry out the proof as follows. Fix an $\varepsilon > 0$. Assumption (j$_2$) assures that some $\delta > 0$ exists to have

$$\mid y \mid^{-2} j(x, y) \geq -\varepsilon, \ \forall x \in \Omega, \ y \in \mathbb{R}, \ \mid y \mid \leq \delta. \qquad (4.5.15)$$

Lebourg's mean value theorem and the growth condition (j$_1$) imply

$$\mid j(x, y) \mid = \mid j(x, y) - j(x, 0) \mid \leq c(1 + \mid y \mid^{p-1}) \mid y \mid, \ \forall (x, y) \in \Omega \times \mathbb{R}.$$

In particular, we see that

$$\mid j(x, y) \mid \leq c(\delta^{-p+1} + 1) \mid y \mid^p, \ \forall x \in \Omega, \ y \in \mathbb{R}, \ \mid y \mid \geq \delta.$$

Combining it with (4.5.15) we find

$$-j(x, y) \leq \varepsilon \mid y \mid^2 + c(\delta^{-p+1} + 1) \mid y \mid^p, \ \forall (x, y) \in \Omega \times \mathbb{R}.$$

Thus we obtain the next estimate for the functional J in (4.5.12)

$$-J(u) \leq \varepsilon \mid u \mid_{0,2}^2 + c(\delta^{-p+1} + 1) \mid u \mid_{0,p}^p, \ \forall u \in L^p(\Omega) \qquad (4.5.16)$$

and thus

$$-J_{|_{H_0^1(\Omega)}}(u) \leq \varepsilon B \parallel u \parallel_{H_0^1}^2, \ \forall u \in H_0^1(\Omega), \qquad (4.5.17)$$

provided $\parallel u \parallel_{H_0^1}$ is sufficiently small, where $B > 0$ denotes a constant independent of ε. Here we have used the continuity of the embedding $H_0^1(\Omega) \hookrightarrow L^p(\Omega)$ for the parameter p as defined in (j_1). Then (4.5.9),(4.5.11) and (4.5.17) allow to write for $u \in K$ that

$$I(u) = \Phi(u) \geq \frac{1}{2}(1 - \lambda\lambda_1^{-1} - 2\varepsilon B) \parallel u \parallel_{H_0^1}^2$$

if $\parallel u \parallel_{H_0^1}$ is small enough. Since $\varepsilon > 0$ can be chosen arbitrarily small, property (i) in Corollary 4.3.5 is verified. We pass to the proof of condition (ii) in Corollary 4.3.5. For this we need the formula below involving the generalized gradient ∂_t with respect to $t \in \mathbb{R}$

$$\partial_t(t^{-\mu}j(x, ty)) = \mu t^{-1-\mu}(\mu^{-1}ty\partial_y j(x, ty) - j(x, ty))$$

for all $x \in \Omega$, $y \in \mathbb{R}$ and $t > 0$. By Lebourg's mean value theorem and the previous relation we see that

$$t^{-\mu}j(x, ty) - j(x, y) \leq \mu\tau^{-1-\mu}(\mu^{-1}j_y^0(x, \tau y; \tau y) - j(x, \tau y))(t - 1)$$

for all $x \in \Omega$, $y \in \mathbb{R}$, $t > 1$, with some $\tau \in (1, t)$. Then assumption (j_3) implies

$$j(x, ty) \leq t^\mu j(x, y) \quad \text{for all } x \in \Omega, \ y \geq 0, \ t > 1 \ . \qquad (4.5.18)$$

Now from (4.5.18) and assumption (j_4) we deduce

$$I(te) \ = \ \Phi(te) \leq \frac{t^2}{2}(\parallel e \parallel_{H_0^1}^2 - \lambda \mid e \mid_{0,2}^2) + t^\mu \int_\Omega j(x, e(x))dx$$

$$-t\int_\Omega g(x)e(x)dx \longrightarrow -\infty \quad \text{as } t \to +\infty$$

because $\mu > 2$. It follows that (ii) in Corollary 4.3.5 is satisfied.

Applying Corollary 4.3.5 we obtain a nontrivial critical point $u \in K$ of $I = \Phi + \Psi$. It remains only to justify that the given function $u \in K$ solves (4.5.10). By Definition 4.1.1 we know that $u \in K$ fulfills

$$\int_\Omega \nabla u(\nabla v - \nabla u)dx + J_{|_{H_0^1(\Omega)}}^0(u; v - u)$$

$$\geq \lambda \int_\Omega u(v - u)dx + \int_\Omega g(v - u)dx, \ \forall \, v \in K, \qquad (4.5.19)$$

where $\lambda < \lambda_1$. On the other hand assumption (j_1) insures

$$J^0_{|_{H^1_0(\Omega)}}(u; w) \leq J^0(\gamma(u); \gamma(w))$$

$$\leq \int_\Omega j^0_y(x, u(x); w(x))dx, \ \forall u, \ w \ \in \ H^1_0(\Omega). \tag{4.5.20}$$

Combining (4.5.19) and (4.5.20) we arrive at (4.5.10). This completes the proof. ∎

Corollary 4.5.4   Under the same assumptions as in Theorem 4.5.3 excepting that (j_3) is replaced by the stronger formulation

(j_3') $\mu^{-1}j^0_y(x, y; y) \leq j(x, y) - a \mid y \mid^q, \ \forall \, x \ \in \ \Omega, \ y > 0$, with constants $\mu > 2, \ a > 0$ and $q > 2$,

the conclusion of Theorem 4.5.3 holds for every $\lambda \in \mathbb{R}$.

Proof.   The first difference from the proof of Theorem 4.5.3 lies in the fact that (j_3') allows to deduce the boundedness of the sequence $\{u_n\} \subset K$ in (4.5.13) for every $\lambda \in \mathbb{R}$, so condition (PS) holds whenever $\lambda \in \mathbb{R}$. Indeed, here we have

$$\Phi(u_n) - \frac{1}{\mu}\Phi^0(u_n; u_n)$$

$$\geq \ (\frac{1}{2} - \frac{1}{\mu}) \parallel u_n \parallel^2_{H^1_0} + \lambda(\frac{1}{\mu} - \frac{1}{2}) \int_\Omega \mid u_n \mid^2 dx$$

$$+ (\frac{1}{\mu} - 1) \int_\Omega g u_n dx + a \int_\Omega \mid u_n \mid^q dx$$

$$\geq \ (\frac{1}{2} - \frac{1}{\mu}) \parallel u_n \parallel^2_{H^1_0} - C,$$

for some constant $C > 0$. The second difference from the proof of Theorem 4.5.3 consists in the argument to establish condition (i) of Corollary 4.3.5. Here we carry out the proof arguing by contradiction. Assume there exists a sequence $\{u_n\}$ in $K\setminus\{0\}$ such that $u_n \to 0$ in $H^1_0(\Omega)$ and

$$I(u_n) = \Phi(u_n) \leq \frac{1}{n} \parallel u_n \parallel^2_{H^1_0} \ .$$

Without loss of generality we may suppose $p > 2$ (see (j_1)). Denoting $w_n = u_n / \parallel u_n \parallel_{H^1_0}$ one has

$$\Phi(u_n) \parallel u_n \parallel^{-2}_{H^1_0} = \frac{1}{2} - \frac{1}{2}\lambda \mid w_n \mid^2_{0,2}$$

$$+ \parallel u_n \parallel_{H_0^1}^{-2} \int_\Omega j(x, u_n(x))dx - \parallel u_n \parallel_{H_0^1}^{-1} \int_\Omega g w_n dx \le \frac{1}{n}. \qquad (4.5.21)$$

Admit for moment that

$$\liminf_{n\to\infty} \parallel u_n \parallel_{H_0^1}^{-2} \int_\Omega j(x, u_n(x))dx \ge 0 . \qquad (4.5.22)$$

Passing to a subsequence if necessary, it is allowed to assume that $w_n \rightharpoonup w$ in $H_0^1(\Omega)$ and $\gamma(w_n) \to \gamma(w)$ in $L^2(\Omega)$ for some $w \in K$. Notice that $w \neq 0$ because otherwise (4.5.21), (4.5.22) and (4.5.9) would imply

$$\liminf_{n\to\infty} \Phi(u_n) \parallel u_n \parallel_{H_0^1}^{-2} \ge \frac{1}{2}$$

which contradicts the choice of the sequence $\{u_n\}$. Then, applying again (4.5.9), we find

$$\int_\Omega g w dx < 0 .$$

Passing to limit in (4.5.21) for $n \to \infty$ and taking into account (4.5.22) we are led to the conclusion

$$\lim_{n\to\infty} \Phi(u_n) \parallel u_n \parallel_{H_0^1}^{-2} = +\infty .$$

This contradicts the choice of $\{u_n\}$. To complete the proof of (i) in Corollary 4.3.5 we have to prove (4.5.22). At this point we utilize (4.5.17), which was established on the basis of (j$_2$), to obtain

$$\liminf_{\substack{u\to 0 \\ u\in H_0^1(\Omega)}} J_{|_{H_0^1(\Omega)}}(u) \parallel u \parallel_{H_0^1}^{-2} \ge -B\varepsilon .$$

Since $\varepsilon > 0$ is arbitrary, inequality (4.5.22) follows. Condition (ii) of Corollary 4.3.5 can be checked exactly as in the proof of Theorem 4.5.3. Then Corollary 4.3.5 implies directly the desired conclusion. ∎

A relevant example of locally Lipschitz function $j : \mathbb{R} \to \mathbb{R}$ satisfying conditions (j$_1$)-(j$_4$) is the following

$$j(t) = \max\{-\frac{1}{p} \mid t \mid^p, -\frac{1}{r} \mid t \mid^r\}, \ \forall \, t \in \mathbb{R},$$

with constants $p, r \in (2, \frac{2N}{N-2})$ if $N \ge 3$ and $p, r > 2$ if $N = 1$ or $N = 2$.

Note that the strict inequality condition (4.5.9) is only used to prove Corollary 4.5.4 and the condition $g \le 0$ a.e. in Ω is sufficient to prove Theorem 4.5.3.

Application 4.5.5   An adhesive contact problem.

Let us here consider the adhesive contact problem described in Section 2.11.3 and formulated through the hemivariational inequality (2.11.48). We set (the notations and requirements of Section 2.11.3 are here again used)

$$X = U_{ad},$$

$$\langle Au, v \rangle = \sum_{m=1}^{l} a(u^{(m)}, u^{(m)}), \ \forall \, u, \, v \, \in \, X,$$

$$\langle f, v \rangle = \sum_{m=1}^{l} \int_{\Omega^{(m)}} f^{(m)}.u^{(m)} dx + \sum_{m=1}^{l} \int_{\Gamma_F^{(m)}} F^{(m)}.\gamma_{\Gamma_F^{(m)}} u^{(m)} ds, \ \forall \, u \in \, X,$$

$$H_{N,q}(z) = \int_{\Gamma_q} j_{N(q)}(z) ds, \forall z \in L^1(\Gamma_q),$$

and

$$H_{T,q}(z) = \int_{\Gamma_q} j_{T(q)}(z) ds, \forall z \in L^1(\Gamma_q).$$

We also set

$$J_q(z) = H_{N,q}(\gamma_{\Gamma_q,N}(z)) + H_{T,q}(\gamma_{\Gamma_q,T}(z)), \ \forall \, z \, \in \, U_{ad}^{(q)},$$

$$J(u) = \sum_{q=1}^{k} J_q(u^{(q)}), \ \forall \, u \, \in \, X$$

and

$$\Phi(u) = \frac{1}{2}\langle Au, u \rangle - \langle f, u \rangle, \ \forall \, u \, \in \, X.$$

Let us first note that for $u, v \in X$ we have the relations (See Section 1.2)

$$J^0(u; v) \leq \sum_{q=1}^{k} J_q^0(u^{(q)}; v^{(q)})$$

$$\leq \sum_{q=1}^{k} H_{N,q}^0(\gamma_{\Gamma_q,N}(u^{(q)}); \gamma_{\Gamma_q,N}(v^{(q)})) + H_{T,q}^0(\gamma_{\Gamma_q,T}(u^{(q)}); \gamma_{\Gamma_q,T}(v^{(q)}))$$

$$\leq \int_{\Gamma_q} j_{N(q)}^0(\gamma_{N|\Gamma_q}(u^{(q)}); \gamma_{N|\Gamma_q}(v^{(q)})) + \int_{\Gamma_q} j_{T(q)}^0(\gamma_{T|\Gamma_q(u^{(q)})}; \gamma_{T|\Gamma_q}(v^{(q)})).$$

It results that if u is a critical point of $I = \Phi + J$ then u solves (2.11.48). In addition to the requirements of Section 2.11.3 we assume that ($q = 1, \cdots, k$)

$$\mid j_{N(q)}(t) \mid \leq C_{N(q)} \mid t \mid + D_{N(q)}, \ \forall \, t \, \in \, \mathbb{R}$$

and

$$| j_{T(q)}(t) | \leq C_{T(q)} | t | + D_{T(q)}, \ \forall \, t \ \in \ \mathbb{R}^3,$$

where $D_{N(q)}, D_{T(q)} \in \mathbb{R}$ and $C_{N(q)}, C_{T(q)}$ are positive constants. We claim that the conditions of Theorem 4.1.6 are satisfied. It is clear that I satisfies (H_1). Let us now check that I satisfies the (HPS) condition. Let $\{u_n\}$ be a sequence such that $I(u_n) \to c$ and satisfying (4.1.7). For n great enough, we have $I(u_n) \leq c + n^{-1}$ and thus

$$C_0 \parallel u_n \parallel_X^2 \leq C_1 \parallel u_n \parallel_X + C_2,$$

for some positive constants C_0, C_1 and $C_2 \in \mathbb{R}$. Here we have used our assumptions on the functions $j_{N(q)}, j_{T(q)}$ and the fact that A is coercive on X in the sense that

$$C_0 \parallel u_n \parallel_X^2 \leq \langle Au, u \rangle, \forall \, u \ \in \ X,$$

as a consequence of the Korn's first inequality (see Remark 2.11.2). It results that the sequence u_n is bounded. For a subsequence $u_n \rightharpoonup u$ and

$$C_0 \parallel u_n - u \parallel^2 \leq \langle Au_n - Au, u_n - u \rangle$$

$$\leq J^0(u_n; u - u_n) + \varepsilon_n \parallel u - u_n \parallel - \langle Au, u_n - u \rangle + \langle f, u_n - u \rangle,$$

$$\leq \sum_{q=1}^{k} H_{N,q}^0(\gamma_{\Gamma_q, N}(u_n^{(q)}); \gamma_{\Gamma_q, N}(u^{(q)} - u_n^{(q)}))$$

$$+ H_{T,q}^0(\gamma_{\Gamma_q, T}(u_n^{(q)}); \gamma_{\Gamma_q, T}(u^{(q)} - u_n^{(q)}))$$

$$+ \varepsilon_n \parallel u - u_n \parallel - \langle Au, u_n - u \rangle + \langle f, u_n - u \rangle.$$

Taking the limit is this last inequality we obtain that $u_n \to u$. The coercivity of A together with our assumptions on the functions $j_{N(q)}, j_{T(q)}$ entail that I is bounded from below. The existence of at least one critical point of I follows from Theorem 4.1.6.

Chapter 5

TOPOLOGICAL METHODS FOR INEQUALITY PROBLEMS

The aim of this Chapter is to discuss an approach based on the use of topological tools (Leray-Schauder degree and continuation results) in a way that is suitable in the setting of unilateral analysis. Particular attention is paid to some nice and fundamental theorems. The material developed in this Chapter will be used later in various directions. In particular, the topological methods constitute powerful devices for the study of unilateral eigenvalue problems (see Chapter 10). In preparing this Chapter we have followed the works of Goeleven [176], Goeleven and Théra [161], Goeleven and Motreanu [164], [172], Lloyd [254], Quittner [366], [367], Rabinowitz [368] and Szulkin [404].

5.1 TOPOLOGICAL TOOLS

5.1.1 TOPOLOGICAL DEGREE OF LERAY - SCHAUDER

Let $\mathcal{D}$ be an open subset of some Banach space X, p a point of X and $f : \mathcal{D} \to X$ a continuous mapping from $\mathcal{D}$ into X. One way to think the degree of f at p relative to $\mathcal{D}$ is as a number which measures the number of solutions of the equation $f(x) = p$ which depends continuously of f as well on p, i.e. for small perturbations of f this number remains unchanged in a small neighborhood V of p.

We start with the definition of degree for C^1-functions defined on $\mathbb{R}^n$ ($n \in \mathbb{N}, n \geq 1$). Let $\mathcal{D} \subset \mathbb{R}^n$ be an open bounded set, $\Phi : \bar{\mathcal{D}} \to \mathbb{R}^n$

a mapping and $p \in \mathbb{R}^n \setminus \Phi(\partial \mathcal{D})$. Suppose that Φ is a $C^1(\bar{\mathcal{D}})$-mapping. Let $\mathcal{N}$ be the set of critical points in $\bar{\mathcal{D}}$, i.e.

$$\mathcal{N} = \{x \in \bar{\mathcal{D}} : J_\Phi(x) = 0\} \tag{5.1.1}$$

where $J_\Phi(x)$ is the Jacobian of Φ at x, that is $J_\Phi(x) := det \, D\Phi(x)$, where $D\Phi$ is the Jacobian matrix of Φ at x, i.e.

$$D\Phi(x) := \begin{bmatrix} \frac{\partial \Phi_1}{\partial x_1}(x) & \cdots & \frac{\partial \Phi_n}{\partial x_1}(x) \\ \cdot & & \cdot \\ \cdot & & \cdot \\ \cdot & & \cdot \\ \frac{\partial \Phi_1}{\partial x_n}(x) & \cdots & \frac{\partial \Phi_n}{\partial x_n}(x) \end{bmatrix} .$$

We also set here

$$\Phi^{-1}(p) = \{x \in \bar{\mathcal{D}} : \Phi(x) = p\}. \tag{5.1.2}$$

If $\Phi^{-1}(p) \cap \mathcal{N} = \emptyset$ then we set

$$deg(\Phi, \mathcal{D}, p) = \sum_{x \in \Phi^{-1}(p)} sign \, J_\Phi(x) \tag{5.1.3}$$

and we call this number the degree of Φ at p relative to $\mathcal{D}$. Note that if $r(x)$ denotes the number of real negative eigenvalues of the matrix $D\Phi(x)$ then (5.1.3) can be written equivalently as follows:

$$deg(\Phi, \mathcal{D}, p) = \sum_{x \in \Phi^{-1}(p)} (-1)^{r(x)} . \tag{5.1.4}$$

Example 5.1.1   Let $\mathcal{D} = (a, b)$ be an open bounded interval and let f be a $C^1(\mathcal{D}) \cap C^0(\bar{\mathcal{D}})$-function on the closure of the interval with values in $\mathbb{R}$. Then

$$deg(f, (a, b), 0) = \begin{cases} 0 & \text{if} & f(a)f(b) > 0 \\ +1 & \text{if} & f(a) < 0 \text{ and } f(b) > 0 \\ -1 & \text{if} & f(a) > 0 \text{ and } f(b) < 0. \end{cases}$$

Example 5.1.2   Let $\mathcal{D}$ be a bounded domain in $\mathbb{C}$ which we identify with $\mathbb{R}^2$. Let $\Phi = \Phi_1 + i\Phi_2$ be a holomorphic function defined on

$\bar{\mathcal{D}}$. Then $deg(\Phi, \mathcal{D}, p)$ is the number of p-points of Φ in $\mathcal{D}$, counting multiplicity, that is also

$$deg(\Phi, \mathcal{D}, p) = \frac{1}{2\pi i} \int_{\partial \mathcal{D}} \frac{\Phi'(z)}{\Phi(z) - p} dz$$

for simply connected $\mathcal{D} \subset \mathbb{C}$.

The Sard's theorem [382] ensures that

$$\mathcal{L}_n(\Phi(\mathcal{N})) = 0. \tag{5.1.5}$$

Therefore we may view the elements $p \in \Phi(\mathcal{N})$ as limits of sequences of elements $p_n \notin \Phi(\mathcal{N})$ and we define the degree as

$$deg(\Phi, \mathcal{D}, p) = \lim_{n \to \infty} deg(\Phi, \mathcal{D}, p_n). \tag{5.1.6}$$

It can be proved that the definition is independent of the particular sequences $\{p_n\}$. Since $C^1(\mathcal{D}) \cap C^0(\bar{\mathcal{D}})$ is a dense subset of $C^0(\bar{\mathcal{D}}; \mathbb{R}^n)$ we can define the degree of a map $\Phi \in C^0(\bar{\mathcal{D}}, \mathbb{R}^n)$ as

$$deg(\Phi, \mathcal{D}, p) = \lim_{n \to \infty} deg(\Phi_n, \mathcal{D}, p) \tag{5.1.7}$$

where $\{\Phi_n\}$ is a sequence of $C^1(\mathcal{D}) \cap C^0(\bar{\mathcal{D}})$ - mappings converging uniformly to Φ. The independence of the degree on the sequence $\{\Phi_n\}$ chosen can also be proved. The degree of a continuous mapping in $\mathbb{R}^n$ is called the Browder degree. The properties of this degree are listed and proved in [254].

Many problems in mathematics can be reduced to a fixed point problem like

$$x = \Phi(x), \quad x \in X \tag{5.1.8}$$

where Φ is a mapping defined on an infinite dimensional Banach space X.

We now extend the degree so as to consider this kind of problems. Let X be a Banach space and let $\mathcal{D} \subset X$ be an open bounded subset of X. Let $\Phi : \bar{\mathcal{D}} \to X$ be a compact mapping, that is Φ is continuous and $\overline{\Phi(\bar{\mathcal{D}})}$ is compact.

The technique to define a degree theory so as to study problem (5.1.8) consists to approximate Φ by mappings of finite rank. Indeed it is known that for every $\varepsilon > 0$, there exists a continuous mapping $\Phi_\varepsilon : \bar{\mathcal{D}} \to X_\varepsilon$, $X_\varepsilon \subset X$ being a subspace and $dim\, X_\varepsilon < \infty$, such that

$$\| \Phi_\varepsilon(x) - \Phi(x) \| \leq \varepsilon, \ \forall\, x \in \mathcal{D}.$$

Let

$$d = d(0, (id_X - \Phi)(\partial \mathcal{D})) > 0.$$

If we choose $\varepsilon < \frac{d}{2}$ then we obtain

$$
\begin{aligned}
\| x - \Phi_\varepsilon(x) \| &= \| x - \Phi(x) + \Phi(x) - \Phi_\varepsilon(x) \| \\
&\geq \| x - \Phi(x) \| - \| \Phi(x) - \Phi_\varepsilon(x) \| \geq \frac{d}{2} > 0,
\end{aligned}
$$

for all $x \in \partial \mathcal{D}$.

If we consider the restriction

$$(id_X - \Phi_\varepsilon)_{|\bar{\mathcal{D}} \cap X_\varepsilon} : \bar{\mathcal{D}} \cap X_\varepsilon \to X_\varepsilon$$

then $deg((id_X - \Phi_\varepsilon)_{|\bar{\mathcal{D}} \cap X_\varepsilon}, \mathcal{D} \cap X_\varepsilon, 0)$ is well-defined. We set

$$deg(id_X - \Phi, \mathcal{D}, 0) = \lim_{\varepsilon \to 0} deg((id_X - \Phi_\varepsilon)_{|\bar{\mathcal{D}} \cap X_\varepsilon}, \mathcal{D} \cap X_\varepsilon, 0). \qquad (5.1.9)$$

It is known that this limit exists, it is finite, and it does not depend on the choice of Φ_ε.

The degree defined by (5.1.9) is called the Leray-Schauder degree. We end this section with the main properties of the Leray-Schauder degree (see e.g. [97], [203], [245], [254], [302]).

Proposition 5.1.3

1. Normalized Property. If $0 \in \mathcal{D}$ then $deg(id_X, \mathcal{D}, 0) = 1$. If $0 \notin \bar{\mathcal{D}}$ then $deg(id_X, \mathcal{D}, 0) = 0$.

2. Solution Property. If $deg(id_X - \Phi, \mathcal{D}, 0) \neq 0$ then there is $x \in \mathcal{D}$ such that $x = \Phi(x)$.

3. Homotopy Property. If f, g are two compact mappings on $\bar{\mathcal{D}}$, and for each $t \in [0, 1]$ there exists f_t compact with $f_0 = f$, $f_1 = g, t \to f_t$ continuous, $0 \notin (id_X - f_t)(\partial \mathcal{D})$, then

$$deg(id_X - f, \mathcal{D}, 0) = deg(id_X - g, \mathcal{D}, 0).$$

4. Boundary Value Dependence. If $\Phi = \Psi$ on $\partial \mathcal{D}$. Then if $0 \notin (id_X - \Phi)(\partial \mathcal{D})$ then

$$deg(id_X - \Phi, \mathcal{D}, 0) = deg(id_X - \Psi, \mathcal{D}, 0).$$

5. **Additivity Property.** If $0 \notin (id_X - \Phi)(\partial \mathcal{D})$ and $\mathcal{D}$ is the disjoint union of finitely many open sets $\mathcal{D}_i$ $(i = 1, 2, \cdots)$ then

$$deg(id_X - \Phi, \mathcal{D}, 0) = \sum_i deg(id_X - \Phi_i, \mathcal{D}_i, 0).$$

6. **Excision Property.** If $K \subset \bar{\mathcal{D}}$ is closed and $0 \notin (id_X - \Phi)(K \cup \partial \mathcal{D})$ then

$$deg(id_X - \Phi, \mathcal{D}, 0) = deg(id_X - \Phi, \mathcal{D} \backslash K, 0).$$

7. **Odd Mappings.** Suppose that $\mathcal{D}$ is an open bounded and symmetric subset of X with $0 \in \mathcal{D}$. If

$$\frac{x - \Phi(x)}{\| x - \Phi(x) \|} \neq -\frac{(x + \Phi(-x))}{\| x + \Phi(-x) \|}, \quad x \in \partial \mathcal{D}$$

then $deg(id_X - \Phi, \mathcal{D}, 0)$ is odd.

8. **Homeomorphism.** Suppose that $id - \Phi$ is a one to one mapping of $\bar{\mathcal{D}}$ onto $(id_X - \Phi)(\bar{\mathcal{D}})$. If $0 \in (id_X - \Phi)(\mathcal{D})$ then $deg(id_X - \Phi, \mathcal{D}, 0) = \pm 1$.

In the case of continuous mappings from $\mathcal{D} \in \mathbb{R}^n$ onto $\mathbb{R}^n$ the previous properties hold true for the Brouwer degree [68]. In this last case, the structure of the considered mapping, let us say f, is arbitrary (i.e. not necessarily of the form $f(x) = x - \Phi(x)$). Many results in mathematics have been proved thanks to the degree. One of them and which has been used in Section 4.4 is the famous Borsuk - Ulam theorem (see e.g. [254]).

Theorem 5.1.4 Let $\mathcal{D}$ be a bounded, open and symmetric subset of $\mathbb{R}^n$. There is no odd continuous mapping from ∂D to $\mathbb{R}^{n-1} \backslash \{0\}$ $(n \in \mathbb{N}, n \geq 1)$.

Proof. Suppose that there exists an odd and continuous mapping $\eta : \partial D \to \mathbb{R}^{n-1} \backslash \{0\}$. There exists an odd and continuous mapping $\tilde{\eta} : \bar{D} \to \mathbb{R}^{n-1}$ such that $\tilde{\eta}(x) = \eta(x), \forall x \in \partial D$. We set

$$\Phi(t, x) = t(\tilde{\eta}(x), 0) + (1 - t)(0, \cdots, 0, -1).$$

It is easy to see that

$$\Phi(t, x) \neq 0, \forall x \in \partial D, \forall t \in [0, 1].$$

We have

$$deg(\Phi(1,\cdot),\mathcal{D},0) = deg(\Phi(0,\cdot),\mathcal{D},0).$$

However, $\Phi(0,\cdot)$ is odd and thus $deg(\Phi(0,\cdot),\mathcal{D},0)$ is odd too, and we get a contradiction. ∎

5.1.2   CONTINUATION RESULTS

Let E be a metric space, by a continuum we mean a closed and connected subset in E. We will make use of the following basic results in the context of topological continuation.

Lemma 5.1.5   Let K be a compact metric space and A, B closed, nonempty subsets of K. Then either A, B are separated in K, i.e. there exist compact subsets K_A, K_B of K such that

$$K = K_A \cup K_B, \ A \subset K_A, \ B \subset K_B, \ K_A \cap K_B = \emptyset$$

or A and B are connected in K, i.e., there exists a continuum $\mathcal{C} \subset K$ such that

$$\mathcal{C} \cap A \neq \emptyset \text{ and } \mathcal{C} \cap B \neq \emptyset.$$

Lemma 5.1.6   Let X be a real Banach space and let $h : \mathbb{R} \times X \to X$ be a completely continuous mapping such that

$$h(0,u) = 0, \ \forall\, u \in X.$$

Then the set

$$S = \{\lambda, u) \in \mathbb{R} \times X : u = h(\lambda, u)\} \tag{5.1.10}$$

contains a pair of unbounded subcontinua emanating from $(0,0) \in \mathbb{R} \times X$ and lying respectively in $\mathbb{R}_+ \times X$ and $\mathbb{R}_- \times X$.

Set

$$B(0,R) = \{u \in X : \| u \| < R\}. \tag{5.1.11}$$

Lemma 5.1.7   Let X be a real Banach space and let $h : [a,b] \times X \to X$ be a completely continuous mapping. Assume that there is an $R > 0$

such that for all $r \in (0, R]$ one has $h(a, u) \neq u \neq h(b, u)$ for all u with $\| u \| = r$. Furthermore, assume that for $r \in (0, R]$:

$$deg(id_X - h(a,.), B(0,r), 0) \neq deg(id_X - h(b,.), B(0,r), 0).$$

Then the set

$$\mathcal{S}_{a,b} = \{(\lambda, u) \in [a,b] \times X : u = h(\lambda, u)\}$$

contains an unbounded continuum.

Remark 5.1.8   A proof of Lemma 5.1.5 can be found in [425]. Lemma 5.1.6 is due to Rabinowitz [368]. Lemma 5.1.7 [360] will be used later in Chapter 10. The following result constitutes an additional important property of the Leray-Schauder degree (see e.g. [254]).

Lemma 5.1.9   Let X be a real Banach space and $\mathcal{D} \subset [a,b] \times X$ an open and bounded subset. Let $h : \overline{\mathcal{D}} \to X$ be a completely continuous mapping such that $0 \notin (id_X - h)(\partial \mathcal{D})$. Then

$$deg(id_X - h(\lambda,.), \mathcal{D}_\lambda, 0) = \text{ constant}.$$

Here

$$\mathcal{D}_\lambda = \{u \in X : (\lambda, u) \in \mathcal{D}\}.$$

5.2   FIXED POINT FORMULATIONS

In this Section we discuss some fixed point formulations of variational inequalities leading to possible use of degree theory and continuation theorems.

The following result is of particular interest.

Lemma 5.2.1   Let X be a real reflexive Banach space, $A : X \to X^*$ a strongly monotone, i.e.

$$\langle Ax - Ay, x - y \rangle \geq \alpha \| x - y \|^2, (\alpha > 0), \ \forall \, x, \, y \in X$$

and hemicontinuous operator, $\Phi : X \to \mathbb{R} \cup \{+\infty\}$ a proper l.s.c. and convex function. Then the mapping $\Pi_{A,\Phi} : X^* \to D(\Phi)$ which associates to $f \in X^*$ the unique solution of the variational inequality

$$u \in X : \langle Au - f, v - u \rangle + \Phi(v) - \Phi(u) \geq 0, \ \forall \, v \, \in \, X \qquad (5.2.1)$$

is well-defined and satisfies

$$\| \, \Pi_{A,\Phi}(f) - \Pi_{A,\Phi}(f') \, \| \leq \alpha^{-1} \, \| \, f - f' \, \|_*, \ \forall \, f, \, f' \, \in \, X^*. \qquad (5.2.2)$$

Proof. The fact that $\Pi_{A,\Phi}(f) \neq \emptyset$ for any $f \in X^*$ is a consequence of Theorem 3.11.2 ($C \equiv X, j \equiv 0$) and the strong monotonicity of A which entails (3.11.3) for any $x_0 \in D(\Phi)$. Let $z_f \in \Pi_{A,\Phi}(f)$ and $z_{f'} \in \Pi_{A,\Phi}(f')$, we have

$$\langle Az_f - f, z_f - z_{f'} \rangle - \Phi(z_{f'}) + \Phi(z_f) \leq 0$$

and

$$-\langle Az_{f'} - f', z_f - z_{f'} \rangle - \Phi(z_f) + \Phi(z_{f'}) \leq 0.$$

From these two last inequalities we deduce that

$$\alpha \, \| \, z_f - z_{f'} \, \| \leq \| \, f - f' \, \|_* \, .$$

It results that the solution of (5.2.1) is unique. Indeed, if $f = f'$ then we obtain $z_f = z_{f'}$. It results also that the mapping $f \to \Pi_{A,C}(f)$ is a well-defined single-valued mapping satisfying the condition (5.2.2). ∎

If Φ is the indicator of some nonempty closed convex subset C of X, i.e. $\Phi \equiv \Psi_C$ then

$$\Pi_{A,\Psi_C} \equiv \Pi_{A,C}, \qquad (5.2.3)$$

with $\Pi_{A,C}$ as considered in Remark 3.3.2 (see also Section 3.2).

Let $F : X \to X^*$ be a possibly nonlinear completely continuous operator. We consider the inequality problem: Find $u \in X$ such that

$$\langle Au - F(u), v - u \rangle + \Phi(v) - \Phi(u) \geq 0, \ \forall \, v \, \in \, X. \qquad (5.2.4)$$

This last inequality problem is equivalent to the fixed point problem: Find $u \in X$ such that

$$u = \Pi_{A,\Phi}(F(u)). \qquad (5.2.5)$$

Provided that the inequality (5.2.4) does not admit solution such that $\| \, u \, \| = R$, for some $R > 0$, then the integer

$$deg(id_X - \Pi_{A,\Phi}(F(.)), B(0, R), 0)$$

is well-defined. Indeed, the mapping $\Pi_{A,\Phi}(F(.))$ is completely continuous as the composite of a completely continuous mapping $(F(.))$ and a continuous one $(\Pi_{A,\Phi}(.);$ see (5.2.2)).

The approach will be of particular interest for the study of unilateral eigenvalue problems. Let $F : \mathbb{R} \times X \to X^*$ be a completely continuous mapping. We consider the unilateral eigenvalue problem: Find $(\lambda, u) \in \mathbb{R} \times X$ such that

$$\langle Au - F(\lambda, u), v - u \rangle + \Phi(v) - \Phi(u) \geq 0, \ \forall \, v \, \in \, X. \tag{5.2.6}$$

For $\lambda \in \mathbb{R}$ fixed, the inequality (5.2.6) is equivalent to the fixed point problem

$$u(\lambda) = \Pi_{A,\Phi}(F(u(\lambda)))$$

and the integer

$$deg(id_X - \Pi_{A,\Phi}(F(\lambda, .)), B(0, R), 0)$$

is well-defined provided that (for the parameter λ considered) the inequality (5.2.6) does not admit solution $u(\lambda)$ such that $\| \, u(\lambda) \, \| = R$.

Lemma 5.2.1 has been established for a strongly monotone operator. We will also use, later in this book, this last result together with Leray-Schauder degree theory and the fixed point formulations exposed in this Section so as to deal with several kinds of elliptic variational inequalities and eigenvalue problems for variational inequalities (see Chapters 6 and 10). We now state a similar result for a maximal monotone operator. We know already from Section 3.9 that this class of operators is of particular interest for the study of parabolic variational inequalities.

Lemma 5.2.2   Let H be a real Hilbert space, $\varphi : H \to \mathbb{R} \cup \{+\infty\}$ a proper, convex and lower semi-continuous function and $A : D(A) \to H$ a maximal monotone operator. We suppose that $int\{D(\partial\varphi)\} \cap D(A) \neq \emptyset$. Then for each $\varepsilon > 0, g \in H$, there exists a unique $u_\varepsilon \in D(A)$ satisfying the variational inequality

$$(\varepsilon u_\varepsilon + Au_\varepsilon - g, v - u_\varepsilon) + \varphi(v) - \varphi(u_\varepsilon) \geq 0, \ \forall \, v \, \in \, H. \tag{5.2.7}$$

Moreover, the mapping $P_{A,\varphi,\varepsilon} : H \to D(A) : g \to P_{A,\varphi,\varepsilon}(g)$ (where for a given $g \in H$, the element $P_{A,\varphi,\varepsilon}(g)$ is defined as the unique solution of (5.2.7)) is continuous.

Proof. Problem (5.2.7) is equivalent to the set-valued problem: Find $u_\varepsilon \in D(A) \cap D(\partial\varphi)$ such that

$$0 \in \varepsilon u_\varepsilon + Au_\varepsilon + \partial\varphi(u_\varepsilon) - g. \tag{5.2.8}$$

The map $A + \partial\varphi$ is maximal monotone since $int\{D(\varphi)\} \cap D(A) \neq \emptyset$ (see e.g. [64], [432]). It results that $(\varepsilon id_H + A + \partial\varphi)^{-1}$ is single-valued and problem (5.2.8) has a unique solution $u_\varepsilon = P_{A,\varphi,\varepsilon}(g)$. It remains to prove that $P_{A,\varphi,\varepsilon}$ is continuous. Indeed, let $g_n \to g$. We set $u_n := P_{A,\varphi,\varepsilon}(g_n)$ and $u = P_{A,\varphi,\varepsilon}(g)$. From (5.2.7) we get

$$(\varepsilon u_n + A u_n - g_n, u_n - u) + \varphi(u_n) - \varphi(u) \leq 0$$

and

$$-(\varepsilon u + A u - g, u_n - u) + \varphi(u) - \varphi(u_n) \leq 0.$$

Thus

$$\varepsilon \parallel u_n - u \parallel^2 \leq \parallel g - g_n \parallel \parallel u_n - u \parallel$$

and therefore

$$\parallel u_n - u \parallel \leq \frac{1}{\varepsilon} \parallel g - g_n \parallel.$$

Thus if $g_n \to g$ then $u_n \to u$ which means that the mapping $P_{A,\varphi,\varepsilon}$ is continuous. $\blacksquare$

With some additional conditions on A and φ we may prove that $P_{A,\varphi,\varepsilon}$ is completely continuous.

Lemma 5.2.3   Suppose that the conditions in the statement of Lemma 5.2.2 are satisfied. If in addition we suppose that φ is Lipschitz continuous on bounded sets and the sets $\{u \in D(A) : \parallel u \parallel \leq R$ and $\parallel Au \parallel \leq R\}$ are compact for each $R > 0$ then the map $P_{A,\varphi,\varepsilon}$ is completely continuous.

Proof. We have (with the same notations that the ones used in Lemma 5.2.2)

$$\parallel u_n - u \parallel \leq \frac{1}{\varepsilon} \parallel g - g_n \parallel.$$

If $g_n \rightharpoonup g$ then $\{u_n\}$ is bounded. We have

$$\varepsilon u_n + A u_n - g_n + \zeta_n = 0,$$

for some $\zeta_n \in \partial\varphi(u_n)$. The Lipschitz continuity of φ on bounded sets ensures that $\{\zeta_n\}$ is bounded and thus $\{Au_n\}$ is bounded too. We can now take a subsequence of $\{u_n\}$ which converges strongly to u. Moreover u is uniquely defined as the solution of (5.2.7) and thus the whole sequence $\{u_n\}$ converges strongly to u. $\blacksquare$

We may now consider the inequality problem: Find $u_\varepsilon \in D(A) \cap D(\varphi)$ such that

$$(\varepsilon u_\varepsilon + A u_\varepsilon - F(u_\varepsilon), v - u_\varepsilon) + \varphi(v) - \varphi(u_\varepsilon), \ \forall\, v \in H, \qquad (5.2.9)$$

where $F : H \to H$ is a (possibly nonlinear) continuous mapping. The inequality problem (5.2.9) is equivalent to the fixed point problem: Find $u_\varepsilon \in X$ such that

$$u_\varepsilon = P_{A,\varphi,\varepsilon}(F(u_\varepsilon)).$$

If we suppose that the conditions of Lemma 5.2.3 are satisfied and if (5.2.9) does not admit solution such that $\| u \| = R$ $(R > 0)$ then the integer

$$deg(id_H - P_{A,\varphi,\varepsilon}(F(.)), B(0,R), 0)$$

is well-defined.

5.3   AN ALTERNATIVE THEOREM

Let X be a real reflexive Banach space, $A : X \to X^*$ a strongly monotone and hemicontinuous operator, $\Phi : X \to \mathbb{R} \cup \{+\infty\}$ a proper, convex and l.s.c. function. Let $F : \mathbb{R} \times X \to X^*$ be a completely continuous mapping. We consider the unilateral eigenvalue problem: Find $(\lambda, u) \in \mathbb{R} \times X$ such that

$$\langle Au - F(\lambda, u), v - u \rangle + \Phi(v) - \Phi(u) \geq 0, \ \forall \, v \in X. \qquad (5.3.1)$$

Moreover, we assume that

$$\begin{aligned} A(0) &= 0, \\ F(\lambda, 0) &= 0, \ \forall \, \lambda \in \mathbb{R}, \end{aligned}$$

and

$$0 \in \partial\Phi(0)$$

so that the curve

$$S_0 := \{(\lambda, 0) : \lambda \in \mathbb{R}\}$$

constitutes the curve of trivial solutions of (5.3.1).

Let us also set

$$\mathcal{S} := \overline{\{(\lambda, u) \in \mathbb{R} \times X \backslash \{0\} : (\lambda, u) \ \text{solves (5.3.1)}\}}. \qquad (5.3.2)$$

We say that $(\mu, 0)$ is a bifurcation point of (5.3.1) if there exist a sequence $(\lambda_n, u_n) \in \mathbb{R} \times X$ solution of (5.3.1) and such that $\lambda_n \to \mu$, $u_n \neq 0$ and $u_n \to 0$.

Theorem 5.3.1   Let $a, b \in \mathbb{R}$ $(a < b)$ be such that $u = 0$ is an isolated solution of (5.3.1) for $\lambda = a$ and $\lambda = b$ and suppose that $(0, a), (0, b)$ are not bifurcation points of (5.3.1). If

$$deg(id_X - \Pi_{A,\Phi}(F(a,.)), B(0,r), 0)$$

$$\neq deg(id_X - \Pi_{A,\Phi}(F(b,.)), B(0,r), 0) \tag{5.3.3}$$

for all $r > 0$, r small enough, then there exists a subcontinuum $\mathcal{C}$ of $\mathcal{S} \cup \{[a,b] \times \{0\}\}$ containing $[a,b] \times \{0\}$ such that either

(i) $\mathcal{C}$ is unbounded in $X \times \mathbb{R}$,

or

(ii) $\mathcal{C} \cap ((\mathbb{R} \setminus [a,b]) \times \{0\}) \neq \emptyset$.

Proof. Suppose by contradiction that there does not exist a subcontinuum of $\mathcal{S} \cup \{[a,b] \times \{0\}\}$ containing $[a,b] \times \{0\}$ satisfying either (i) or (ii). Let $\mathcal{C}$ be the connected component (i.e. the maximal connected subset) of $\mathcal{S} \cup \{[a,b] \times \{0\}\}$ containing $[a,b] \times \{0\}$. One can say that $\mathcal{C}$ is weakly compact. Indeed, let us first remark that $\mathcal{C}$ is bounded (from (i) which cannot be satisfied). Let now $\{(\lambda_n, u_n)\} \subset \mathbb{R} \times X$ be a sequence contained in $\mathcal{C} \subset \mathcal{S} \cup \{[a,b] \times \{0\}\}$.

Then
$$u_n = \Pi_{A,\Phi}(F(\lambda_n, u_n)), \ \forall \, n \, \in \, \mathbb{N}. \tag{5.3.4}$$

Indeed, the result is trivial if $(\lambda_n, u_n) \in [a,b] \times \{0\}$. If $(\lambda_n, u_n) \in \mathcal{S}$ then there exist sequences $u_n^j \to u$ in X and $\lambda_n^j \to \lambda_n$ in $\mathbb{R}$ such that

$$u_n^j = \Pi_{A,\Phi}(F(\lambda_n^j, u_n^j)), \ \forall \, j \, \in \, \mathbb{N}.$$

Taking the limit as $j \to +\infty$, we obtain (5.3.4).

Since $\mathcal{C}$ is bounded, we may find a subsequence (again denoted by (λ_n, u_n)) such that $\lambda_n \to \lambda$ in $\mathbb{R}$ and $u_n \rightharpoonup u$ in X. Taking now the limit as $n \to +\infty$ in (5.3.4) we see that

$$u_n = \Pi_{A,\Phi}(F(\lambda_n, u_n)) \to \Pi_{A,\Phi}(F(\lambda, u)) = u.$$

Let U_δ be a δ-neighborhood ($\delta > 0$) of $\mathcal{C}$. From (ii) which is not satisfied we may assume that $U_{2\delta}$ contains no solutions $(\lambda, 0)$ of (5.3.1) for $\lambda \in \mathbb{R} \setminus [a,b]$. Set $A = \mathcal{C}$, $B = \partial U_\delta$ and $K = \overline{U}_\delta \cap \mathcal{S}$. We have $A \cap B = \emptyset$ by construction. Using Lemma 5.1.5, we get the existence of two disjoints weakly compact subsets K_A, K_B of K such that $K = K_A \cup K_B, \mathcal{C} \subset K_A$ and $\partial U_\delta \subset K_B$. Let $\mathcal{D}$ be any open ε-neighborhood in $\mathbb{R} \times X$ of K_A where ε is less than $\min\{d_H(A,B), 2\delta\}$. We have $\mathcal{D} \subset \mathbb{R} \times X, \partial \mathcal{D} \cap K = \emptyset$ and $\mathcal{D}$ contains no trivial solutions other than those in $[a,b]$.

We set
$$\mathcal{D}_\lambda := \{u \in X : (\lambda, u) \in \mathcal{D}\}.$$

By Lemma 5.1.9, we have

$$deg(id_X - \Pi_{A,\Phi}(F(\lambda,.)), \mathcal{D}_\lambda, 0) = \text{const} \tag{5.3.5}$$

for all $\lambda \in [a, b]$. We have

$$deg(id_X - \Pi_{A,\Phi}(F(a,.)), \mathcal{D}_a, 0)$$
$$= \ deg(id_X - \Pi_{A,\Phi}(F(a,.)), \mathcal{D}_a \backslash \overline{B(0,r)}, 0)$$
$$+ deg(id_X - \Pi_{A,\Phi}(F(a,.)), B(0,r), 0)$$

for $r > 0$ sufficiently small to have

$$\overline{\mathcal{D}_a} \cap \overline{B(0,r)} = \emptyset.$$

Such r exists thanks to our assumptions on a. For $\mu \geq a$, we have also

$$deg(id_X - \Pi_{A,\Phi}(F(a,.)), \mathcal{D}_a \backslash \overline{B(0,r)}, 0)$$
$$= \ deg(id_X - \Pi_{A,\Phi}(F(\mu,.)), \mathcal{D}_\mu \backslash \overline{B(0,r)}, 0).$$

For μ large enough, $\mathcal{D}_\mu \backslash \overline{B(0,r)} = \emptyset$ and thus

$$deg(id_X - \Pi_{A,\Phi}(F(a,.)), \mathcal{D}_a \backslash \overline{B(0,r)}, 0) = 0.$$

Similarly, by choosing $r > 0$ possibly smaller to have

$$\overline{\mathcal{D}_b} \cap \overline{B(0,r)} = \emptyset,$$

we check that

$$deg(id_X - \Pi_{A,\Phi}(F(b,.)), \mathcal{D}_b \backslash \overline{B(0,r)}, 0) = 0$$

and thus

$$deg(id_X - \Pi_{A,\Phi}(F(b,.)), \mathcal{D}_b, 0)$$
$$= \ deg(id_X - \Pi_{A,\Phi}(F(b,.)), B(0,r), 0).$$

The contradiction follows from (5.3.5) and our assumption (5.3.3).   ∎

Applications of the previous result will be discussed later in Chapter 10.

5.4   EXISTENCE OF GLOBAL CONTINUA

Let X be a real reflexive Banach space, $A : X \to X^*$ a strongly monotone and hemicontinuous operator. Let K be a nonempty closed and convex subset of X. With a given completely continuous map $F : \mathbb{R} \times X \to X^*$ we consider the problem: Find $\lambda \in \mathbb{R}, u \in K$ such that

$$\langle Au - F(\lambda, u), v - u \rangle \geq 0, \ \forall v \in K. \tag{5.4.1}$$

The following result ensures the existence of global continua for the problem (5.4.1).

Theorem 5.4.1   Assume that A, K, and F have the properties above with

$$0 \in K, \ A(0) = 0, \tag{5.4.2}$$

and

$$F(0, v) = 0, \ \forall v \in K. \tag{5.4.3}$$

Then the solution set Σ of (5.4.1), i.e.,

$$\Sigma = \{(\lambda, u) \in \mathbb{R} \times K : (\lambda, u) \text{ solves } (5.4.1)\} \tag{5.4.4}$$

contains a pair of unbounded subcontinua Γ^+, Γ^- emanating from $(0,0) \in \mathbb{R} \times K$, lying in $\mathbb{R}_+ \times K, \mathbb{R}_- \times K$ respectively. Moreover Σ is locally compact in $\mathbb{R} \times X$.

Proof.   In view of the definition of $\Pi_{A,K} : X^* \to K$ entering Lemma 5.2.1 and 5.2.3 it follows that $(\lambda, u) \in \mathbb{R} \times K$ solves (5.4.1) if and only if

$$T(\lambda, u) := \Pi_{A,K}(F(\lambda, u)) = u. \tag{5.4.5}$$

By the continuity of $\Pi_{A,K}$ (cf. Lemma 5.2.1) and the complete continuity of F we know that the mapping $T : \mathbb{R} \times K \to X$ is completely continuous. We have also

$$T(0, u) = \Pi_{A,K}(F(0, u)) = \Pi_{A,K}(0), \ \forall u \in X, \tag{5.4.6}$$

where (5.4.3) has been used. Lemma 5.2.1, with $f = 0$ in (5.2.2) (and $\Phi = \Psi_K$) guarantees that

$$\Pi_{A,K}(0) = 0. \tag{5.4.7}$$

It is so because by (5.4.2) and (5.2.1) with $f = 0, v = 0$ it is allowed to write

$$\alpha \parallel u \parallel^2 \leq \langle Au, u \rangle \leq 0,$$

with $\alpha > 0$. Combining (5.4.6) and (5.4.7) we obtain that $T(0, u) = 0$, $\forall u \in X$. Thus we checked that the requirements in Lemma 5.1.6 are fulfilled for the map T introduced by (5.4.5). Thus Lemma 5.1.6 yields the first part of the conclusion of Theorem 5.4.1.

For showing that Σ is locally compact let $(\mu_n, y_n) \in \Sigma$ be a bounded sequence. By virtue of the reflexivity of X we can find a subsequence of (μ_n, y_n) denoted again (μ_n, y_n) such that $\mu_n \to \mu$ in $\mathbb{R}$ and $y_n \rightharpoonup y$ in X as $n \to \infty$ for $(\mu, y) \in \mathbb{R} \times X$. Since $(\mu_n, y_n) \in \Sigma$ the equality below holds

$$y_n = T(\mu_n, y_n), \ \forall\, n \geq 1. \tag{5.4.8}$$

The complete continuity of T allows to conclude that $\{y_n\}$ is strongly convergent in X. Then by passing to the limit in (5.4.8), we deduce that $(\mu, y) \in \Sigma$. The proof is thus complete. $\blacksquare$

Remark 5.4.2   More generally, Theorem 5.4.1 holds if the relation $A(0) = 0$ in (5.4.2) is replaced by the assumption that the variational inequality

$$\langle Au, v - u \rangle \geq 0, \ \forall\, v \in K, \tag{5.4.9}$$

has the unique solution $u = 0$. That is also $A(0) \in K^*$ where we recall that K^* denotes the dual set of K. As a further extension, if $u = u_0$ is the unique solution of (5.4.9), then Theorem 5.4.1 can be generalized to obtain two unbounded locally compact subcontinua emanating from $(0, u_0) \in \mathbb{R} \times K$ and lying respectively in $\mathbb{R}_+ \times K$ and $\mathbb{R}_- \times K$.

Remark 5.4.3   Clearly, if $\cup_{\lambda \in R_+} \{u : (\lambda, u) \in \Sigma\}$ is bounded (or the corresponding on R_-) then the projection of $\Gamma^+(\Gamma^-)$ coincides with $\mathbb{R}_+(\mathbb{R}_-)$.

Example 5.4.4   Let $X = H_0^1(\Omega)$ for an open bounded set of class $C^{0,1}$ in $\mathbb{R}^n$ ($n \in \mathbb{N}, n \geq 1$). Suppose that $A : X \to X^*$ is the differential operator described by

$$Au = - \sum_{i,j=1}^{n} \partial_{x_j}(a_{ij}(x, u, \nabla u)\partial_{x_i} u) + c(x, u, \nabla u)u,$$

with $a_{ij}, c \in L^\infty(\Omega), c \geq 0$ and there exist $\beta > 0$ such that

$$\sum_{i,j=1}^{n} a_{ij}(x, y, z)\tau_i\tau_j \geq \beta \mid \tau \mid^2, \ \forall\, \tau \in \mathbb{R}^n, (\forall x \in \Omega, y \in \mathbb{R}, z \in \mathbb{R}^n).$$

Let $f \in C^0(\overline{\Omega} \times \mathbb{R}; \mathbb{R})$ satisfy the growth condition

$$\mid f(x,t) \mid \leq c_1 + c_2 \mid t \mid^\tau, \ \forall \, (\alpha,t) \, \in \, \overline{\Omega} \, \times \, \mathbb{R},$$

with $c_1, c_2 \geq 0$ and $0 \leq \tau < 2n/(n-2)$ if $n \geq 3, 1 \leq \tau < +\infty$ if $n = 1, 2$. The mapping $F : X \to X^*$ defined by

$$\langle Fu, v \rangle = \int_\Omega f(x, u(x))v(x)dx, \ \forall \, u, \ v \, \in \, X,$$

is completely continuous (see e.g. [358], [432]). Let K be a nonempty closed convex subset of X with $0 \in K$. Then for every continuous function $g : \mathbb{R} \to \mathbb{R}$ such that the formula below is valid

$$g(0)F(v) = 0, \ \forall \, v \, \in \, K,$$

Theorem 5.4.1 applies to the variational inequality

$$\lambda \in \mathbb{R}, u \in K : \langle Au - g(\lambda)F(u), v - u \rangle \geq 0, \ \forall \, v \, \in \, K.$$

5.5   A TOPOLOGICAL APPROACH FOR NONCOERCIVE EVOLUTION VARIATIONAL INEQUALITIES

Let H be a real Hilbert space, let $\varphi : H \to \mathbb{R} \cup \{+\infty\}$ be a proper convex and lower semi-continuous function, let $A : D(A) \to H$ be a maximal monotone operator such that $\mathrm{int}D(\partial\varphi) \cap D(A) \neq \emptyset$ and let $B : H \to H$ be a completely continuous operator (this assumption will be relaxed later as soon as we will require some additional conditions on A and φ (see Remark 5.5.2). For $f \in H$ be given, we consider in this Section the nonlinear variational inequality: Find $u \in D(A)$ such that

$$(Au + Bu - f, v - u) + \varphi(v) - \varphi(u) \geq 0, \ \forall \, v \, \in \, H. \qquad (5.5.1)$$

We begin by considering the regularized one: Find $u_\varepsilon \in D(A)$ such that:

$$(\varepsilon u_\varepsilon + Au_\varepsilon + Bu_\varepsilon - f, v - u_\varepsilon) + \varphi(v) - \varphi(u_\varepsilon) \geq 0, \ \forall \, v \, \in \, H. \qquad (5.5.2)$$

This last variational inequality is equivalent to the fixed point problem: Find $u_\varepsilon \in H$ such that

$$u_\varepsilon = P_{A,\varphi,\varepsilon}(f - Bu_\varepsilon). \qquad (5.5.3)$$

The map $z \to P_{A,\varphi,\varepsilon}(z)$ is completely continuous since B is completely continuous and $P_{A,\varphi,\varepsilon}$ is continuous by Lemma 5.2.2. We now assume that

$$(Bu, u) \geq c_1 \parallel Bu \parallel^2 - c_2 \parallel u \parallel - c_3, \ \forall \, u \, \in \, H \tag{5.5.4}$$

for some positive constants c_1, c_2 and c_3. It is known that (see Theorem 1.1.11)

$$\varphi(u) \geq -C_0 - C_1 \parallel u \parallel, \ \forall \, v \, \in \, H, \tag{5.5.5}$$

for some constants $C_0 \in \mathbb{R}, C_1 \geq 0$, and moreover we suppose

$$int\{D(\partial\varphi)\} \cap D(A) \neq \emptyset \tag{5.5.6}$$

and

$$A(0) = 0. \tag{5.5.7}$$

We first prove the following lemma.

Lemma 5.5.1   For each $\varepsilon > 0, f \in H$, problem (5.5.2) has at least one solution.

Proof. We set

$$\mathcal{H}(t, u) := P_{A,\varphi,\varepsilon}(t(f - Bu)).$$

We claim that there exists $R > 0$ such that for each $t \in [0, 1]$ and for each $u(t)$ solution of the fixed point

$$u(t) = \mathcal{H}(t, u(t)), \tag{5.5.8}$$

we have $\parallel u(t) \parallel \leq R$. Indeed, let $(t, u \equiv u(t))$ be a solution of (5.5.8). Then

$$(\varepsilon u + Au + tBu - tf, v - u) + \varphi(v) - \varphi(u) \geq 0, \ \forall \, v \, \in \, H. \tag{5.5.9}$$

Setting $v = 0$ in (5.5.9), we obtain

$$\varepsilon \parallel u \parallel^2 + (Au, u) + t(Bu, u) \leq t(f, u) - \varphi(u) + \varphi(0).$$

Therefore

$$\varepsilon \parallel u \parallel^2 + c_1 t \parallel Bu \parallel^2 + (Au, u)$$
$$\leq \ tc_2 \parallel u \parallel + c_3 t + t \parallel f \parallel \parallel u \parallel - \varphi(u) + \varphi(0).$$

The monotonicity of A and $A(0) = 0$ ensure that $(Au, u) \geq 0$. Using also property (5.5.5) and the fact that $t \leq 1$, we obtain

$$\varepsilon \parallel u \parallel^2 \leq c' \parallel u \parallel + c", \tag{5.5.10}$$

for some positive constants c' and c''. That means that the solution set of (5.5.8) on $[0,1] \times H$ is bounded. Indeed, if we suppose the contrary then we can find a sequence $\{u_n\}$ such that $\| u_n \| \to +\infty$ and satisfying (5.5.10) which is a contradiction. Thus for $\tau > \max\{R, \| P_{A,\varphi,\varepsilon}(0) \|\}$, problem (5.5.8) has no solution on $\partial B(0,\tau)$. Thus

$$deg(id_H - \mathcal{H}(t,.), B(0,\tau), 0)$$

is well-defined and

$$deg(id_H - P_{A,\varphi,\varepsilon}(f - Bu), B(0,\tau), 0)$$

$$= deg(id_H - P_{A,\varphi,\varepsilon}(0), B(0,\tau), 0) = 1.$$

Thus problem (5.5.3) or equivalently problem (5.5.2) has at least one solution. $\blacksquare$

Remark 5.5.2   If the sets $\{u \in D(A) : \| u \| \leq R, \| Au \| \leq R\}$ are compact and φ is Lipschitz continuous on bounded sets then the map $P_{A,\varphi,\varepsilon}$ is completely continuous (see Lemma 5.2.3) and the assumption requiring B completely continuous can be relaxed either by "B weakly continuous", i.e. $u_n \rightharpoonup u \Rightarrow Bu_n \rightharpoonup Bu$, or by "$B$ demi-continuous", i.e. $u_n \to u \Rightarrow Bu_n \rightharpoonup Bu$ and bounded (in the sense that it maps bounded sets into bounded sets). In each case the mapping $P_{A,\varphi,\varepsilon}$ is compact in the sense that it is continuous and it maps bounded sets into relatively compact sets and we can deal with the Leray-Schauder degree as above.

We now prove the main theorem of this Section. Let $A : D(A) \subset H \to H$ be a closed linear operator with dense domain. Moreover, we suppose that

$$N(A) = N(A^*), \tag{5.5.11}$$

$$dim\{N(A)\} < +\infty, \tag{5.5.12}$$

$$A^{-1} : R(A) \to R(A) \text{ is compact.} \tag{5.5.13}$$

Note that assumption (5.5.11) ensures that A is a one-one map of $D(A) \cap R(A)$ onto $R(A)$ so that the inverse in (5.5.13) is well-defined. Assumptions (5.5.11)-(5.5.13) imply also that

$$\{u \in D(A) : \| u \| \leq 1 \text{ and } \| Au \| \leq 1\} \text{ is compact.} \tag{5.5.14}$$

Note that (5.5.14) together with (5.5.11) imply (5.5.12) and (5.5.13) (for a closed linear operator A with dense domain).

Remark 5.5.3   Further discussions on operators satisfying (5.5.14) can be found in [42] and [66].

Theorem 5.5.4   Let H be a real Hilbert space, let $A : D(A) \subset H \to H$ be a closed linear operator with dense domain satisfying conditions (5.5.11)-(5.5.13), let $B : H \to H$ be a bounded and demi-continuous operator satisfying inequality (5.5.4), let $\varphi : H \to \mathbb{R}$ be a convex and Lipschitz continuous function. If

$$(f, v) < \underline{r}_B(v) + \varphi_\infty(v), \ \forall \, v \, \in \, N(A)\backslash\{0\} \tag{5.5.15}$$

then problem (5.5.1) has at least one solution.

Proof.   The operator A is maximal monotone and $A(0) = 0$. We may apply Lemma 5.5.1 and Remark 5.5.2 so as to find for each $\varepsilon > 0$ an element $u_\varepsilon \in D(A)$ such that

$$(\varepsilon u_\varepsilon + A u_\varepsilon + B u_\varepsilon - f, v - u_\varepsilon) + \varphi(v) - \varphi(u_\varepsilon) \geq 0, \ \forall \, v \, \in \, H. \tag{5.5.16}$$

Let $\{\varepsilon_n\}$ be a sequence of positive real numbers satisfying $\varepsilon_n \to 0$ as $n \to +\infty$. We denote by $u_n \in D(A)$ a solution of (5.5.16) for $\varepsilon := \varepsilon_n$. We claim that the sequence $\{u_n\}$ is bounded. If we suppose the contrary then along a subsequence we may assume that $y_n := u_n / \parallel u_n \parallel \rightharpoonup y$ and $\parallel u_n \parallel \to +\infty$. We set $v = 0$ in (5.5.16) to get

$$\varepsilon_n \parallel u_n \parallel^2 + (A u_n, u_n) + (B u_n, u_n) + \varphi(u_n) \leq \varphi(0) + (f, u_n). \tag{5.5.17}$$

Thus

$$\begin{aligned}
c_1 \parallel B u_n \parallel^2 \ &\leq \ c_2 \parallel u_n \parallel + c_3 + \varphi(0) - \varphi(u_n) + \parallel f \parallel \parallel u_n \parallel \\
&\leq \ c_2 \parallel u_n \parallel + c_3 + C_1 \parallel u_n \parallel + \parallel f \parallel \parallel u_n \parallel + C_0 \\
&\leq \ \alpha_0 + \alpha_1 \parallel u_n \parallel
\end{aligned}$$

for some positive constants α_0 and α_1. We have

$$\frac{\parallel B u_n \parallel}{\parallel u_n \parallel} \leq [c_1^{-1}(\frac{\alpha_0}{\parallel u_n \parallel^2} + \frac{\alpha_1}{\parallel u_n \parallel})]^{\frac{1}{2}}$$

Thus $\frac{\parallel B u_n \parallel}{\parallel u_n \parallel} \to 0$ as $n \to +\infty$. There exists $\zeta_n \in \partial \varphi(u_n)$ such that

$$\varepsilon_n u_n + A u_n + B u_n - f + \zeta_n = 0, \ \forall \, n \, \in \, \mathbb{N}.$$

Thus

$$\parallel A u_n \parallel \leq \varepsilon_n \parallel u_n \parallel + \parallel B u_n \parallel + \parallel f \parallel + \parallel \zeta_n \parallel, \ \forall \, n \, \in \, \mathbb{N}. \tag{5.5.18}$$

We have $\parallel \zeta_n \parallel \leq K, \ \forall n \, \in \, \mathbb{N}$, since φ is Lipschitz continuous with a constant $K > 0$. Dividing (5.5.18) by $\parallel u_n \parallel$, we obtain

$$\parallel A y_n \parallel \leq \varepsilon_n + \frac{\parallel B u_n \parallel}{\parallel u_n \parallel} + \frac{\parallel f \parallel}{\parallel u_n \parallel} + \frac{K}{\parallel u_n \parallel}. \tag{5.5.19}$$

Thus there exists $R > 0$ such that the sequence $\{y_n\}$ belongs to the set $\{u \in D(A) : \| u \| \le R$ and $\| Au \| \le R\}$ which is compact. Thus along a subsequence we may assume that $y_n \to y \in D(A), y \ne 0$. Using (5.5.19) again we obtain $y \in N(A)$. Inequality (5.5.17) implies that

$$\langle B(t_n y_n), y_n \rangle + \frac{\varphi(t_n y_n)}{t_n} \le \frac{\varphi(0)}{t_n} + (f, y_n)$$

where $t_n := \| u_n \|$. Thus

$$\underline{r}_B(y) + \varphi_\infty(y) \le (f, y).$$

This is a contradiction to condition (5.5.15) since $y \in N(A)\backslash\{0\}$. Thus the sequence $\{u_n\}$ remains bounded. We have

$$\| Bu_n \|^2 \le \frac{\alpha_0}{c_1} + \frac{\alpha_1}{c_1} \| u_n \|$$

and thus the sequence $\{Bu_n\}$ is bounded too. We have also

$$\varepsilon_n u_n + Au_n + Bu_n - f + \zeta_n = 0$$

and since $\| \zeta_n \| \le K$ we obtain the boundedness of $\{Au_n\}$. Thus we can take a subsequence (again denoted by $\{u_n\}$) such that $u_n \to u \in D(A), Au_n \rightharpoonup y$. The operator A is closed and thus $y = Au$. For $v \in H$, we have

$$(\varepsilon_n u_n + Au_n + Bu_n - f, v - u_n) + \varphi(v) - \varphi(u_n) \ge 0.$$

Taking the limit as $n \to +\infty$, we obtain

$$(Au + Bu - f, v - u) + \varphi(v) - \varphi(u) \ge 0.$$

Since v is arbitrary we may conclude.   ∎

We will see later in Chapter 7 that Theorem 5.5.4 can be used to study nonlinear and noncoercive parabolic variational inequalities.

5.6   THE ASYMPTOTIC RELAXATION PRINCIPLE

We know that the mathematical formulation of some unilateral contact problems in structural mechanics lead to a linear variational inequality $LVI(M, q, K)$ of the form:

$$u \in K : (Mu)^T (v - u) - q^T (v - u) \ge 0, \ \forall v \in K, \qquad (5.6.1)$$

where $K \subset \mathbb{R}^N$ is a nonempty closed convex set, $q \in \mathbb{R}^N$ is a vector and $M \in \mathbb{R}^{N \times N}$ is a real matrix. For instance, in the case of some engineering structure containing members subjected to noninterpenetration conditions, the set K denotes the set of admissible displacements u, M is the stiffness matrix and q is the loading vector (see Section 2.11.11).

If the set K is a closed convex cone, i.e. $0 \in K$, $\alpha K \subset K$, $\forall \alpha > 0$ and $K + K \subset K$, then the variational inequality (5.6.1) turns out to be equivalent to the linear complementarity problem $LCP(M, q, K)$ (see Section 1.1.9):

$$u \in K, \tag{5.6.2}$$

$$u^T M u - q^T u = 0, \tag{5.6.3}$$

$$v^T (Mu - q) \geq 0, \ \forall \, v \in K. \tag{5.6.4}$$

The following Theorem is the basic result for this Section. It reduces the study of a LVI to $LCPs$ involving the recession cone K_∞ and the relaxations of the matrix M by taking the convex combinations of M and the identity matrix I.

Theorem 5.6.1   If for all $t \in [0, 1]$, 0 is the unique solution of the problem $LCP((1 - t)I + tM, 0, K_\infty)$ then for each $q \in \mathbb{R}^N$, the solution set of $LVI(M, q, K)$ is nonempty and compact.

Proof. Problem $LVI(M, q, K)$ is equivalent to the fixed point problem (see Section 1.1.6):

$$u = P_K(u - (Mu - q)), \tag{5.6.5}$$

where P_K is the projection operator on K. We set

$$H(t, u) = P_K(tu - t(Mu - q)),$$

for each $t \in [0, 1]$. We claim that there exists $R > 0$ such that if $t \in [0, 1]$ and $u(t) \in \mathbb{R}^N$ is solution of the fixed point problem

$$u(t) = H(t, u(t))$$

then $\| u(t) \| \leq R$. Indeed, if we suppose the contrary then we can find sequences $\{t_n\} \subset [0, 1]$ and $\{u_n\} \subset \mathbb{R}^N$ satisfying

$$\| u_n \| \to +\infty$$

and

$$u_n = H(t_n, u_n). \tag{5.6.6}$$

We have $tu - t(Mu - q) = u - (t(Mu - q) + (1 - t)u)$ and thus problem (5.6.6) is equivalent to the variational inequality:

$$u_n \in K : ((1 - t_n)u_n + t_n(Mu_n - q))^T(v - u_n) \geq 0, \ \forall \ v \ \in K. \quad (5.6.7)$$

We can assume along a subsequence that $w_n := u_n / \| u_n \| \to w \neq 0$ and $t_n \to t \in [0, 1]$. Let $\mu \geq 0$ and $x \in K$ be fixed. For n great enough,

$$u_n \frac{\mu}{\| u_n \|} + (1 - \frac{\mu}{\| u_n \|})x \in K.$$

Taking the limit as $n \to +\infty$, we obtain

$$\mu w + x \in K.$$

Since $\mu \geq 0$ and $x \in K$ are arbitrary, it results that $w \in K_\infty \backslash \{0\}$. Setting now $v := v_0$ where v_0 is fixed in K, we obtain

$$((1 - t_n)u_n + t_n(Mu_n - q))^T u_n \leq ((1 - t_n)u_n + t_n(Mu_n - q))^T v_0. \quad (5.6.8)$$

Dividing (5.6.8) by $\| u_n \|^2$ and taking the limit as $n \to +\infty$, we obtain

$$((1 - t)w + tMw)^T w \leq 0. \quad (5.6.9)$$

If $e \in K_\infty$ then $u_n + e \in K$ and thus using (5.6.7) again, we obtain

$$((1 - t_n)u_n + t_n(Mu_n - q))^T e \geq 0, \ \forall \ e \ \in \ K_\infty. \quad (5.6.10)$$

Dividing this last relation by $\| u_n \|$ and taking the limit as $n \to +\infty$, we obtain

$$((1 - t)w + tMw)^T e \geq 0,$$

and since e is arbitrary, we obtain

$$((1 - t)w + tMw)^T e \geq 0, \ \forall \ e \ \in \ K_\infty. \quad (5.6.11)$$

Thus

$$((1 - t)w + tMw)^T w = 0$$

as a consequence of (5.6.9), (5.6.11) and the fact that $w \in K_\infty$. Thus w is a non-zero solution of $LCP((1-t)I + tM, 0, K_\infty)$. This is a contradiction.

This means in particular that the solution set of $LVI(M, q, K)$ is bounded. It is also easy to see that it is closed and thus compact. Moreover, for R great enough the Brouwer degree of the map $u - H(t, u)$ is well-defined for all $t \in [0, 1]$ and using the homotopy invariance property of the degree, we obtain (Recall that $B(0, R) \ = \ \{u \in \mathbb{R}^N : \| u \| < R\}$)

$$deg(I - P_K(I - (M. - q)), B(0, R), 0)$$
$$= \ deg(I - H(1, .), B(0, R), 0)$$
$$= \ deg(I - H(0, .), B(0, R), 0)$$
$$= \ deg(I - P_K(0), B(0, R), 0) = 1,$$

(the last equality follows from the fact that for R great enough the element $P_K(0)$ belongs to $B(0, R)$) and thus problem (5.6.5) or equivalently the variational inequality $LVI(M, q, K)$ has at least one solution.

∎

Example 5.6.2   Variational Inequalities involving P-matrices

One says that M is a P-matrix if all its principal minors are positive. The class of P-matrices includes the positive definite matrices, the non-singular **M**-matrices, that are the matrices M for which there exist a nonnegative matrix B ($b_{ij} \geq 0$, $\forall i, j = 1, \cdots, N$) and a real number $s > \rho(B)$ ($\rho(B)$ denotes the spectral radius of B) such that $M = sI - B$, and the class of diagonally stable matrices, that are the matrices M for which there exists a positive diagonal matrix D such that $MD + DM^T$ is positive definite (see e.g. [49]).

Let us denote by $\{e^1, \cdots, e^N\}$ the canonical basis of $\mathbb{R}^N$. We say that a cone C is stable with respect to $\{e^1, \cdots, e^N\}$ if the following property

$$w \in C \Rightarrow (w^T e^j)e^j \in C, \ \forall\, j = 1, \cdots N$$

is satisfied. In other words, if $w \in C$ then its projection onto the space $X_j := \{x \in \mathbb{R}^N : x_k = 0, \ \forall k \neq j\}$ remains in C (for each j in $\{1, \cdots, N\}$). That is for instance the case of the cones $\mathbb{R}^N_+, \mathbb{R}^{N-1} \times \mathbb{R}_+$, $\mathbb{R}^{N-2} \times \mathbb{R} \times \mathbb{R}_-$, etc. If K is a rectangle of $\mathbb{R}^N$, i.e. $K = [a_1, b_1] \times [a_2, b_2] \times \cdots \times [a_N, b_N]$, $a_i \in \mathbb{R} \cup \{-\infty\}$, $b_i \geq a_i, b_i \in \mathbb{R} \cup \{+\infty\}, i = 1, \cdots, N$, then K_∞ is stable with respect to the canonical basis of $\mathbb{R}^N$.

We obtain the following result.

Corollary 5.6.3   Let M be a $N \times N$ P-matrix and let K be a non-empty closed convex set such that its recession cone K_∞ is stable with respect to $\{e^1, \cdots, e^N\}$. Then for each $q \in \mathbb{R}^N$ the solution set of problem $LVI(M, q, K)$ is nonempty and compact.

Proof. The matrix M is a P-matrix and thus for each nonzero $w \in \mathbb{R}^N$, some entry of the Hadamard product $(w) \circ (Mw)$ is positive, that is there is some $\alpha \in \{1, \cdots, N\}$ such that

$$w_\alpha(Mw)_\alpha > 0$$

(see [192]; property 2.5.6.2, p. 120). Thus for all $t \in (0, 1]$, we have also

$$(1 - t)w_\alpha w_\alpha + t w_\alpha(Mw)_\alpha > 0.$$

If $t = 0$ then there exists also some index $\beta \in \{1, \cdots, N\}$ such that $w_\beta^2 > 0$ since $w \neq 0$. Let $t \in [0, 1]$ and suppose that there exists $w \neq 0$ solution of $LCP((1 - t)I + tM, 0, K_\infty)$. Then for some $k \in \{1, \cdots, N\}$ we have thus $(1 - t)w_k w_k + tw_k(Mw)_k > 0$. Moreover, we have also

$$((1 - t)w + tMw)^T w = 0 \qquad (5.6.12)$$

and

$$((1 - t)w + tMw)^T v \geq 0, \ \forall \, v \in K_\infty. \qquad (5.6.13)$$

We set $v = (w^T e^j)e^j$ in (5.6.13) to obtain

$$(1 - t)w_j^2 + t(Mw)_j w_j \geq 0.$$

We have

$$((1 - t)w + tMw)^T w = \sum_{j=1,\cdots N} (1 - t)w_j^2 + t(Mw)_j w_j$$

$$= \sum_{j \neq k}((1 - t)w + tMw)_j w_j + ((1 - t)w + tMw)_k w_k > 0,$$

which is a contradiction to (5.6.12). Thus for all $t \in [0, 1]$, 0 is the unique solution of $LCP((1 - t)I + tM, 0, K_\infty)$ and we may apply Theorem 5.6.1 to conclude. ∎

Note that if K itself is a stable cone with respect to the canonical basis of $\mathbb{R}^N$ then the uniqueness can also be proved. Indeed, suppose that problem $LVI(M, q, K) \equiv LCP(M, q, K)$ admits two different solutions u^1 and u^2. Then set $w^1 = Mu^1 - q$ and $w^2 = Mu^2 - q$. We have

$$(w^\alpha)^T v \geq 0, \ \forall \, v \in K(\alpha = 1, 2)$$

and thus if we set $v := ((u^\beta)^T e^i)e^i (i = 1, \cdots, N)$ then we obtain the inequalities

$$(w^\alpha)_i(u^\beta)_i \geq 0, \ \forall \, \alpha, \beta \in \{1, 2\}, \ i \in \{1, \cdots N\}.$$

We have also

$$(w^\alpha)^T u^\alpha = 0, \ \forall \, \alpha \in \{1, 2\}.$$

Therefore

$$[(u^2)_i - (u^1)_i][(w^2)_i - (w^1)_i] \leq 0$$

and thus $[(u^2)_i - (u^1)_i][(M(u^2 - u^1))_i] \leq 0, \ \forall i \in \{1, \cdots N\}$, which is a contradiction to the property of a P-matrix already used here above to prove the existence of at least one solution.

Example 5.6.4   Variational inequalities involving positive stable matrices.

Let A be a given square matrix. We denote by $sp(A)$ the spectrum of A and by $sp_r(A) = sp(A) \cap \mathbb{R}$ the set of real eigenvalues of A. Let us now consider the generalized eigenvalue problem which consists to find $\lambda \in \mathbb{R}$ and $u \in K_\infty \backslash \{0\}$ solution of $CP(\lambda I - A, 0, K_\infty)$, that is

$$\lambda u^T(v - u) - (Au)^T(v - u) \geq 0, \ \forall \ v \ \in \ K_\infty. \tag{5.6.14}$$

We denote by $\sigma(A, K_\infty)$ the set of eigenvalues of the complementarity problem (5.6.14), i.e. the set of $\lambda \in \mathbb{R}$ for which there exists at least one $u \in K_\infty \backslash \{0\}$ solution of (5.6.14). If $K_\infty = \mathbb{R}^N$ then $\sigma(A, K_\infty) = sp_r(A)$. Let us now give some conditions on the generalized spectrum of the matrix $-M$ ensuring the solvability of $LVI(M, q, K)$ for all $q \in \mathbb{R}^N$.

Corollary 5.6.5   If $\sigma(-M, K_\infty) \subset (-\infty, 0)$ then for each $q \in \mathbb{R}^N$ the solution set of problem $LVI(M, q, K)$ is nonempty and compact.

Proof.   Let $t \in (0, 1]$. The inequality

$$((1 - t)u + tMu)^T(v - u) \geq 0,$$

is equivalent to the inequality

$$(\lambda u + Mu)^T(v - u) \geq 0,$$

with $\lambda = (1 - t)/t$. Thus if $\sigma(-M, K_\infty) \subset (-\infty, 0)$ then it is clear that 0 is the unique solution of $CP((1 - t)I + tM, 0, K_\infty)$ for all $t \in [0, 1]$. We conclude by using Theorem 5.6.1.   ∎

Corollary 5.6.6   Let us assume that all real eigenvalues of M are positive and suppose that K_∞ has the following form

$$K_\infty = \{x \in \mathbb{R}^N : x^T d \geq 0\},$$

with some $d \in \mathbb{R}^N \backslash \{0\}$. Then if

$$\{\lambda \in \mathbb{R} \backslash sp_r(-M) : d^T(\lambda I + M)^{-1}d = 0\} \subset (-\infty, 0) \tag{5.6.15}$$

then for each $q \in \mathbb{R}^N$ the solution set of problem $LVI(M, q, K)$ is nonempty and compact.

Proof.   The dual cone of K_∞ is given by $K_\infty^* = \{\alpha d; \alpha \geq 0\}$. Suppose that $\lambda \in \mathbb{R} \backslash sp_r(-M)$ and set

$$R(\lambda, M) := (\lambda I + M)^{-1}.$$

Such λ belongs to $\sigma(-M, K_\infty)$ if and only if $d^T R(\lambda, M)d = 0$. Indeed, suppose that $d^T R(\lambda, M)d = 0$ and set

$$u = R(\lambda, M)d.$$

It is clear that $u \in K_\infty$ since $d^T u = d^T R(\lambda, M)d = 0$. Moreover $u \neq 0$ since $\lambda \notin sp_r(-M)$. We have also

$$(\lambda I + M)(u) = d \in K_\infty^*$$

and

$$(\lambda u + Mu)^T u = d^T u = 0.$$

Conversely, if $\lambda \in \sigma(-M, K_\infty)$ then there exists $u \in K_\infty \backslash \{0\}$ such that

$$(\lambda u + Mu)^T u = 0 \tag{5.6.16}$$

and

$$\lambda u + Mu \in K_\infty^*. \tag{5.6.17}$$

Using (5.6.17) we get the existence of $\alpha \geq 0$ such that $\lambda u + Mu = \alpha d$ and thus (5.6.16) entails

$$\alpha^2 d^T R(\lambda, M)d = 0.$$

Note that $\alpha > 0$ since $\lambda \notin sp_r(-M)$ and thus $d^T R(\lambda, M)d = 0$. Thus $\sigma(-M, K_\infty) \subset sp_r(-M) \cup \{\lambda \in \mathbb{R} \backslash sp_r(-M) : d^T R(\lambda, M)d = 0\}$. The conclusion follows from Corollary 5.6.5 since $\{\lambda \in \mathbb{R} \backslash sp_r(-M) : d^T R(\lambda, M)d = 0\} \subset (-\infty, 0)$ and all the real eigenvalues of M are positive. ∎

One says that a matrix A is positive stable if $re(\lambda) > 0$, $\forall \lambda \in sp(A)$. It is clear that Corollary 5.6.6 apply for this matrix class for which there is no subset-superset relationship with the one of P-matrices. Note that a matrix whose all principal minors are nonnegative with at least one positive has all its real eigenvalues positive.

Remark 5.6.7   i) To compute $sp(-M)$ we look for the zeros of the characteristic polynomial $p(\lambda) = det(\lambda I + M)$ while to compute $\sigma(-M, K_\infty) \backslash sp_r(-M)$ we look for the zeros of $R(\lambda, M) := d^T (\lambda I + M)^{-1}d$ (in the framework of Corollary 5.6.6). ii) For example if we take $N = 2$, $M_{11} = M_{22} = 2$, $M_{12} := 10, M_{21} := -20$ and $d := (1, 0)^T$ then $sp(-M) = \{-2 + 10\sqrt{2}i, -2 - 10\sqrt{2}i\}$ and $\sigma(-M, K_\infty) = \{-2\}$. Thus M is positive stable, not positive semidefinite and $\sigma(-M, K_\infty) \subset (-\infty, 0)$ so that for each $q \in \mathbb{R}^2$, problem $LVI(M, q, K)$ has at least one solution.

5.7   A DEGREE THEORETIC APPROACH FOR VARIATIONAL-HEMIVARIATIONAL INEQUALITIES

We have seen that generalized projection operators can be used to formulate fixed point problems which are equivalent to some variational inequalities. We have also seen that the fixed point problem constitutes a suitable formulation for the use of topological tools. Hemivariational and variational-hemivariational inequalities are more delicate from this point of view. Indeed, the nonconvex term involved in these last models prevents the formulation of equivalent fixed point problems. Here we show that provided that some "strong conditions" are imposed on the nonconvex term, then an intermediate fixed point problem can be used to study the inequality problem via the Leray-Schauder degree. Further problems using the present approach will be discussed in Chapter 10.

Let X be a real reflexive Banach space which is densely and compactly embedded in $L^2(\Omega)$ for a nonempty open bounded set Ω of class $C^{0,1}$ in $\mathbb{R}^n$ ($n \geq 1; n \in \mathbb{N}$). Let $A : X \to X^*$ be a strongly monotone and hemicontinuous operator and $\Phi : X \to \mathbb{R} \cup \{+\infty\}$ a proper, convex and l.s.c. function. Let us state here a variant of the mapping $\Pi_{A,\Phi}$ on $L^2(\Omega)$ by saying that for $f \in L^2(\Omega)$, $\Pi_{A,\Phi}(f)$ is the unique solution of the variational inequality

$$u \in X : \langle Au, v - u \rangle + \Phi(v) - \Phi(u)$$

$$\geq \int_\Omega f(x)(v(x) - u(x))dx, \ \forall \, v \in X. \tag{5.7.1}$$

By following similar computations as the one detailed in Lemma 5.2.1, we obtain that

$$\| \, \Pi_{A,\Phi}(f) - \Pi_{A,\Phi}(f') \, \| \leq \, | \, f - f' \, |_{0,2}, \ \forall f, f' \in L^2(\Omega). \tag{5.7.2}$$

Let us now consider the variational-hemivariational inequality: Find $u \in X$ such that

$$\langle Au, v - u \rangle + \int_\Omega j_y^0(x, u(x); v(x) - u(x))dx$$

$$+\Phi(v) - \Phi(u) \geq 0, \ \forall \, v \in X. \tag{5.7.3}$$

We suppose that $j : \Omega \times \mathbb{R} \to \mathbb{R}$ satisfies the conditions:

$$j(.,y) : \Omega \to \mathbb{R} \text{ is measurable, } \forall \, y \in \mathbb{R},$$

$$j(.,0) \in L^1(\Omega),$$

$$j(x,.) : \mathbb{R} \to \mathbb{R} \text{ is locally Lipschitz, } \forall\, x \,\in\, \Omega.$$

In order to study the inequality (5.7.3) through a fixed point formulation, we suppose in addition that there exists a continuous mapping $W : L^2(\Omega) \to L^2(\Omega)$ such that $W(v)(x) \in \partial_y j(x, v(x))$, $\forall v \in V$, a.e. $x \in \Omega$.

Suppose that the problem: Find $u \in X, w \in L^2(\Omega)$ such that

$$w(x) \in \partial_y j(x, u(x)) \text{ for a.e. } x \in \Omega$$

and

$$\langle Au, v - u \rangle + \Phi(v) - \Phi(u) \geq - \int_\Omega w(x)(v(x) - u(x))dx, \ \forall\, v \,\in\, X$$

has no solution such that $\parallel u \parallel = r$ $(r > 0)$. Then we may consider the integer

$$deg(id_X - \Pi_{A,\Phi}(-W(.)), B(0,r), 0).$$

The mapping $\Pi_{A,\Phi}$ is indeed completely continuous since if $u_n \rightharpoonup u$ in X then $u_n \to u$ in $L^2(\Omega)$, $W(u_n) \to W(u)$ in $L^2(\Omega)$ and thus $\Pi_{A,\Phi}(-W(u_n)) \to \Pi_{A,\Phi}(-W(u))$ in X. If

$$deg(id_X - \Pi_{A,\Phi}(-W(.)), B(0,r), 0) \neq 0$$

then there exists $u \in X$ such that

$$\langle Au, v - u \rangle + \Phi(v) - \Phi(u) + \int_\Omega W(u)(x)(v(x) - u(x))dx \geq 0, \ \forall\, v \,\in\, X$$

and since (Proposition 1.2.7)

$$\int_\Omega j_y^0(x, u(x); v(x) - u(x))dx \geq \int_\Omega W(u)(x)(v(x) - u(x))dx$$

we obtain that u solves (5.7.3).

Remark 5.7.1   i) Let $J : L^2(\Omega) \to \mathbb{R}$ be defined by

$$J(u) = \int_\Omega j(x, u(x))dx.$$

If ∂J is continuous as a multivalued map from $L^2(\Omega)$ to $L^2(\Omega)$, then there exists $W \in C^0(L^2(\Omega); L^2(\Omega))$ such that $W(v)(x) \in \partial_y j(x, v(x))$, $\forall v \in L^2(\Omega)$, a.e. $x \in \Omega$.

Indeed, since the values of the multivalued map ∂J are closed and convex sets (see Proposition 1.2.7), the continuity of ∂J ensures the existence of a continuous mapping W with $W(v) \in \partial J(v)$ for each $v \in L^2(\Omega)$ (see Aubin and Cellina [28]). We conclude by using Theorem 1.2.20. ii) Remark i) is in particular true if the integrand $j : \Omega \times \mathbb{R} \to \mathbb{R}$ satisfies

$$| z_1 - z_2 | \leq k(x) \, | y_1 - y_2 |$$

whenever $z_i \in \partial_y j(x, y_i), i = 1, 2$, a.e. $x \in \Omega$, for some $k \in L^\infty(\Omega)$.

References

[1] Abraham R. and Robbin J. (1967) Transversal Mappings and Flows. *Benjamin, New York.*

[2] Adams R. (1975) Sobolev Spaces. *Academic Press, New York.*

[3] Adly S., Goeleven D. and Théra M. (1995) Recession Methods in Monotone Variational Hemivariational Inequalities. *Topological Methods in Nonlinear Analysis*, Vol. 5, pp. 397-409.

[4] Adly S. and Goeleven D. (1995) Homoclinic Orbits for a Class of Hemivariational Inequalities. *Applicable Analysis*, Vol. 58, pp. 229-240.

[5] Adly S. (1995) Analyse Variationnelle Appliquée aux Problèmes Unilatéraux: Analyse de Récession, méthodes Variationnelles et Algorithmiques. *Thèse de l'Université de Limoges, France.*

[6] Adly S., Goeleven D. and Théra M. (1996) Recession Mappings and Noncoercive Variational Inequalities. *Nonlinear Analysis, TMA*, Vol. 26, pp. 1573-1603.

[7] Al-Fahed, A.M. Stavroulakis G.E. and Panagiotopoulos P.D. (1991) Hard and Soft Fingered Robot Grippers. The Linear Complementarity Approach. *Z.A.M.M.*, Vol. 71, pp. 257-265.

[8] Al-Fahed A.M., Stavroulakis G.E. and Panagiotopoulos P.D. (1992) A Linear Complementarity Approach to the Frictionless Gripper. *Int. J. of Robot. Res.*, Vol. 11 pp. 112-122.

[9] Al-Fahed A.M., Stavroulakis G.E. (2000) A complementarity problem formulation of the frictional grasping problem. *Computer Methods in Applied Mechanics and Engineering*, Vol. 190, pp. 941-952.

[10] Al-Abed M.R., C.D. Bisbos and Panagiotopoulos P.D. (1996) The Debonding Effect in Saddles: Comparison of the Performance of Flügge's and Sanders' Shell Theories. *Int. J. Pres. Ves. & Piping*, Vol. 65, pp 53-74.

[11] Ambrosetti A. and Rabinowitz P. H. (1973) Dual Variational Methods in Critical Point Theory and Applications. *J. Func. Anal.* Vol. 14, pp. 349-381.

[12] Ambrosetti A. and Turner R.E.L. (1988) Some Discontinuous Variational Problems. *Differential and Integral Equations*, Vol. 1, pp. 341-349.

[13] Ambrosetti A. and Badiale M. (1989) The Dual Variational Principle and Elliptic Problems with Discontinuous Nonlinearities. *J. Math. Anal. Appl.*, Vol. 140, pp. 363-373.

[14] Ambrosetti A., Calahorrano M. and Dobardo F. (1990) Global Branching for Discontinuous Problems. *Comment. Math. Univ. Carolinae*, Vol. 32, pp. 213-222.

[15] Ambrosetti A. (1992) Critical Points and Nonlinear Variational Problems. *Suppl. Bull Math. France*, Vol. 120(fasc. 2).

[16] Andersson L.E. (1989) A Global Existence Result for Quasistatic Contact Problem with Friction. *LITH-MAT-R-89-00, Linköping Institute of Technology, Linköping, Sweden.*

[17] Andrews K. T., Klarbring A., Wright S. and Shillor M. (1997) A dynamic contact problem with friction and wear. *Int. J. Egng. Sci.*, Vol. 35, pp. 1291-1309.

[18] Ang D.D. and Le V.K. (1995) Contact of a Plate and an Elastic Body. *Z.A.M.M.*, Vol. 75, pp. 115-126.

[19] Antes H. and Panagiotopoulos P.D. (1992) The Boundary Integral Approach to Static and Dynamic Contact Problems. Equality and Inequality Methods. *Birkhäuser Verlag, Basel, Boston.*

[20] Anzellotti G. (1984) Elasticity with Unilateral Constraints on the Stresses. *Univ. Trento, Dip. Matematica, U.T.M. 158.*

[21] Argyris J.H. (1965) Three-Dimensional Anisotropic and Inhomogeneous Elastic Media. Matrix Analysis for Small and Large Displacements. *Ing. Archiv*, Vol 34, pp. 33-55.

[22] Argyris J.H. (1966) Continua and Discontinua, Proc. 1st Conf. Matrix Meth. Struct. Mech. *Wright Patterson Air Force Base, Dayton, Ohio 1965, AFFDL TR*, pp. 66-80.

[23] Aribert J.M. and Abdel Aziz U. (1985) Calcul des Poutres Mixtes jusqu'à l' Etat Ultime avec un Effet de Soulèvement à l' Interface Acier-Béton. *Construction Métallique*, Vol. 4, pp. 3-36.

[24] Aubin J.P. (1977) Applied Abstract Analysis. *J. Wiley and Sons, New York*

[25] Aubin J.P. (1979) Applied Functional Analysis. *J. Wiley and Sons, New York.*

[26] Aubin J.P. and Clarke F.H. (1979) Shadow Prices and Duality for a Class of Optimal Control Problems. *SIAM J. Control Optimization*, Vol. 17, pp. 567-586.

[27] Aubin J.P. and Ekeland I. (1984) Applied Nonlinear Analysis. *Wiley-Inter-Science, N. York.*

[28] Aubin J.P. and Cellina A. (1984) Differential inclusions. *Springer-Verlag, Berlin.*

[29] Aubin J.P. and Frankovska H. (1990) Set-Valued Analysis. *Birkhäuser Verlag, Basel, Boston.*

[30] Awbi B., Rochdi M. and Sofonea M. (2000) Abstract evolution equations related to viscoelastic frictional contact problems. *Z. angew. Math. Phys.*, Vol. 51 pp. 218-235.

[31] Baiocchi C. and Capelo A. (1984) Variational and Quasivariational Inequalities, Applications to Free-Boundary Problems. *John Wiley and Sons, New York.*

[32] Baiocchi C., Gastaldi F. and Tomarelli F. (1986) Some Existence Results on Noncoercive Variational Inequalities. *Annali Scuola Normale Superiore - Pisa, Classe di Scienze*, Serie IV - Vol. XIII, pp. 617-659.

[33] Baiocchi C., Buttazzo G., Gastaldi F. and Tomarelli F. (1988) General Existence Theorems for Unilateral Problems in Continuum Mechanics. *Arch. Rat. Mech. Anal.*, Vol. 100, pp. 149-180.

[34] Baniotopoulos C.C. (1985) Analysis of Structures for "Complete" Constitutive Laws, Doctoral Dissertation. *Scientific Annual of the*

Faculty of Technology of the Aristotle University, Nr. 27 of the Θ' Issue, Thessaloniki.

[35] Baniotopoulos C.C. and Panagiotopoulos P.D. (1990) A Hemivariational Inequality Approach to the Rock Interface Problem. *Engineering Fracture Mechanics*, Vol. 35, pp. 399-407.

[36] Baniotopoulos C.C. (1990) Friction and Contact Effects on Pipelines. *In: Proceedings of the Pipeline Technology Congress, Oostende, Royal Flemish Society of Engineers, Antwerpen, Belgium*

[37] Baniotopoulos C.C. (1991) A Contribution to the Analysis of the Corner Cracking Problem in Masonry Veneer Walls by Means of the Theory of Hemivariational Inequalities. *J. of Theoretical and Applied Mechanics*, Vol. XXII(2), pp. 91-100.

[38] Baniotopoulos C.C. (1995) On the Separation Process in Bolted Steel Splice Plates. *J. Constr. Steel Research*, Vol. 32, pp. 15-35.

[39] Barbu V. (1976) Nonlinear Semigroups and Differential Equations in Banach Spaces. *Noordhoff, Leiden.*

[40] Barbu V. (1982) Necessary Conditions for Multiple Integral Problem in the Calculus of Variations. *Math. Ann.*, Vol. 260, pp. 175-189.

[41] Barbu V. (1989) Mathematical Methods in Optimization of Differential Systems (in Romanian). *Ed. Acad., Bucuresti.*

[42] Baras P. and Hassan J.C. (1977) Compacité de l'Operateur Définissant la Solution d'une Equation d'Evolution Non Homogène. *C.R.A.S.*, t. 284, Ser. A, pp. 799-802.

[43] Bardzokas D, Parton V.Z. and Theocaris P.S. (1989) The Plane Problem of the Theory of Elasticity for an Orthotropic Field with a Defect. *Dokl. Akad. Nauk USSR*, Vol. 309, pp. 1072-1077.

[44] Bardzokas D, Parton V.Z. and Theocaris P.S. (1989) The General Case of the Plane Problem of the Theory of Elasticity for Multiply-Connected Fields. *PMM* Vol. 53, pp. 485-496.

[45] Becker E. and Bürger W. (1975) Kontinuumsmechanik. *B.G. Teubner, Stuttgart.*

[46] Berger M.S. (1967) On von Kármán's Equations and the Buckling of a Thin Elastic Plate. I. The Clamped Plate. *Comm. Pure Appl. math.*, Vol. XX, pp. 687-719.

[47] Berger M.X. and Fife P.C. (1968) Von Kármán's Equations and the Buckling of a Thin Elastic Plate. II. Plate with General Edge Conditions. *Comm. Pure Appl. Math.* Vol. XXI, pp. 227-241.

[48] Berge C. (1959) Espaces Topologiques, Fonctions Multivoques. *Dunod, Paris.*

[49] Berman A. and Plemmons R.J. (1979) Nonnegative Matrices in the Mathematical Sciences. *Academic Press, New York.*

[50] Bershchanskii Y.M. and Meerov M.V. (1983) The Complementarity Problem : Theory and Methods of Solution. *Avtomat. Remote Control,* Vol. 44, pp. 687-710.

[51] Bezine G., Cimetière A., Gelbert J.P. (1985) Unilateral Buckling of Thin Elastic Plates by the Boundary Equation Method. *International Journal for Numerical Methods in Engineering,* Vol. 21, pp. 2189-2199.

[52] Birkhoff G. (1967) Lattice Theory. *American Mathematical Society, New York.*

[53] Bohl P. (1904) Über die Bewegung eines Mechanischen Systems in der Nähe einer Gleichgewichtslage. *J. Reine Angew. Math.,* Vol. 127, pp. 179-276.

[54] Boieri P., Gastaldi F. and Kinderlehrer D. (1987) Existence, Uniqueness and Regularity Results for the Two Bodies Contact Problem. *Appl. Math. Optim.,* Vol. 15, pp. 251-277.

[55] Borwein J.M. (1984) Generalized Linear Complementarity Problems Treated Without Fixed-Point Theory. *Journal of Opt. Theory and Appl.* Vol. 43, pp. 343-357.

[56] Borwein J.M. and Dempster M.A.H. (1989) The Ordered Linear Complementarity Problem. *Mathematics of Operation Research,* Vol. 14, pp. 534-558.

[57] Bouguima S.M. and Boucherif A. (1993) A Discontinuous Semilinear Elliptic Problem without a Growth Condition. *Dyn. Syst. Applic.,* Vol. 2, pp. 183-188.

[58] Bouguima S.M. (1995) A Quasilinear Elliptic Problem with a Discontinuous Nonlinearity. *Nonlinear Analysis, TMA,* Vol. 25 , pp. 1115-1121.

[59] Bourbaki N. (1966) Eléments de Mathématiques - Espaces vectoriels Topologiques. *Hermann, Paris.*

[60] Bressan A. (1980) On Equilibrium in the Presence of Friction I. *Rend. del Circ. Mat. di Palermo*, Vol. XXIX, Serie II, pp. 435-449.

[61] Bressan A. (1981) On Equilibrium in the Presence of Friction II. *Rend. del Circ. Mat. di Palermo*, Vol. XXX, Serie II, pp. 148-156.

[62] Brézis H. (1968) Equations et Inéquations Non Linéaires dans les Espaces Vectoriels en Dualité. *Ann. Inst. Fourier, Grenoble*, Vol. 18, pp. 115-175.

[63] Brézis H. (1972) Problèmes Unilatéraux. *J. Math. Pures et Appl.* Vol. 51, pp. 1-168.

[64] Brézis H. (1973) Opérateurs Maximaux Monotones et Semigroupes de Contractions dans les Espaces de Hilbert. *North-Holland Publ. Co., Amsterdam and American Elsevier Publ. Co., New York.*

[65] Brézis H. and Haraux, A., (1976) Images d'une Somme d'Opérateurs Monotones et Applications. *Israël Journal of Mathematics*, Vol. 23, N°2, pp. 165-186.

[66] Brézis H. and Nirenberg L. (1978) Characterizations of the Ranges of Some Nonlinear Operators and Applications to Boundary Value Problems. *Ann. Scuola Normale Superiore Pisa, Classe di Scienze*, Serie IV, vol. V, N°2, pp. 225-326.

[67] Brézis H. (1992) Analyse Fonctionnelle. Théorie et Applications. *Masson, Paris.*

[68] Brouwer L.E.J. (1912) Uber Abbildung von Mannigfaltigkeiten. *Math. Ann.*, Vol. 71, pp. 97-115.

[69] Browder F.E. (1965) Nonlinear Monotone Operators and Convex Sets. *Bull. Amer. Math. Soc.* Vol. 71, pp. 780-785.

[70] Browder F.E. (1970) Pseudo-Monotone Operators and the Direct Method of the Calculus of Variations. *Arch. Rat. Mech. Anal.*, Vol. 38, pp. 268-277.

[71] Browder F.E. (1976) Nonlinear Operators and Nonlinear Equations of Evolution in Banach Spaces. *Proceedings of Symposia in Pure Mathematics, Vol. XVIII, Part 2, American Mathematical Society.*

[72] Budiansky B. (1994) Theory of Buckling and Postbuckling Behaviour of Elastic Structures. *In: Advances in Applied Mechanics (ed. by Chia-Shun Yih), Academic Press, London, pp. 1-65.*

[73] Burkill J.C. (1963) The Lebesgue Integral. *Cambridge Tracts in Mathematics and Mathematical Physics, N⁰ 40, Cambridge University Press, Cambridge.*

[74] Buttazzo G. (1989) Semicontinuity, Relaxation and Integral Representation in the Calculus of Variations. *Pitman Research Notes in Mathematics Series, N° 207, Longman.*

[75] Buttazzo G. and Tomarelli F. (1991) Compatibility Conditions for Nonlinear Neumann Problems. *Advances in Mathematics,* Vol. 89, pp. 127-143.

[76] Cadivel M., Goeleven D. and Shillor M., (2000) Study of a Unilateral Oscillator with Friction, *Mathematical and Computer Modelling,* Vol. 32, pp. 381-391.

[77] Calvert B.D. and Gupta C.P. (1978) Nonlinear Elliptic Boundary Value Problems in L^p-Spaces and Sums of Ranges of Accretive Operators. *Nonlinear Analysis, TMA,* Vol. 2, pp. 1-26.

[78] Carlsson L.A. and Pipes R.B. (1987) Experimental Characterization of Advanced Composite Materials. *Prentice-Hall Inc., Englewood Cliff, New Jersey.*

[79] Chang K.C. (1976) On the Multiple Solutions of the Elliptic Differential Equations with Discontinuous Nonlinear Terms. *Scientia Sinica,* Vol. XXI, pp. 139-158.

[80] Chang K.C. (1981) Variational Methods for Non-Differentiable Functionals and their Applications to Partial Differential Equations. *J. Math. Anal. Appl.* Vol. 80, pp. 102-129.

[81] Cholet C. (1994) Chocs de Solides Rigides. *Rapport de D.E.A. Laboratoire des Matériaux et des Structures du Génie Civil, UMR 113 LCPC-CNRS, France.*

[82] Ciarlet P.P. and Rabier P. (1980) Les Equations de von Kármán. *Lectures Notes in Mathematics 836, Springer Verlag.*

[83] Ciarlet P.P. and Nečas J. (1985) Unilateral Problems in Nonlinear Three-Dimensional Elasticity. *Arch. Rational Mech. Anal.,* Vol. 87, pp. 319-338.

[84] Ciarlet P.P. and Nečas J. (1987) Injectivity and Self-Contact in Nonlinear Elasticity. *Arch. Rational Mech. Anal.,* Vol. 97, pp. 171-188.

[85] Cimetière A. (1980) Un Problème de Flambement Unilatéral en Théorie des Plaques. *Journal de Mécanique*, Vol. 19, pp. 183-202.

[86] Cimetière A. (1993) Un Résultat de Différentiabilité dans un Problème d'Obstacle pour des Poutres en Flexions. C.R. Acad. Sc. Paris, t. 316, Série I, pp. 749-754.

[87] Clark D.C. (1972) A variant of the Ljusternik-Schnirelmann Theory. *Indiana Univ. Math. J.*, Vol. 22, pp. 65-74.

[88] Clarke F.H. (1973) Necessary Conditions for Nonsmooth Problems in Optimal Control and the Calculus of Variations. *Ph.D. Thesis, University of Washington, Seattle.*

[89] Clarke F.H. (1981) Generalized Gradients of Lipschitz Functionals. *Advances in Math.* Vol. 40, pp. 52-67.

[90] Clarke F.H. (1983) Optimization and Nonsmooth Analysis. *Wiley, New York.*

[91] Cocu M. (1984) Existence of Solutions of Signorini Problems with Friction. *Int. J. Engng. Sci.* Vol. 22, pp. 567-575.

[92] Coffman C.V. (1969) A Minimum-Maximum Principle for a Class of Nonlinear Integral Equations. *J. Analyse Math.*, Vol. 22, pp. 391-419.

[93] Comninou J. (1977) Interface Crack with Friction in the Contact Zone. *J. Appl. Mech.* Vol. 44, pp. 780-781.

[94] Cottreli B. and Rice J.R. (1980) Slightly Curved or Kinked Cracks. *Int. J. Fracture*, Vol. 16, pp. 155-169.

[95] Crisfield M.A. (1986) Snap-Through and Snap-Back Response in Concrete Structures and the Dangers of Under-Integration. *Int. J. Num. Meth. in Eng.* Vol. 22, pp. 751-767.

[96] Crisfield M.A. and Wills J. (1988) Solution Strategies and Softening Materials. *Comp. Meth. Appl. Mech. Eng.* Vol. 66, pp. 267-289.

[97] Cronin J. (1964) Fixed Points and Topological Degree in Nonlinear Analysis. *American Mathematical Society, Providence, Rhode Island.*

[98] Curnier A. (1986) A Theory of Friction. *Int. J. Solids Struct.* Vol. 20, pp. 637-647.

[99] Dautray, R. and Lions, J.L. (1988) Analyse Mathématique et Calcul Numérique pour les Sciences et les Techniques. *Masson Paris.*

[100] Davies P. and Benzeggagh M.L. (1975) Interlaminar Mode-I Fracture Testing. *In: Application of Fracture Mechanics ot Composite Materials (ed. K.Friedrich). Elsevier Publishers B.V, Amsterdam.*

[101] Davison J.B., Kirby P.A. and Nethercot D.A. (1987) Rotational Stiffness Characteristics of Steel Beam to Column Connections. *J. Constr. Steel Res.* Vol. 8, pp. 17- 54.

[102] Demkowicz L. and Oden J.T. (1982) On some Existence and Uniqueness Results in Contact Problems with Nonlocal Friction. *Nonl. Anal.* , Vol. 6, pp. 1075-1093.

[103] Delassus E. (1917) Théorie des Liaisons Finies Unilatérales. *Annales Scientifiques de l'Ecole Normale Supérieure, Paris*

[104] Del Piero G. and Maceri F. (1985) On the Delamination Problem of Two-Layer Plates. *In: Proc. 2nd Meeting on Unilateral Problems in Struct. Analysis, Springer Verlag, Wien, New York, pp. 1-15.*

[105] Demyanov V.F., Stavroulakis G.E., Polyakova L.N. and Panagiotopoulos P.D. (1996) Quasidifferentiability and Nonsmooth Modelling in Mechanics, Engineering and Economics. *Dordrecht, Kluwer Academic.*

[106] Dieudonne J. (1968) Eléments d'Analyse. *Gauthier-Villars.*

[107] Dinca G., Panagiotopoulos P.D. and Pop G. (1995) Inégalités Hémi-variationnelles Semi-Coercives sur des Ensembles Convexes. *C. R. Acad. Sci. Paris*, t. 320, Série I, pp. 1183-1186.

[108] Dinca G., Panagiotopoulos P.D. and Pop G. (1997) An Existence Result on Noncoercive Hemivariational Inequalities, *Ann. Fac. Sci. Toulouse Math. 6, pp. 609-632.*

[109] Do C. (1975) Problèmes de Valeurs Propres pour une Inéquation Variationnelle sur un Cône et Application au Flambement Unilateral d'une Plaque Mince. *C.R. Acad. Sc. Paris*, t. 280, pp. 45-48.

[110] Do C. (1977) Bifurcation Theory for Elastic Plates Subjected to Unilateral Conditions. *Journal of Mathematical Analysis and Applications*, Vol. 60, pp. 435-448.

[111] Du Y. (1991) A Deformation Lemma and Some Critical Point Theorems. *Bull. Austral. Math.*, Vol. 43, pp. 161-168.

[112] Dubourg M.C., Mouwakeh M. and Villechaise B. (1988) Interaction Fissure-Contact. Etude Théorique et Expérimentale. *J. de Mécanique Théor. et Appl.* Vol. 7, pp. 623-643.

[113] Dubourg M.C. (1989) Le Contact Unilateral avec Frottement le long de Fissures de Fatigue dans les Liaisons Mécaniques. *Doct. Dissertation, Institut Nat. Sc. Appl. Lyon*, Nr.891 SAL 0088.

[114] Dugill J.W. (1976) On Stable Progressively Fracturing Solids. *J. of Applied Math. and Phys.* Vol. 27, pp. 423-437.

[115] Dugill J.W. and Rida M.A.M. (1980) Further Considerations of Progressively Fracturing Solids. *ASCE Eng. Mech. Div.* Vol. 106.

[116] Dumont Y., D. Goeleven D., Rochdi M. and Shillor M. (2000) Frictional Contact of a Nonlinear Spring, *Mathematical and Computer Modelling*, Vol. 31, pp. 83-97.

[117] Dumont Y., Goeleven D., Kuttler K.L., Rochdi M. and Shillor M. (2000) A Dynamic Model with Friction and Adhesion with Applications to Rocks *Journal of Mathematical Analysis and Applications*, Vol. 247, pp. 87-109.

[118] Dundurs J. and Comninou M. (1979) Some Consequences of the Inequality Conditions in Contact and Crack Problems. *J. of Elast.* Vol. 9, pp. 131-137.

[119] Dunford N. and Schwartz J.T. (1966) Linear Operators Part I: General Theory. *Interscience Publishers, New York.*

[120] Duvaut G. and Lions J.L. (1971) Un Probléme d'élasticité avec Frottement, *J. de Mécanique* Vol. 10, pp. 409-420.

[121] Duvaut G. and Lions J.L. (1972) Les Inéquations en Mécanique et en Physique. *Dunod, Paris.*

[122] Duvaut. G. (1980) Equilibre d' un Solide Elastique avec Contact Unilateral et Frottement de Coulomb. *C.R. Acad. Sc. Paris*, Vol. 290, pp. 263-265.

[123] Ekeland I. (1974) On the Variational principle. *J. Math. Anal. Appl.*, Vol. 47, pp. 324-353.

[124] Ekeland I. and Teman R. (1976) Convex Analysis and Variational Problems, *Studies in Mathematics and its Applications* . American *Elsevier Publishing Company, INC.-New York.*

[125] Ekeland I. (1979) Nonconvex Minimization Problems. *Bull. (New Series) Amer. Math. Soc.*, Vol. 1, pp. 443-474.

[126] Ekeland I. and Szulkin A. (1989) Minimax Results of Lusternik-Schnirelman Type and Applications. *Les Presses de l'Université de Montréal 107, Montréal.*

[127] Ellaia R. and Hassouni A. (1991) Characterization of Nonsmooth Functions through their Generalized Gradients. *Optimization,* Vol. 22, pp. 401-416.

[128] Evans L.C. and Gariepy R.F. (1992) Measure Theory and Fine Properties of Functions. *Studies in Advanced Mathematics, CRC Press, Boca Raton.*

[129] Fan K. (1972) A Minimax Inequality Theorem and Applications. *In: Inequalities III (O. Shisha, ed.), Academic Press, New York.*

[130] Fichera G. (1963) The Signorini Elastostatics Problem with Ambiguous Boundary Conditions. *Proc. Int. Conf. Application of the Theory of Functions in Continuum Mechanics,* Vol. I, Tbilisi.

[131] Fichera G. (1972) Boundary Value Problems in Elasticity with Unilateral Constraints. *In: Encyclopedia of Physics (ed. by S.Flügge), Springer-Verlag, Berlin* Vol. VI a/2.

[132] Filippov A.F. (1967) Classical Solutions of Differential Equations with Multivalued Right-Hand Side. *SIAM J. Control,* Vol. 5, pp. 609-621.

[133] Filippov A.F. (1988) Differential Equations with Discontinuous Righthand Sides. *Kluwer Academic Publishers, The Netherlands.*

[134] Floegl H. and Mang H.A. (1982) Tension Stiffening Concept Based on Bond Slip. *ASCE (ST 12)* Vol. 108, pp. 2681-2701.

[135] Francu J. (1971) On Signorini Problem for von Kármán Equations. *Aplicace Matematiky,* Vol. 24, pp. 355-371.

[136] Frémond M. (1982) Adhérence des Solides. *C. R. Acad. Sc. Paris,* Série II, t. 295, pp. 769-772.

[137] Frémond M. (1983) Conditions Unilatérales et Non Linéarité en Calcul à la Rupture. *Matem. Aplicadá e Computational Brazil,* Vol. 2, pp. 237- 256.

[138] Frémond M. (1984) Sur la Fissuration. *C. R. Acad. Sc. Paris Série II,* Vol. 299, pp. 487-490.

[139] Frémond M. (1987) Contact Unilatéral avec Adhérence: Une Théorie du Premier Gradient. *Unilateral Problems in Structural Analysis - 2, (ed. by Del Pierro G. and Maceri F.), CISM Courses and Lectures, Springer-Verlag, N.York, Wien*, Vol. 304, pp. 33-45.

[140] Frémond M. (1988) Contact with Adhesion. *In Topics in Nonsmooth Mechanics (ed by J.J.Moreau, P.D.Panagiotopoulos, G. Strang), Birkhäuser Verlag, Boston, Basel, Berlin*, pp. 157-185.

[141] Frémond M. (1993) Endommagement et Principe des Puissances Virtuelles. *C.R. Acad. Sci. Paris*, t. 317, Série II, pp. 857-864.

[142] Frémond M. (1995) Rigid Bodies Collisions. *Physics Letters A*, Vol. 204, pp. 33-41.

[143] Friedman A. (1982) Variational Principles and Free-Boundary Problems. *Interscience, New York*.

[144] Fundo M. (1997) An Existence Result on a General Type of Hemivariational Inequalities. *Appl. Anal., 65, pp. 373-394.*

[145] Fundo M. (1998) Hemivariational Inequalities in Subspaces of $L^p(\Omega)$ $(p \geq 3)$. *Nonlinear Anal., 33, pp. 341-350.*

[146] Gastaldi F. and Tomarelli F. (1987) Some Remarks on Nonlinear and Noncoercive Variational Inequalities. *Bollettino U.M.I.*, Vol. 1-B (7), pp. 143-165.

[147] Gastaldi F. (1988) Remarks on a Noncoercive Contact Problem with Friction in Elastostatics. *Istituto Anal. Num. C.N.R, Pavia Publ.*, N° 649.

[148] Gastaldi F. and Martins J.A.C. (1988) A Noncoercive Steady-Sliding Problem with Friction. *Istituto Anal. Num. C.N.R. Pavia Publ.*, N° 650.

[149] Gattesco N. (1990) Long-Span Steel and Concrete Beams with Partial Shear Connection. *Studi e Ricerche* Vol. 12, pp. 243-266.

[150] Germain P. (1973) Cours de Mécanique des Milieux Continus I. *Masson, Paris*.

[151] Germain P. (1973) La Méthode de Puissances Virtuelles en Mécanique des Milieux Continus, 1ère Partie. Théorie du Second Gradient. *J. de Mécanique*, Vol. 12, pp. 235-274.

[152] Germain P. (1973) The Method of Virtual Power in Continuum Mechanics. Part 2: Microstructure. *SIAM J. Appl. Math.* Vol. 25, pp. 556-575.

[153] Gerstle K.H. (1988) Effect of Connections on Frames. *J. Constr. Steel Res.* Vol. 10, pp. 241-268.

[154] Gianessi F. (1991) Complementarity Systems and some Applications in the Fields of Structural Engineering and of Equilibrium on a Network. *Preprint, Un. of Pisa.*

[155] Girkmann K. (1963) Flächentragwerke. *Springer Verlag, Wien.*

[156] Glowinski R., Lions J.L. and Tremolieres R. (1981) Numerical Analysis of Variational Inequalities. *Studies in Mathematics and its Applications, Vol. 8, North-Holland-Elsevier, Amsterdam, New-York.*

[157] Goeleven D. (1993) On the Solvability of Linear Noncoercive Variational Inequalities in Separable Hilbert Spaces. *Journal of Optimization Theory and Applications,* Vol. 79, pp. 493-511.

[158] Goeleven D. (1993) A Note on Palais-Smale Condition in the Sense of Szulkin. *Differential and Integral Equations,* Vol. 6, pp. 1041-1043.

[159] Goeleven D., Nguyen V.H. and Willem M. (1994) Existence and Multiplicity Results for Semicoercive Unilateral Problems. *Bull. of the Australian Math. Society,* Vol. 49, pp. 489-498.

[160] Goeleven D. and Nguyen V.H. (1994) On the One-Dimensional Nonlinear Elastohydrodynamic Lubrication Problem. *Bull. of the Australian Mathematical Society,* Vol. 50, pp. 353-372.

[161] Goeleven D. and Théra M. (1994) Some Global Results for Nonlinear Eigenvalue Problems Governed by a Variational Inequality. *Rend. del Sem. Matematico e Fisico di Milano,* Vol. LXI, pp. 185-198.

[162] Goeleven D. and Théra M. (1995) Semicoercive Variational Hemivariational Inequalities. *Journal of Global Optimization,* Vol. 6, pp. 367-381.

[163] Goeleven D. (1996) On a Class of Hemivariational Inequalities Involving Hemicontinuous Monotone Operators. *Numerical Functional Analysis and Optimization,* Vol. 17, pp. 77-92.

[164] Goeleven D. and Motreanu D. (1996) Eigenvalue and Dynamic Problems for Variational and Hemivariational Inequalities. *Commun. Appl. Nonlinear Anal.,* Vol. 3 , pp. 1-21.

[165] Goeleven D. (1996) Noncoercive Variational Problems and Related Results. *Pitman Research Notes in Mathematics Series, N⁰357, Longman.*

[166] Goeleven D. (1996) A Uniqueness Theorem for the Generalized Order Linear Complementarity Problem. *Linear Algebra and its Applications*, Vol. 235, pp. 221-227.

[167] Goeleven D. (1996) Noncoercive Hemivariational Inequalities and its Applications in Nonconvex Unilateral Mechanics. *Applications of Mathematics*, Vol. 41, pp. 203-229.

[168] Goeleven D. (1996) Noncoercive Hemivariational Inequality Approach to Constrained Problems for Star-Shaped Admissible Sets. *Journal of Global Optimization*, Vol. 9, pp. 121-140.

[169] Goeleven D., Stavroulakis G.E. and Panagiotopoulos P.D. (1996) Solvability Theory for a Class of Hemivariational Inequalities Involving Copositive Plus Matrices, Applications in Robotics. *Mathematical Programming*, Vol. 75, pp. 441-465.

[170] Goeleven D. and Motreanu D. (1996) Szulkin's Type Minimax Methods in Unilateral Problems. *Advances in Functional Analysis (In the Honour of Late Professor P.K. Kamthan) (ed. P.K. Jain), John Wiley & Sons.*

[171] Goeleven D., Panagiotopoulos P.D., Lebeau C. and Plotnikova G. (1997) Inequality Forms of d'Alembert Principle in Mechanics of Systems with Holonomic Unilateral Constraints. *Z.A.M.M.*, Vol. 77, pp. 483-501.

[172] Goeleven D. and Motreanu D. (1997) A Degree Theoretic Approach for the Study of Problems in Variational-Hemivariational Inequalities. *Differential and Integral Equations*, Vol. 10, pp. 893-904.

[173] Goeleven D., Stavroulakis G.E, Salmon G. and Panagiotopoulos P.D. (1997) Solvability Theory and Projection Methods for a Class of Singular Variational Inequalities. Elastostatic Unilateral Contact Applications. *Journal of Optimization Theory and Applications*, Vol. 95, pp. 263-293.

[174] (1997) Goeleven D. and Panagiotopoulos P.D. On a Class of Noncoercive Hemivariational Inequalities Arising in Nonlinear Elasticity. *Applicable Analysis*, Vol. 66, pp. 1-37.

[175] Goeleven D. and Motreanu D. (1999) A Mathematical Approach to the Rock Interface Problem. *Journal of Elasticity* Vol. 55 pp. 79-97.

[176] Goeleven D. and Motreanu D. (2000) On the Solvability of Linear Variational Inequalities via Relaxed Complementarity Problems. *Communications in Applied Analysis*, Vol. 4, pp. 533-546.

[177] Gol'dshtein R.V. and Spector A.A. (1984) Variational Method of Investigation of Three-Dimensional Mixed Problems of a Plane Cut in an Elastic Medium in the Presence of Slip and Adhesion of its Surfaces. *PMM U.S.S.R.* , Vol. 47, pp. 232-239.

[178] Gol'dshtein R.V. and Spector A.A. (1988) Variational Methods of Solution and Analysis of Spatial Contact and Mixed Problems with Friction. *In: Mechanics of Deformable Solids (ed. by A. Yu. Ishlinski) Allerton Press, N.York.*

[179] Gowda M.S. and Seidman J. (1990) Generalized Linear Complementarity Problems. *Mathematical Programming*, Vol. 46, pp. 329-340.

[180] Graves L.M. (1950) Some Mappings Theorems. *Duke Math. J.*, Vol 17, pp. 111-114.

[181] Granas A. (1985) Méthodes Topologiques en Analyse Non Linéaire. *Les Presses de l'Université de Montréal, Montréal.*

[182] Green A.E. and Naghdi P.M. (1965) A General Theory of an Elastic-Plastic Continuum. *Arch. Rat. Mech. Anal.*, Vol. 18, pp. 251-281.

[183] Grossinho M.R. and Tersian S. (In Press) Existence Results for Semilinear Fourth Order Equations with Discontinuous Nonlinearities, To appear in Z.A.M.M..

[184] Halphen B. and Son N.Q. (1974) Plastic and Viscoplastic Materials with Generalized Potential. *Mech. Res. Comm.* Vol. 1, pp. 43-47.

[185] Halphen B. and Son N.Q. (1975) Sur les Matériaux Standards Généralisés. *J. de Mécanique,* Vol. 14, pp. 39-63.

[186] Hartman P. and Stampacchia G. (1966) On Some Non-Linear Elliptic Differential-Functional Equations. *Acta Math.*, Vol. 115, pp. 271-310.

[187] Haslinger J. and Panagiotopoulos P.D. (1984) The Reciprocal Variational Approach to the Signorini Problem with Friction. Approximation Results. *Proc. Royal Soc. of Edinburgh,* Vol. 98A, pp. 365-383.

[188] Hess P. (1974) On Semi-Coercive Nonlinear Problems. *Indiana University Mathematical Journal,* Vol. 23, pp. 645-654.

[189] Hlavacek I., Haslinger J., Nečas J. and Lovisek J. (1982) Solutions of Variational Inequalities in Mechanics. *Applied Mathematical Sciences 66, Springer Verlag, New York* .

[190] Hoffmann K.H. and Sprekels J. (eds). (1990) Free Boundary Problems: Theory and Applications I,II. *Pitman Res. Notes in Math. Vol. 185, Longman Scient. and Techn., Essex.*

[191] Hoffmann K.H. and Sprekels J. (eds). (1990) Free Boundary Value Problems. *ISNM Vol 95, Birkhäuser Verlag, Basel.*

[192] Horn R.A. and Johnson C.R. (1991) Topics in Matrix Analysis. *Cambridge University Press, Cambridge.*

[193] Hu L.T. (1959) Homotopy Theory. *Academic Press, New York.*

[194] Hu B. (1990) A Quasi-Variational Inequality Arising in Elastohydrodynamics. *SIAM J. Math. anal.,* Vol. 21, pp. 18-36.

[195] Hult J. and Travniček L. (1983) Carrying capacity of Fibre Bundles with Varying Strength and Stiffness. *J. Méc. Théor. et Appl.,* Vol. 2, pp. 643-657.

[196] Ioffe A.D. and Levin V.L. (1972) Subdifferentials of Convex Functions. *Trans. Moscow Math. Soc.,* Vol 26, pp. 1-72.

[197] Isac G. (1985) Problème de Complémentarité (en Dimension Infinie). *Publications du Département de Mathématiques et Informatique de l'Université de Limoges.*

[198] Isac G. (1990) A Special Variational Inequality and the Implicit Complementary Problem. *Journal of the Faculty of Science, the University of Tokyo,* Vol. 37, (Sec. IA), pp. 109-127.

[199] Isac G. and Kostreva M. (1991) The Generalized Order Complementarity Problem. *J.O.T.A.,* Vol. 71, pp. 517-534.

[200] Isac G. and Goeleven D. (1993) Existence Theorems for the Implicit Complementarity Problem. *Internat. J. Math. Sci.,* Vol. 16, $N^0$1, pp. 67-74.

[201] Isac G. and Goeleven D. (1993) The Implicit General Order Complementarity Problem, Models and Iterative Methods. *Annals of Operation Research,* Vol. 44, pp. 63-92.

[202] Isac G. (1993) Complementarity Problems. *Lecture Notes in Mathematics 1528, Springer-Verlag, Heildelberg*

[203] Istrătescu V. I. (1981) Fixed Point Theory. An Introduction. *Reidel Publ. Co., Dordrecht.*

[204] Jarusek J. (1983) Contact Problems with Bounded Friction. Coercive Case. *Czech. Math. J.*, Vol. 33, pp. 254-278.

[205] Jarusek J. Dynamic Contact Problems with Given Friction for Viscoelastic Bodies. *Czech. Math. J.*, Vol. 46, pp.475-487.

[206] Jean M. and Pratt E. (1985) A System of Rigid Bodies with Dry Friction. *Int. J. Eng. Sci.*, Vol. 23, pp. 497-513.

[207] Jean M. and Moreau J.J. (1987) Dynamics in the Presence of Unilateral Contacts and Dry Friction; a Numerical Approach. *In: Unilateral Problems in Structural Analysis 2 (ed. G. Del Piero and F.Maceri), CISM Courses and Lectures N^0 304, pp. 151-196, Springer-Verlag, Wien*

[208] John O. (1977) On Signorini Problem for von Kármán Equations. *Aplikace Matematiky*, Vol. 22, pp. 52-68.

[209] Johnson R.P. and May I.M. (1975) Partial-Interaction Design of Composite Beams. *The Struct. Eng.* Vol. 53, pp. 305-311.

[210] Jonsson A. and Wallin H. (1984) Function Spaces on Subsets of $\mathbb{R}^3$. *Math. Rep. Vol. 2, Harwood Acad. Publ., Chur, London.*

[211] Jones R. (1975) Mechanics of Composite Materials. *McGraw Hill, New York.*

[212] Kalker J.J. (1988) The Quasistatic Contact Problem with Friction for Three Dimensional Elastic Bodies. *J. Méc. Theor. et Appl.*, Vol. 7 (Special issue), pp. 55-66.

[213] Kalker J.J. (1988) Contact Mechanical Algorithms. *Comm. in Applied Num. Methods*, Vol. 4, pp. 25-32.

[214] Kalker J.J. (1990) Three Dimensional Elastic Bodies in Rolling Contact. *Kluwer Acad. Publ., Dordrecht.*

[215] Kaplan A.A. and Tichatschke R. (1994) Stable Methods for Ill-Posed Variational Problems. *Academic Verlag, Berlin.*

[216] Karamanlis I. (1991) Buckling Problems in Composite von Kármán Plates. *Doct. Thesis, Aristotle University Dept. of Civil Eng.*

[217] Karamanlis H.N. and Panagiotopoulos P.D. (1993) The Eigenvalue Problems in Hemivariational Inequalities and its Applications to Composite Plates. *Journal Mech. Behaviour of Materials, Vol. 3, pp.151-175.*

[218] Kavian O. (1993) Introduction à la Théorie des Points Critiques et Applications aux Problèmes Elliptiques, *Springer-Verlag, Paris.*

[219] Kelley J.L. Topologia General. *Eudeba, Buenos Aires (English ed. (1955), D. Van Nostrand, Princeton).*

[220] Kikuchi N. and Oden J.T. (1988) Contact Problems in Elasticity. A Study of Variational Inequalities and Finite Element Methods. *SIAM Publ., Philadelphia.*

[221] Kim S.J. and Oden J.T. (1984) Generalized Potentials in Finite Elastoplasticity. *Int. J. Engng. Sci., Vol. 22, pp. 1235-1257.*

[222] Kim S.J. and Oden J.T. (1985) Generalized Flow Potentials in Finite Elastoplasticity - II. Examples. *Int. J. Engng. Sci. Vol. 23, pp. 515-530.*

[223] Kinderleher D. and Stampacchia G. (1980) An Introduction to Variational Inequalities. *Academic Press, New York.*

[224] Kinderlehrer D. (1981) Remarks About Signorini's Problem in Linear Elasticity. *Ann. Scuola Norm. Sup. Pisa Cl. Sci., Vol. IV (8), pp. 605-645.*

[225] Klarbring A. (1984) Contact Problems with Friction Using a Finite - Dimensional Description and the Theory of Linear Complementarity. *Linköping Studies in Science and Technology, Thesis No. 20, Linköping Institute of Technology, Linköping, Sweden.*

[226] Klarbring A. (1987) Contact Problems with Friction by Linear Complementarity. *In: Unilateral Problems in Structural Analysis, 2 (ed. by Del Piero G., Maceri F.). CISM Courses and Lectures, N^0 304, Springer Verlag, Wien, N.York.*

[227] Knaster B., Kuratowski C. and Mazurkiewicz S. (1929) Ein Beweis des Fixpunktktsatzes für n-dimensionales Simplexe. *Fund. Math., Vol. 14, pp. 132-137.*

[228] Koltsakis E.K. (1991) Theoretical and Numerical Study of Structures with Nonmonotone Boundary Conditions. *Application to Adhesion Joints, Doct. Dissertation, Dept. of Civil Eng., Aristotle University, Thessaloniki.*

[229] Köthe G. (1969) Topological Vector Spaces I. *Springer-Verlag.*

[230] Kozlov V.V. and Treshchëv D.V. (1991) Billiards. A Genetic Introduction to the Dynamics of Systems with Impacts. *Amer. Math. Soc. (Transl. of Math. Monographs Vol. 89) Providence.*

[231] Krasnoselskii M.A. (1964) Topological Methods in the Theory of Nonlinear Integral Equations. *Pergamon Press, Oxford*

[232] Kubrusly R.S. and Oden J.T. (1981) Nonlinear Eigenvalue Problems Characterized by Variational Inequalities with Applications to the Postbuckling Analysis of Unilaterally-Supported Plates. *Nonlinear Analysis, TMA*, Vol. 5, pp. 1265-1284.

[233] Kubrusly R.S. (1982) On the Existence of Post-Buckling Solutions of Shallow Shells Under A Certain Unilateral Constraint. *Int. Egng Sci.*, Vol. 20, pp. 93-99.

[234] Kuczma M.S. and Stein E. (1994) On Nonconvex Problems in the Theory of Plasticity. *Arch. Mech.*, Vol. 46, pp. 505-529.

[235] Kufner A., John O. and Fučik S. (1977) Function Spaces. *Noordhoff International Publ, Leyden.*

[236] Kwak B.M. and Lee S.S. (1988) A Complementarity Problem Formulation for Two-Dimensional Frictional Contact Problem. *Comp. Struct.*, Vol. 28, pp. 469-480.

[237] Kwak B.M. (1991) Complementarity Problem Formulation of Three Dimensional Frictional Contact. *ASME J. Applied Mech.*, Vol. 58, pp. 134-140.

[238] Lanczos C. (1966) The Variational Principles of Mechanics. *University of Toronto Press, Toronto.*

[239] Lebourg G. (1975) Valeur Moyenne pour Gradient Généralisé. *C. R. Acad. Sci. Paris*, Vol. 281, pp. 795-797.

[240] Lebeau C. (1996) Une Etude en Théorie des Percussions et Liaisons Unilatérales. *Mémoire de D.E.A., U.C.L., Belgium*

[241] Lefter C. Critical Point Theorems for Lower Semicontinuous Functions. *Unpublished Manuscript, University of Iasi.*

[242] Lefter C. and Motreanu D. (1992) Critical Point Methods in Nonlinear Eigenvalue Problems with Discontinuities. *In: International Series of Numerical Mathematics, Vol. 107, Birkhäuser Verlag.*

[243] Léné F. (1973) Sur les Matériaux Elastiques à Energie de Déformation Non Quadratique. *Thèse de 3ème cycle, Université Paris VI.*

[244] Léné F. (1974) Sur les Matériaux Elastiques à Energie de Déformation Non Quadratique. *J. de Mécanique,* Vol. 13, pp. 499-534.

[245] Leray J. and Schauder J. (1934) Topologie et Equations Fonctionnelles. *Ann. Sci. Ecole Norm. Sup. Sér.,* Vol. 3, pp. 45-78.

[246] Liolios A.A. (1986) A Linear Complementarity Approach for the Signorini Problem with Friction. *Z.A.M.M.,* Vol. 66, pp. 349-352.

[247] Liolios A.A. (1987) Upper and Lower Solution Estimates in Unilateral Viscoelastodynamics. *Acta Mech.,* Vol. 66, pp. 275-278.

[248] Liolios A.A. (1988) Seismic Interaction Between Adjacent Structures: A Linear Complementarity Approach for the Unilateral Elastoplastic Softening Contact with Friction. *In: Structural Dynamics and Earthquake Engineering, (ed. by Kounadis A.N. and Krätzig W.B.) Athens.*

[249] Liolios A.A. (1989) A Linear Complementarity Approach for the Non-Convex Dynamic Problem of Unilateral Contact with Friction Between Adjacent Structures. *Z.A.M.M.,* Vol. 69, pp. 420-422.

[250] Lions J.L. and Stampacchia G. (1967) Variational Inequalities. *Comm. Pure Applied Math.,* Vol. XX, pp. 493-519.

[251] Lions J.L. (1969) Quelques Méthodes de Résolution des Problèmes aux Limites Non Linéaires. *Dunod/Gauthier-Villars, Paris.*

[252] Lions J.L. and Magenes E. (1972) Non-homogenous Boundary Value Problems and Applications. *Spinger-Verlag, New York.*

[253] Ljusternik L. and Schnirelmann L. (1934) Méthodes Topologiques dans les Problèmes Variationnels. *Herman and Cie, Paris.*

[254] Lloyd N.G. (1978) Degree Theory. *Cambridge Tracts in Mathematics 73, Cambridge University Press, Cambridge.*

[255] Lötstedt P. (1982) Mechanical Systems of Rigid Bodies Subject to Unilateral Constraints. *SIAM J. Appl. Math.,* Vol. 42, pp. 281-296.

[256] Lui E.M. and Chen W.F. (1987) Steel Frame Analysis with Flexible Joints. *J. Constr. Steel Res.,* Vol. 8, pp. 161-202.

[257] Lupo D. (1989) A Bifurcation Result for a Dirichlet Problem with Discontinuous Nonlinearity. *Rendicondi del Circolo Matematico di Palermo,* Vol. II(XXXVIII), pp. 305-318.

[258] Maier G. (1968) A Quadratic Programming Approach for Certain Classes of Nonlinear Structural Problems. *Meccanica,* Vol. 3, pp. 121-130.

[259] Maier G. (1973) Mathematical Programming Methods in Structural Analysis. *In Variational Methods in Engineering (ed. by C.A. Brebbia and H. Tottenham), Vol. II. Southampton Univ. Press, Southampton.*

[260] Martins J.A.C. (1986) Dynamic Frictional Contact Problems Involving Metallic Bodies. *Ph.D.Dissertation, University of Texas at Austin.*

[261] Martins J.A.C. and Oden J.T. (1987) Existence and Uniqueness Results for Dynamic Contact Problems with Nonlinear Normal and Friction Interface Laws. *Nonlinear Anal.,* Vol. 11, pp. 407-428.

[262] Marques M.M.D.P. (1988) Inclusões Differenciais e Choques inelasticos. *Doct. Dissertation, Faculty of Sciences, Univ. of Lisbon.*

[263] Marquez M.D.P.M. (1994) An Existence, Uniqueness and Regularity Study of the Dynamics of Systems with One-Dimensional Friction. *Eur. J. Mech., A/Solids,* Vol. 13, pp. 277-306.

[264] Maugin G.A. (1980) The Method of Virtual Power in Continuum Mechanics. Application to Coupled Fields. *Acta Meccanica,* Vol. 35, pp. 1-70.

[265] Mawhin J. and Willem M. (1989) Critical Point Theory and Hamiltonian Systems. *Spinger Verlag, New York.*

[266] Michel P. and Penot J.P. (1992) A Generalized Derivative for Calm and Stable Functions, *Diff. and Int. Equ.,* Vol. 5, pp. 433-454.

[267] Michalowski R. and Mroz Z. (1978) Associated and Non-Associated Sliding Rules in Contact Friction Problems. *Arch. of Mech. (Arch. Mech. Stosowanej),* Vol. 30, pp. 259-276.

[268] Miettinen M. (1993) Approximation of Hemivariational Inequalities and Optimal Control Problems. *Doct. Dissertation, Dept. of Mathematics, Univ. of Jyväskylä, Rep. 59, Jyväskyla, Finland.*

[269] Minty J.G. (1967) On the Generalization of a Direct Method of the Calculus of Variations. *Bull. Amer. Math. Soc.,* Vol. 73, pp. 315-321.

[270] Mironescu P. and Radulescu V. (1995) A Multiplicity Theorem for Locally Lipschitz Periodic Functionals. *J. Math. Anal. Appl.,* Vol. 195, pp. 621-637.

[271] Mistakidis E. (1992) Theoretical and Numerical Study of Structures with Nonmonotone Boundary and Constitutive laws. Algorithms and Applications. *Doct. Dissertation, Dept. of Civil Eng. Aristotle Univ. Thessaloniki.*

[272] Mistakidis E., Panagiotopoulos P.D. and Panagouli O.K. (1993) Fractal Surfaces and Interfaces in Structures. *Methods and Algorithms. Chaos, Solitons and Fractals*, Vol. 2, pp. 551-574.

[273] Mitsopoulou E. (1983) Unilateral Contact, Dynamic Analysis of Beams by a Time-Stepping Quadratic Programming Procedure. *Meccanica*, Vol. 18, pp. 254-265.

[274] Mitsopoulou E.N. and Doudoumis I.N. (1987) A Contribution to the Analysis of Unilateral Contact Problems with Friction. *Solid Mech. Arch.*, Vol. 12, pp. 165-186.

[275] Mitsopoulou E., Panagiotopoulos P.D. and Zervas P.A. (1991) Dynamic Boundary Integral "Equation" Method for Unilateral Contact Problems. *Eng. Anal. with Bound. Elements*, Vol. 8, pp. 192-199.

[276] Moffat U.R and Dowling P.J. (1978) The Longitudinal Bending Behaviour of Composite Box Girder Bridges having Incomplete Interaction. *The Structural Engineer* Vol. 56B, pp. 53-60.

[277] Moré J.J. (1974) Classes of Functions and Feasibility Conditions in Nonlinear Complementarity Problems. *SIAM Review*, Vol. 17, pp. 1-16.

[278] Moreau J.J. (1967) Fonctionnelles Convexes. Séminaire sur les Equations aux Dérivées Partielles. *Collège de France, Paris.*

[279] Moreau J.J. (1968) La Notion du Surpotentiel et les Liaisons Unilatérales on Elastostatique. *C.R. Acad. Sci. Paris*, Vol. 167A, pp. 954-957.

[280] Moreau J.J. (1970) Sur les Lois de Frottement, de Plasticité et de Viscosité. *C.R. Acad. Sc. Paris*, Vol. 271A, pp. 608-611.

[281] Moreau J.J. (1971) Mécanique Classique. *Masson & Cie, Paris.*

[282] Moreau J.J. (1983) Liaisons Unilatérales sans Frottement et Chocs Inélastiques. *C. R. Acad. Sc. Paris*, t. 296, pp. 1473-1476.

[283] Moreau J.J. (1986) Une Formulation du Contact Frottement Sec. Application au Calcul Numérique. *C.R. Acad. Sci. Paris*, Sér. II, Vol. 302, pp. 799-801.

[284] Moreau J.J. and Panagiotopoulos P.D. (eds) (1988) Nonsmooth Mechanics and Applications. *CISM, Vol. 302, Springer Verlag, Wien.*

[285] Moreau J.J., Panagiotopoulos P.D. and Strang G. (eds) (1988) Topics in Nonsmooth Mechanics. *Birkhäuser Verlag, Basel, Boston.*

[286] Moreau J.J. (1988) Unilateral Contact and Dry Friction in Finite Freedom Dynamics. *In: Nonsmooth Mechanics and Applications (ed. by J.J.Moreau and P.D.Panagiotopoulos), CISM, Vol. 302, Springer Verlag, Wien, N.York.*

[287] Morosanu G. (1988) Nonlinear Evolution Equation and Applications. *D. Reidel Publishing Company, Dordrecht.*

[288] Mosco U. (1969) Convergence of Convex Sets and of Solutions of Variational Inequalities. *Advances in Mathematics*, Vol. 3, pp. 510-585.

[289] Mosco U. (1976) Implicit Variational Problems and Quasi-Variational Inequalities. *In Nonlinear Operators and the Calculus of Variations, Lecture Notes in Mathematics 543, pp. 85-156, Springer-Verlag, Berlin.*

[290] Moser K. (1992) Faserkunststoffverbund. *VDI Verlag, Düsseldorf.*

[291] Motreanu D. and Pavel N.H. (1982) Quasi-tangent Vectors in Flow-Invariance and Optimization Problems on Banach Manifolds. *J. Math. Anal. Appl.*, Vol. 88, pp. 116-132.

[292] Motreanu D. (1986) Existence for Minimization with Nonconvex Constraints. *J. Math. Anal. Appl.*, Vol. 117, pp. 128-137.

[293] Motreanu D. and Panagiotopoulos P.D. (1993) Hysteresis: The Eigenvalue Problem for Hemivariational Inequalities. *In: Models of Hysteresis (ed. by A. Visintin) Pitman Research Notes in Mathematics, Longman, Harlow.*

[294] Motreanu D. (1995) Existence of Critical Points in a General Setting. *Set-Valued Anal.*, Vol. 3, pp. 295-305.

[295] Motreanu D. and Naniewicz Z. (1996) Discontinuous Semilinear Problems in Vector-Valued Function Spaces. *Differ. Int. Equations*, Vol. 9, pp. 581-598.

[296] Motreanu D. and Panagiotopoulos P.D. (1999) Minimax Theorems and Qualitative Properties of the Solutions of Hemivariational Inequalities. *Kluwer, Boston*

[297] Motreanu D. Nonlinear Eigenvalue Problems with Constraints. (In Press) *To appear in Top. Math. Nonlin. Anal.*

[298] Mroz Z. (1973) Mathematical Models of Inelastic Material Behaviour. *Univ. of Waterloo Press, Waterloo.*

[299] Mroz Z. and Stupkiewicz S. (1992) Constitutive Modelling of Slip and Wear in Elastic, Frictional Contact. *In: Proc. Contact Mech. Int. Symp. (ed. by A. Curnier Presses Polyt. et Univers. Romandes, Lausanne,* pp. 133-156.

[300] Munkres J.R. (1966) Elementary Differential Topology. *Princeton University Press, Princeton.*

[301] Murty K.G. (1988) Linear Complementarity. *Linear and Nonlinear Programming. Heldermann Verlag, Berlin.*

[302] Nagumo M. (1951) A Theory of Degree of Mapping Based on Infinitesimal Analysis. *Amer. J. Math.*, Vol. 73, pp. 485-496.

[303] Naniewicz Z. (1989) On Some Nonconvex Variational Problems Related to Hemivariational Inequalities. *Nonlin. Anal.*, Vol. 13, pp. 87-100.

[304] Naniewicz Z. and Wozniak C.Z. (1989) On the Quasi-Stationary Models of Debonding Processes in Layered Composites. *Ing. Archiv,* Vol. 60, pp. 31-40.

[305] Naniewicz, Z. (1989) On Some Nonmonotone Subdifferential Boundary Conditions in Elastostatics. *Ingenieur-Archiv.,* Vol. 60, pp. 31-40.

[306] Naniewicz, Z. (1992) On the Pseudomonotonicity of Generalized Gradients of Nonconvex Functions. *Applicable Analysis,* Vol. 47, pp. 151-172.

[307] Naniewicz Z. (1992) On Some Noncoercive Problems Related to Delamination in Layered Composites. *In: Nonsmooth Optimization, Methods and Application (ed. F. Giannessi), Gordon and Breach Science Publishers, Switzerland, USA.*

[308] Naniewicz Z. (1993) On the Existence of Solutions to the Continuum Model of Delamination. *Nonl. Anal. TMA,* Vol. 20, pp. 481-507

[309] Naniewicz Z. (1994) Hemivariational Inequality Approach to Constrained Problems for Admissible Sets. *Journal of Optimization, Theory and Applications,* Vol. 83, pp. 97-112.

[310] Naniewicz Z. (1994) Hemivariational Inequalities with Functions Fulfilling Directional Growth Condition. *Applicable Analysis*, Vol. 55, pp. 259-285.

[311] Naniewicz Z. and Panagiotopoulos P.D. (1995) The mathematical Theory of Hemivariational Inequalities and Applications. *Marcel Dekker, N. York.*

[312] Naniewicz Z. Hemivariational Inequalities as Necessary Conditions for Optimality for a Class of Nonsmooth Nonconvex Functionals. *Preprint RW 96-07(19), Institute for Applied Mathematics and Mechanics, Warsaw University 1996.*

[313] Nečas J. (1967) Méthodes Directes en Théorie des Equations Elliptiques. *Masson, Paris.*

[314] Nečas J., Jarusek J. and Haslinger J. (1980) On the Solution of the Variational Inequality to the Signorini Problem with Small Friction. *Bulletino U.M.I.*, Vol. 17B, pp. 796-811.

[315] Nethercot D.A. and Chen W.F. (1988) Effects of Connections on Columns. *J. Constr. Steel Res.*, Vol. 10, pp. 201-240.

[316] Nguyen Q.S. (1995) Stabilité des Structures Elastiques. *Mathématiques & Applications 18, Springer Verlag, Berlin.*

[317] Nguyen V.D. (1989) Constructing Stable Grasps. *Int. J. of Robot. Res.*, Vol. 8, pp. 26-36.

[318] Oden J.T. and Pires E. (1981) Contact Problems in Elastostatics with Non-Local Friction Laws. *TICOM Report 81-12, University of Texas at Austin.*

[319] Oden J.T. and Martins J.A.C. (1985) Models and Computational Methods for Dynamic Friction Phenomena. *Comp. Meth. Appl. Mech. Eng.*, Vol. 52, pp. 527-634.

[320] Oden J.T. and Wu S.R. (1985) Existence of Solutions to the Reynolds Equation of Elastohydrodynamic Lubrication. *Int. J. Engng Sci.*, Vol. 23, pp. 207-215.

[321] Oehlers D.J. and Johnson R.D. (1987) The Strength of Stud Shear Connections in Composite Beams. *The Structural Engineer*, Vol. 65B, pp. 44-48.

[322] Pagano N.J. (1978) Stress Fields in Composite Laminates. *Int. J. Solids Struct.*, Vol. 14, pp. 385-400.

[323] Paipetis S.A. and Papanicolaou G.C. (eds) (1988) Engineering Applications of New Composites. *Omega Scientific, Oxford.*

[324] Palmer A.C., Maier G. and Drucker D.C. (1967) Normality Relations and Convexity of Yield Surfaces for Unstable Materials of Structural Elements. *Trans. ASME,* Vol. 24, pp. 464-470.

[325] Panagiotopoulos P.D. (1975) A Nonlinear Programming Approach to the Unilateral Contact – and Friction – Boundary Value Problem in the Theory of Elasticity. *Ing. Archiv.,* Vol. 44, pp. 421-432.

[326] Panagiotopoulos P.D. (1976) A Variational Inequality Approach to the Inelastic Stress-Unilateral Analysis of Cable Structures. *Comp. and Struct.,* Vol. 6, pp. 133-139.

[327] Panagiotopoulos P.D. (1976) Convex Analysis and Unilateral Static Problems. *Ing. Archiv,* Vol. 45, pp. 55-68.

[328] Panagiotopoulos P.D. and Talaslidis D. (1980) A Linear Analysis Approach to the Solution of Certain Classes of Variational Inequality Problems in Structural Analysis. *Int. J. Solids and Struct.* Vol. 16, pp. 991-1006.

[329] Panagiotopoulos P.D. (1981) Non-Convex Superpotentials in the Sense of F.H. Clarke and Applications. *Mech. Res. Comm.,* Vol. 8, pp. 335-340.

[330] Panagiotopoulos P.D. (1981) Dynamic and Incremental Variational Inequality Principles, Differential Inclusion and their Applications to Co-Existent Phases Problems. *Acta Mechanica,* Vol. 40, pp. 85-107.

[331] Panagiotopoulos P.D. (1982) Non-Convex Energy Functionals. Application to Non-convex Elastoplasticity. *Mech. Res. Comm.,* Vol. 9, pp. 23-29.

[332] Panagiotopoulos P.D. (1983) Nonconvex Energy Functions. Hemivariational Inequalities and Substationarity Principles. *Acta Mechanica,* Vol. 42, pp. 160-183.

[333] Panagiotopoulos P.D. (1983) A Boundary Integral Inclusion Approach to Unilateral B.V.Ps in Elastostatics. *Mech. Res. Comm.,* Vol. 10, pp. 91-96.

[334] Panagiotopoulos P.D. (1983) Une Généralisation Non-Convexe de la Notion du Surpotentiel et ses Applications. *C.R. Acad. Sc. Paris,* Vol. 296II, pp. 1105-1108.

[335] Panagiotopoulos P.D. and Baniotopoulos C.C. (1984) A Hemivariational Inequality and Substationarity Approach to the Interface Problem. *Theory and Prospects of Applications, Engineering Analysis* Vol. 1, pp. 20-31.

[336] Panagiotopoulos P.D. (1985) Inequality Problems in Mechanics and Applications, Convex and Nonconvex Energy Functions. *Birkhaüser, Basel.*

[337] Panagiotopoulos, P.D. (1985) Nonconvex Problems of Semipermeable Media and Related Topics. *Z.A.M.M.*, Vol 65, 1, pp. 29-36.

[338] Panagiotopoulos P.D. (1985) Hemivariational Inequalities and Substationarity in the Static Theory of von Kármán Plates. *Z.A.M.M.*, Vol. 65, 6, pp. 219-229.

[339] Panagiotopoulos P.D. and Baniotopoulos C.C. (1986) Hemivariational Incqualitics and Substationarity Principles in Structural Analysis and Their Applications. *J. Struct. Mech.*, Vol. 14, pp. 77-103.

[340] Panagiotopoulos P.D. and Koltsakis E.K. (1987) Interlayer Slip and Delamination Effect: A Hemivariational Inequality Approach. *Canadian Society for Mech. Engineering,* Vol. 11 , pp. 43-52.

[341] Panagiotopoulos P.D. and Koltsakis E.K. (1987) Hemivariational Inequalities for Linear and Nonlinear Elastic Materials. *Meccanica,* Vol. 22 , pp. 65-75.

[342] Panagiotopoulos P.D. (1987) Multivalued Boundary Integral Equations for Inequality Problems. The Convex Case. *Acta Mechanica,* Vol. 70, pp. 145-167.

[343] Panagiotopoulos P.D. Hemivariational Inequalities and their Applications. *In: Topics in Nonsmooth Mechanics (ed. by J.J.Moreau, P.D.Panagiotopoulos and G. Strang) Birkhäuser Verlag, Boston, 1980.*

[344] Panagiotopoulos P.D. (1988) Variational Hemivariational Inequalities in Nonlinear Elasticity. The Coercive Case. *Aplikace Matematiky (now Applications of Mathematics),* Vol. 33, pp. 249-268.

[345] Panagiotopoulos P.D. and Liolios A.A. (1988) On the Dynamic of Inelastic Shocks. A New Approach. *J. Proc of the Greek-German Seminar in Structural Dynamics and Earthquake Engineering, (ed. by A.N. Kounadis and W.B. Krätzig) Publ. of the Hellenic Soc. Theor. Appl. Mech., Athens,* pp.12-18.

[346] Panagiotopoulos P.D. (1989) Boundary Integral Equations for Inequality Problems. The Nonconvex Case. *Acta Mechanica*, Vol. 72, pp. 152-168.

[347] Panagiotopoulos P.D. (1989) Semicoercive Hemivariational Inequalities. On the Delamination of Composite Plates. *Quart. of Appl. Math.*, Vol. XLVII , pp. 611-629.

[348] Panagiotopoulos P.D. and Stavroulakis G. (1990) The Delamination Effect in Laminated von Kármán Plates under Unilateral Boundary Conditions. A Variational-Hemivariational Inequality Approach. *J. of Elasticity,* Vol. 23, pp. 69-96.

[349] Panagiotopoulos P.D. and Koltsakis E.K. (1990) The Nonmonotone Skin Effects in Plane Elasticity Obeying to Linear Elastic and subdifferential Material Laws. *Z.A.M.M.,* Vol. 70, pp. 13-21.

[350] Panagiotopoulos, P.D. (1991) Coercive and Semicoercive Hemivariational Inequalities. *Nonlinear Analysis, TMA,* Vol. 16, pp. 209-231.

[351] Panagiotopoulos P.D. and Haslinger J. (1992) On the Dual Reciprocal Variational Approach to the Signorini-Fichera Problem. Convex and Nonconvex Generalizations. *Z.A.M.M.,* Vol. 72, pp. 497-506.

[352] Panagiotopoulos P.D. (1992) Adhesive Joints and Interfaces of Linear Elastic Bodies in Loading and Unloading. Semicoercive Hemivariational Inequalities. *J. of Elasticity,* Vol. 28, pp. 29-54.

[353] Panagiotopoulos P. D. (1993) Hemivariational inequalities. Applications in Mechanics and Engineering. *Springer-Verlag, Berlin, Heidelberg, New York.*

[354] Panagiotopoulos P.D. (1994) Variational Principles for Contact Problems Including Impact Phenomena. *In Contact Mechanics (eds. M. Raons, M. Jean, J.J. Moreau), Plenum Publ., N. York.*

[355] Panagiotopoulos P.D. (1995) Modelling of Nonconvex Nonsmooth Energy Problems, Dynamic Hemivariational Inequalities with Impact Effects. *J. of Comp. and Applied Math.,* Vol. 63, pp. 123-138.

[356] Paoli L. and Schatzman M. (1993) Mouvements à un Nombre Fini de Degrés de Liberté avec Contraintes Unilatérales: Cas avec Perte d'Energie. *Modélisation Mathématique et Analyse Numérique,* Vol. 27, pp. 673-717.

[357] Pardalos P.M. and Panagiotopoulos P.D. (eds) (1995) Nonconvex Energy Functions: Applications in Engineering. *Journal of Global Optim., Vol. 6, Issue 4.*

[358] Pascali D. and Sburlan S. (1978) Nonlinear Mappings of Monotone Type. *Sijthoff and Noordhoff International Publishers, Amsterdam, The Netherlands.*

[359] Pazy A. (1983) Semigroups of Linear Operators and Applications to Partial Differential Equations. *Applied Mathematical Sciences 44, Springer-Verlag, Berlin.*

[360] Peitgen H.O. and Schmitt K. (1984) Global Analysis of Two-Parameter Elliptic Eigenvalue Problems. *Transactions of the American Mathematical Society*, Vol. 283, pp. 57-95.

[361] Potier-Ferry M. (1974) Problèmes Semi-Coercifs. Applications aux Plaques de von Kármán. *J. Math. Pures et Appl.*, Vol. 53, pp. 331-346.

[362] Potier-Ferry M. (1975) Sur un Système d'Equations Rencontré en Théorie des Plaques. *C. R. Acad. Sc. Paris*, t. 280, pp. 1385-1387.

[363] Prager W. (1957) On Ideal-Locking Materials. *Trans. Soc. Rheol.*, Vol 1, pp. 169-175.

[364] Prager W. (1958) Elastic Solids of Limited Compressibility. *Proc. 9th Int. Congress Appl. Mech. Brussels, Vol. 5.*

[365] Przemieniecki J.S. (1968) Theory of Matrix Structural Analysis. *McGraw-Hill Book Co., New York.*

[366] Quittner P. (1986) Spectral Analysis of Variational Inequalities. *Commentationes Mathematicae Universitatis Carolinae*, Vol. 27, pp. 605-629.

[367] Quittner P. (1989) Solvability and Multiplicity Results for Variational Inequalities, *Commentationes Mathematicae Universitatis Carolinae*, Vol. 30, pp. 281-302.

[368] Rabinowitz P.H. (1971) Some Global Results for Nonlinear Eigenvalue Problems. *J. Funct. Anal.*, Vol. 7, pp. 487-513.

[369] Rabinowitz P.H. (1986) Minimax Methods in Critical Point Theory with Applications to Differential Equations. *CBM Reg. Conf. Ser. in Math. N^0 65, Amer. Math. Soc., Providence.*

[370] Rauch J. (1977) Discontinuous Semilinear Differential Equations and Multiple Valued Maps. *Proc. Amer. Math. Soc.*, Vol. 64, pp. 277-282.

[371] Reddy B.D. and Tomarelli F. (1990) The Obstacle Problem for an Elastoplastic Body. *Appl. Math. Optim.*, Vol. 21, pp. 89-110.

[372] Ribarska N., Tsachev T. and Krastanov M. (1995) Deformation Lemma, Ljusternik-Schnirellmann Theory and Mountain Pass Theorem on C^1-Finsler Manifolds. *Serdica Math. J.*, Vol. 21, pp. 239-266.

[373] Rockafellar R.T. (1968) Integrals which are Convex Functionals. *Pacific J. Math.* Vol. 24, pp. 525-539.

[374] Rockafellar R.T. (1970) Convex Analysis. *Princeton Univ. Press, Princeton.*

[375] Rockafellar R.T. (1979) La Théorie des Sous-Gradients et ses Applications à l'optimisation. Fonctions Convexes et Non-convexes. *Les Presses de l' Université de Montréal, Montréal.*

[376] Rockafellar R.T. (1980) Generalized Directional Derivatives and Subgradients of Non-convex Functions. *Can. J. Math.*, Vol. XXXII, pp. 257-280.

[377] Rodriguez J. F. (1987) Obstacle problems in Mathematical Physics. *Mathematics Studies n^0 134, Elsevier Science Publishers B.V.*

[378] Rodrigues J.F. (ed). (1989) Mathematical Models for Phase Change Problems. *ISNM Vol 88, Birkhäuser Verlag, Basel.*

[379] Roman I., Harlet H. and Marom G. (1981) Stress Intensity Factor Measurements in Composite Sandwich Structures. *In: I.H. Marshal, ed., Proc. 1st Conf. on Composite Structures, Applied Science Publishers, London,* pp. 633-645.

[380] Roseau M. (1984) Vibrations des Systèmes Mécaniques : Méthodes Analytiques et Applications. *Masson, Paris.*

[381] Salencon J. and Tristán-Lopez A. (1980) Analyse de la Stabilité des Talus en Sols Cohérents Anisotropes. *C.R. Acad. Sc. Paris*, Vol. 290B, pp. 493-496.

[382] Sard A. (1942) The Measure of the Critical Values of Differentiable Maps. *Bull. Amer. Math. Soc.*, Vol. 48, pp. 883-890.

[383] Schatzman M. (1973) Problèmes aux Limites Non Linéaires, Non Coercifs. *Ann. Scuola Norm. Sup. Pisa, Cl. Sci.*, Vol. 27, pp. 641-686.

[384] Schellekens J.C.J. and De Borst R. (1991) Application of Linear and Nonlinear Fracture Mechanism Options to Free Edge Delamination in Laminated Composites. *Heron,* Vol. 36, pp. 37-48.

[385] Schwartz M.M. (1984) Composite Materials Handbook. *McGraw-Hill, New-York.*

[386] Schwartz L. (1993) Analyse III. Calcul Intégral. *Hermann, Paris.*

[387] Shi P. and Shillor M. (1991) Noncoercive Variational Inequalities with Applications to Friction Problems. *Proceedings of the Royal Society of Edinburgh,* Vol. 117A, pp. 275-293.

[388] Signorini A. (1959) Questioni di Elasticità Nonlinearizzata e Semi-linearizata. *Rend. Mat.,* Vol. 18,

[389] Sipcic S.R., Rabinovich V.L. and Rugg R.A. (1993) Unilateral Contact Problem for Composite Finite Bodies. *AIAA-93-1610-CP,* pp. 2661-2267.

[390] Spanier E.H. (1966) Algebraic Topology. *McGraw-Hill, New York.*

[391] Spector A.A. (1982) Variational Methods of Analysis for Certain Classes of Spatial Problems of Contact Between Elastic Bodies with Friction. *Dokl. Acad. Nauk. SSR,* Vol. 265, pp. 111-117.

[392] Spektor A.A. (1987) Variational Methods in Three-Dimensional Problems of Non-Stationary Interaction of Elastic Bodies with Friction. *PMM U.S.S.R.,* Vol. 51, pp. 56-62..

[393] Stavroulakis G.E. and Panagiotopoulos P.D. (1988) Laminated Orthotropic Plates under Subdifferential Boundary Conditions. A Variational-Hemivariational Inequality Approach. *Z.A.M.M.,* Vol. 6, pp. 213-216.

[394] Stavroulakis G.E. (1991) Analysis of Structures with Interfaces. Formulation and Study of Variational-Hemivariational Inequality Problems. *Doct. Dissertation, Dept. of Civil Eng., Aristotle University, Thessaloniki.*

[395] Stavroulakis G.E., Panagiotopoulos P.D. and Al-Fahed A.M. (1991) On the Rigid Body Displacements and Rotations in Unilateral Contact Problems and Applications. *Comp. and Struct.,* Vol. 40, pp. 599-614.

[396] Stavroulakis G.E. and Panagiotopoulos P.D. (1991) Delamination of Multilayered Plates in Bending under Monotone Boundary and Nonmonotone Interlayer Conditions. A Variational- Hemivariational Inequality Approach. *J. of Theoretical and Applied Mechanics*, Vol. XXII(2), pp. 38-46.

[397] Stavroulakis G.E. (1993) Convex Decomposition for Nonconvex Energy Problems in Elastostatics and Applications. *European J. of Mech. A /Solids*, Vol. 12, pp. 1-20.

[398] Stavroulakis G.E., Goeleven D. and Panagiotopoulos P.D. (1995) New Models for a Class of Adhesive Grippers. The Hemivariational Inequality Approach. *Archive of Applied Mechanics*, Vol. 67, pp. 50-61.

[399] Stavroulakis G.E., Goeleven D. and Panagiotopoulos P.D. (1997) Models of Singular Variational Inequalities and Complementarity Problems Arising in F.E.M. and B.E.M. Unilateral Contact Problems. *In Proceedings of the 8th French-German Conference on Optimization (eds. P. Gritzmann, R. Horst, E. Sachs, R. Tichatschke), Lecture Notes in Economics and Mathematical Systems, Springer Verlag.*

[400] Stavroulakis G.E., Antes H., Panagiotopoulos P.D. Transient elastodynamics around cracks including contact and friction. (1999) *Computer Methods in Applied Mechanics and Engineering, Special Issue: Computational Modeling of Contact and Friction, Eds.: J.A.C. Martins and A. Klarbring*, Vol. 177, pp. 427-440.

[401] Stavroulakis G.E. (2000) Inverse and crack identification problems in engineering mechanics. *Habilitation Thesis, Technical University Braunschweig.*

[402] Stuart C.A. and Toland J.F. (1980) A Variational Method for Boundary Value Problems with Discontinuous Nonlinearities. *J. London Math. Soc.*, Vol. 21, pp. 319-328.

[403] Stuart C.A. and Toland J.F. (1980) A Property of Solutions of Elliptic Differential Equations with Discontinuous Nonlinearities. *J. London Math. Soc.*, Vol. 21, 329-335.

[404] Szulkin A. (1985) Positive Solutions of Variational Inequalities: A Degree-Theoretic Approach. *Journal of Differential Equations*, Vol. 57, pp. 90-111.

[405] Szulkin A. (1986) Minimax Principles for Lower Semicontinuous Functions and Applications to Nonlinear Boundary Value Problems. *Ann. Inst. Henri Poincaré* , Vol 3, pp. 77-109.

[406] Tato Y. (1987) Signorini's Problem with Friction in Linear Elasticity. *Japan J. of Appl. Math.*, Vol. 4, pp. 237-268.

[407] Telega J.J. (1988) Topics on Unilateral Contact Problems of Elasticity and Inelasticity. *In: Nonsmooth Mechanics and Applications (ed. by Moreau J.J. Panagiotopoulos P.D.), CISM Courses and Lectures, N⁰ 302, Springer Verlag, Wien.*

[408] Teman R. (1985) Mathematical Problems in Plasticity. *Gauthier Villars, Paris.*

[409] Theocaris P.S. and Makrakis G.N. (1986) The Kinked Crack Solved by Mellin Transform. *J. Elasticity*, Vol. 16, pp. 393-411.

[410] Theocaris P.S. and Makrakis G.N. (1987) Crack Kinking in Anti-Plane Shear Solved by the Mellin Transform. *Int. J. Fracture*, Vol. 34, pp. 251-262.

[411] Theocaris P.S. and Panagiotopoulos P.D. (1991) On Debonding Effects in Adhesively Bonded Cracks - A Boundary Integral Approach. *Arch. of Appl. Mech.*, Vol. 61, pp. 578-587.

[412] Theocaris P.S. and Panagiotopoulos P.D. (1992) On the Consideration of Unilateral Contact and Friction in Cracks. The Boundary Integral Method. *Int. J. Num. Meth. Eng.*, Vol. 35, pp. 1697-1708.

[413] Thompson J.M.T. and Hunt G.W. (1973) A General Theory of Elastic Stability. *J.Wiley and Sons, London.*

[414] Tomarelli F. (1991) A Quasi-Variational Problem in Nonlinear Elasticity. *Annali di Matematica Pura et Applicata*, Vol. CLVIII(IV), pp. 331-389.

[415] Tomarelli F. (1993) Noncoercive Variational Inequalities for Pseudomonotone Operators. *Rend. Sem. Mat. Fis. Univ. Milano*, Vol. n.83/P.

[416] Tonti E. (1973) A Systematic Approach to the Search for Variational Principles. *In Variational Methods in Engineering (ed. by C.A. Brebbia and H.Tottenham), Vol. I, Southampton Univ. Press, Southampton.*

[417] Troianiello G.M. (1987) Elliptic Differential Equations and Obstacle Problems. *Plenum Press, New York*

[418] Truesdell C. and Noll W. (1965) The Non-linear Field Theories of Mechanics. *In: Encyclopedia of Physics. Vol. III/3 (ed. by S. Flfügge) Springer-Verlag, Berlin.*

[419] Tzaferopoulos M.A. (1991) Numerical Analysis of Structures with Monotone and Nonmonotone, Nonsmooth Material Laws and Boundary Conditions: Algorithms and Applications. *Doct. Dissertation, Aristotle University, Dept. of Civil Eng., Thessaloniki.*

[420] Tzaferopoulos M.A. and Panagiotopoulos P.D. (1991) Analysis of Steel Frames with Nonmonotone Flexible Joints. *In: Proc. 1st Nat. Conf. on Steel Structures (ed. by A.N. Kounadis), Athens, pp. 130-140.*

[421] Vainberg M. M. (1973) Variational Method and Method of Monotone Operators in the Theory of Nonlinear Equations. *John Wiley & Sons, New York.*

[422] Verma R.U. (1997) Nonlinear Variational and Constrained Hemivariational Inequalities Involving Relaxed Operators. *To appear in Z. Angew. Math. Mech., 77, pp. 387-391.*

[423] Visitin A. (ed.) (1993) Models of Hysteresis. *Pitman Research Note in Mathematics Series, 286, Longman, Essex UK.*

[424] Vladimirov V. (1979) Distributions en Physique Mathématique. Edition MIR, Moscow.

[425] Whyburn G.T. (1958) Topological Analysis. Chap. 1, Princeton Univ. Press, Princeton, N.J.

[426] Williams J.G. and Rhodes M.D. (1982) Effect of Resin on Impact Damage Tolerance of Graphite/Epoxy Laminates. *In: Proceedings of the 6th International Conference on Composite Materials, Testing and Design, (ed. by I.M. Daniel). ASTM Special Technical Publication 787, ASTM Philadelphia,* pp.450-480.

[427] Williams J.F., Stouffer D.C., Ilič S. and Jones R. (1986) An Analysis of Delamination Behaviour. *Comp. Struct.,* Vol. 5, pp. 203-216.

[428] Woźniak C. (1984) Materials with Generalized Constraints. *Arch. Mech.,* Vol. 36, 4, pp. 539-551.

[429] Woźniak C. (1987) A Nonstandard Method of Modeling of Thermoelastic Periodic Composites. *Int. J. Eng. Sci.,* Vol. 25, pp. 483-498.

[430] Woźniak C. (1990) Discrete and Continuum Modeling of Delamination Processes. *Ingenieur-Archiv,* Vol. 60, pp. 335-344

[431] Woźniak C. and Kleiber M. Nonlinear Mechanics of Structures. PWN-Polish Sci. Publ. Warszawa, Kluwer Academic Publishers, Dordrecht, London 1991.

[432] Zeidler E. (1990) Nonlinear Functional Analysis and its Applications. *Springer Verlag, Berlin.*

[433] Zervas P.A. (1992) Seismic Behaviour of Frame Structures with Unilateral Contact Conditions. *Doct. Dissertation, Dept. of Civil Eng., Aristotle Univ., Thessaloniki.*

[434] Zhuravlev V.F. and Foufaev N.A. (1993) Mechanics of Systems with Unilateral Constraints (in russian). *Moscow, Nauka.*

Appendix A
List of Notations

We have listed here the main notations which are used throughout
the text and we refer the reader where necessary, to the Section or the
formula, wherein the definition is given. Throughout this book the summation convention with respect to a repeated index is employed, unless
otherwise stated.

$\mathbb{R}^n$	Euclidean n-dimensional space
$\mathbb{R}^n_+$	The set of nonnegative vectors of $\mathbb{R}^n$, i.e. $x \in \mathbb{R}^n_+ \Leftrightarrow x_i \geq 0, \forall i = 1, ..., n$
$\mathbb{R}$	$= \mathbb{R}^1$
$\mathbb{R}_+$	$= \mathbb{R}^1_+$
$x = \{x_i\}_{i=1,...,n}$	Vector of $\mathbb{R}^n$ whose coordinates are $x_1, ..., x_n$
$\mid x \mid$	Length of $x \in \mathbb{R}^n$, i.e. $\mid x \mid = (\sum_{i=1}^n x_i^2)^{\frac{1}{2}}$
$\langle x, y \rangle_n$	Scalar product in $\mathbb{R}^n$
$x^T y$	Scalar product in $\mathbb{R}^n$
$x.y$	Scalar product in $\mathbb{R}^n$
$x \wedge y$	Vector product in $\mathbb{R}^n$

x^+	The positive part of a vector $x \in \mathbb{R}^n$, i.e. $x_i^+ = \max\{0, x_i\}, i = 1, ..., n$
x^-	The negative part of a vector $x \in \mathbb{R}^n$, i.e. $x_i^- = \max\{0, -x_i\}, i = 1, ..., n$
$M = (m_{ij})$	Matrix of $\mathbb{R}^{m \times n}$ with entries $m_{ij}(i = 1, ..., m; j = 1, ..., n)$
M^T	Transpose of M
$det\ M$	Determinant of M
M^{-1}	Inverse of M
$sp(M)$	Spectrum of M
$sp_r(M)$	Set of real eigenvalues of M, i.e. $\mathrm{sp}_r(M) = \mathrm{sp}(M) \cap \mathbb{R}$
I	Identity matrix
X^*	Topological dual of a Banach space X
$X^\perp$	Orthogonal of X
$\oplus$	Direct sum
$dim\ \{X\}$	Dimension of X
$codim\ \{X\}$	Codimension of X
$\langle .,. \rangle$	Duality product
$\langle .,. \rangle_{X^*,X}$	Duality product between X^* and X
$\| \cdot \|$	Norm
$\| \cdot \|_X$	Norm in X
$\| \cdot \|_*$	Dual norm
$\| \cdot \|_{X^*}$	Dual norm in X^*
$(.,.)$	Scalar product
$(.,.)_X$	Scalar product in X

$d(.,.)$	Metric
$d_H(.,.)$	Hausdorff distance
$d(x, K)$	Dstance from $x \in X$ to $K \subset X$, i.e. $d(x, K) := \inf_{y \in K} d(x, y)$
$\mathcal{L}(X, Y)$	Space of linear and continuous mapping from X onto Y
$\to$	Strong convergence
$\rightharpoonup$	Weak convergence
$P_C : X \to C$	Projection operator from X onto C, see Sec. 1.1.6
$J : X \to X^*$	Duality mapping, see Sec. 1.1.6
$\Pi_{A,C}$	See Sec. 3.2
$\Pi_{A,\Phi}$	See Sec. 5.2
$\Pi_{A,\varphi,\varepsilon}$	See Sec. 5.2
$\vee\{x, y\}$	Least upper bound of the pair x, y, see Sec. 1.1.10, 3.6, 6.1
$\wedge\{x, y\}$	Greatest lower bound of the pair x, y, see Sec. 1.1.10, 3.6, 6.1
$A \cup B, A \cap B, A - B$	Union, intersection and difference of sets A and B
$A \times B$	Cartesian product of the sets A and B
$X \otimes Y$	Tensor product of the vector spaces X and Y
$int\{A\}$	Interior of a set A
$\overline{A}$	Closure of A
$conv\{A\}$	Convex hull of A
A^*	Dual cone of A, see Sec. 1.1.2
A^+	Polar cone of A, see Sec. 1.1.2
2^A	Set of all subsets of A

A_∞	Recession cone of A, see Sec. 1.4.1
Ψ_A	Indicator (function) of A, see (1.1.3)
$N_A(x)$	Normal cone of A at x
$N_\delta(A)$	Closed neighborhood of A, see Sec. 4.4
$\gamma(A)$	Genus of Krasonel'skii of A, see Sec. 4.4
$cat_X(A)$	Category of A in X, see Sec. 9.4
$deg(\Phi, \mathcal{D}, p)$	Degree of Φ at p relative to $\mathcal{D}$
$span\{x_1, ..., x_n\}$	Vector space generated by the elements $x_1, ..., x_n$
$\{u_n\}$	Sequence
$B(u_0, R)$	Open ball, i.e. $B(u_0, R) = \{u \in X :\parallel u - u_0 \parallel < R\}$
$\overline{B(u_0, R)}$	Closed ball, i.e. $\overline{B(u_0, R)} = \{u \in X :\parallel u - u_0 \parallel \leq R\}$
B_R	$= \overline{B(0, R)}$
$\mathcal{S}(A, f, C)$	See (3.5.2)
$\mathcal{S}(A, f, C, u_0)$	See (3.5.11)
$\mathcal{S}(A, f, C, \Phi, j)$	See (3.12.4)
$\Lambda_\infty(A, f, C)$	See (3.5.3)
$\Lambda_\infty(A, f, C, u_0)$	See (3.5.12)
$\Lambda_\infty(A, f, C, \Phi, j)$	See (3.12.5)
sin, cos, tg	Sinus, cosinus, tangent
$sign(a)$	Sign of $a \in \mathbb{R}$
$u', \frac{du}{dt}$	First order time-derivative of $u : (a, b) \to X$, see Sec. 2.10.5
$u'', \frac{d^2 u}{dt^2}$	Second order time-derivative of $u : (a, b) \to X$, see Sec. 2.10.5

$\dot{u}, u'$	First order (weak) derivative of $u : (a, b) \to \mathbb{R}^n$
$\ddot{u}, u''$	Second order (weak) derivative of $u : (a, b) \to \mathbb{R}^n$
$u^{(k)}$	(Weak) derivative of order k of $u : (a, b) \to \mathbb{R}^n$
$D^\alpha u(x), \alpha = (\alpha_1, ..., \alpha_n)$	$= \frac{\partial^{|\alpha|} u(x)}{\partial x_1^{\alpha_1} \partial x_2^{\alpha_2} ... \partial x_n^{\alpha_n}}, \| \alpha \| = \alpha_1 + ... + \alpha_n$
u_{x_i}	$= \frac{\partial u}{\partial x_j}$
$u_{i,j}$	$= \frac{\partial u_i}{\partial x_j}$
∇u	$\{\frac{\partial u}{\partial x_i}\}$
$\nabla u \nabla v$	$= (\nabla u)^T (\nabla v) = \frac{\partial u}{\partial x_i} \frac{\partial v}{\partial x_i}$
$t_{ij,k}$	$= \frac{\partial t_{ij}}{\partial w_k}$
$f : X \to \mathbb{R}$	Real-valued function
$\varphi : X \to \mathbb{R} \cup \{+\infty\}$	Extended real-valued function
$L_{loc}(X; \mathbb{R})$	Set of locally Lipschitz functions on X
$\Gamma_0(X; \mathbb{R} \cup \{+\infty\})$	Set of proper, convex and l.s.c. functionals on X
$D(\varphi)$	Domain of φ, see (1.1.5)
$epi(\varphi)$	Epigraph of φ, see (1.1.4)
$\varphi^{-1}(0)$	Set of zeros of φ, i.e. $\varphi^{-1}(0) = \{x \in X : \varphi(x) = 0\}$
$\varphi^{-1}(c)$	$= \{x \in X : \varphi(x) = c\}$
∂f	Generalized gradient of $f \in L_{loc}(X; \mathbb{R})$, see Sec. 1.2.1
$\partial \varphi$	Convex subdifferential of $\varphi \in \Gamma_0(X; \mathbb{R} \cup \{+\infty\})$, see (1.3.1)
$\partial_y f$	Generalized gradient of $f : T \times X \to \mathbb{R}$ with respect to the second variable
$\partial_y \varphi$	Convex subdifferential of $\varphi : T \times X \to \mathbb{R} \cup \{+\infty\}$ with respect to the second variable

$f^0(u; v)$	Generalized directional derivative of f at u in the direction v, see Sec. 1.2.1
$f'(u; v)$	One-sided directional derivative of f at u in the direction v, see Sec. 1.2.1
$f^0_y(x, u; v)$	Generalized directional derivative of $f(x,.)$ at u in the direction v
$f'_y(x, u; v)$	One-sided directional derivative of $f(x,.)$ at u in the direction v
$f^0_{i,y}(x, u; v)$	Generalized directional derivative of $f_i(x,.)$ at u in the direction v
$f'(u)$	Gâteaux derivative of $f : X \rightarrow \mathbb{R}$ at u, i.e. $\langle f'(u), h \rangle = \lim_{\theta \to 0} \frac{f(u+\theta h) - f(u)}{\theta}$
$\nabla f(u)$	Gradient of f at u
$f"(u)$	Second derivative of $f : X \rightarrow \mathbb{R}$ at u, i.e. $f"(u)(h, k) = \langle g'(u), k \rangle$ with $g(x) = \langle f'(x), h \rangle$
$f \circ g$	Composite function i.e. $(f \circ g)(x) = f(g(x))$
$f * g$	Convolution product of f and g
f_c	See (1.1.32)
f^c	See (1.1.33)
$\varphi\infty$	Recession function of $\varphi \in \Gamma_0(X; \mathbb{R} \cup \{+\infty\})$, see (1.4.5)
$\underline{r}_A$	Recession mapping of Brézis and Nirenberg associated to $A : X \rightarrow X^*$, see (1.4.18)
$\underline{r}_{A,u_0}$	See (1.4.19)
$\underline{\gamma}_j$	Asymptotic potential mapping associated to $j \in L_{loc}(X; \mathbb{R})$, see (1.4.23)
h_+, h_-	See Sec. 1.4.4, 1.4.5
$\check{j}, \hat{j}$	See Sec. 1.4.6
$supp\{f\}$	Support of $f : \Omega \rightarrow \mathbb{R}$, i.e. $supp\{f\} = \{x \in \Omega : f(x) \neq 0\}$

$\underline{\beta}$	$\underline{\beta}(t) = \lim_{\delta \to 0} \operatorname{ess\,inf}_{\|\tau - t\| < \delta} \beta(\tau)$, see Sec. 1.2.3
$\overline{\beta}$	$\overline{\beta}(t) = \lim_{\delta \to 0} \operatorname{ess\,sup}_{\|\tau - t\| < \delta} \beta(\tau)$, see Sec. 1.2.3
$\beta(t - 0)$	Limit of $\beta(t)$ as $t \to 0^-$
$\beta(t + 0)$	Limit of $\beta(t)$ as $t \to 0^+$
$L^\infty_{loc}(\mathbb{R})$	Space of locally bounded functions from $\mathbb{R}$ onto $\mathbb{R}$
ess inf, ess sup	Essential infinum, essential supremum
inf, sup	Infinum, supremum
min, max	Minimum, maximum
φ^+	The positive part of a function φ, i.e. $\psi^+(x) = \max\{0, \psi(x)\}, \forall x$
φ^-	The negative part of a function φ, i.e. $\varphi^-(x) = \max\{0, -\varphi(x)\}, \forall x$
$Ker(f)$	$= \{x \in X : \langle f, x \rangle = 0\}$ for $f \in X^*$
$A \hookrightarrow B$ Continuously	$A \subset B$ and the identity map from A to B is continuous
$A \hookrightarrow B$ Compactly	$A \subset B$ and the identity map from A to B is compact
l.s.c.	Lower semi-continuous, see Sec. 1.1.1
u.s.c.	Upper semi-continuous, see Sec. 1.1.1
w.l.s.c.	Weakly l.s.c., see Sec. 1.1.1
w.u.s.c.	Weakly u.s.c., see Sec. 1.1.1
KKM	Knaster-Kuratowski-Mozurkiewicz, see Sec. 1.1.7
a-compact	See Sec. 3.5
(PS)	See (4.1.2)
(PSC)	See (4.1.5)
(HPS)	See (4.1.6)

(HPS')	See (4.1.7)
(VPS)	See (4.1.11)
(VPS')	See (4.1.12)
a.e.	Almost everywhere
μ-a.e.	Almost everywhere for the measure μ
Ω	Regular open bounded set
$\partial\Omega, \Gamma$	Boundary of Ω
Γ	$= \partial\Omega$
$\mathcal{L}_n$	n-dimensional Lebesgue measure
$\mathcal{H}_{n-1}$	$(n-1)$-dimensional Hausdorff measure
(T, τ, μ)	Positive complete measure space
dx	Formal notation for $d\mathcal{L}_n$
ds	Formal notation for $d\mathcal{H}_{n-1}$
$C^m(\Omega; X)$	Space of functions $f : \Omega \to X$ with continuous derivatives up to order m
$C^m(\Omega)$	$= C^m(\Omega; \mathbb{R})$
$C^\infty(\Omega)$	$= \cap_{m=0}^\infty C^m(\Omega)$
$C_0^\infty(\Omega)$	$= \{v \in C^m(\Omega) : supp\{v\}$ is compact in $\Omega\}$
$D(\Omega)$	$= C_0^\infty(\Omega)$
$D^*(\Omega)$	Space of distributions and the topological dual of $D(\Omega)$
$C^k(\overline{\Omega})$	The space of functions $\varphi \in C^m(\Omega)$ such that $D^\alpha\varphi$ ($\mid \alpha \mid \leq m$) possesses a unique continuous extension to $\overline{\Omega}$
$C^{m,n}(\overline{\Omega})$	The space of functions $\varphi \in C^m(\overline{\Omega})$ such that $D^\alpha\varphi$ ($\mid \alpha \mid \leq m$) satisfies the Hölder condition $\mid D^\alpha\varphi(x) - D^\alpha\varphi(y) \mid \leq c \mid x - y \mid^n$, ($c$ const. > 0) $\forall x, y \in \Omega$

$L^p(T; X)$	The space of equivalence classes of measurable functions $f : T \to X$ for which $\int_\Omega \| f(x) \|_X^p \, d\mu < +\infty$, Lebesgue integration being implied
$L^p(T)$	$= L^p(T; \mathbb{R})$
$W^{k,p}(\Omega), W^{k,p}(\Omega; \mathbb{R}^n)$	Sobolev spaces, see Sec. 2.10.1
$H^k(\Omega), H^{k,p}(\Omega; \mathbb{R}^n)$	Sobolev spaces, see Sec. 2.10.1
$\langle .,. \rangle_{k,p}, \| \cdot \|_{k,p}, \| \cdot \|_{0,p},$	see Sec. 2.10.1
$\| \cdot \|_{0,\infty}, \| \cdot \|_\infty,$	See Sec. 2.10.1
$(.,.)_{k,2}$	See Sec. 2.10.1
$\gamma, \bar{\gamma}, \gamma_N, \gamma_T$	Trace operators, see Sec. 2.10.2
$\Omega \in C^{k,\mu}$	See Sec. 2.10.1
$W^{k,p}(a,b; \mathcal{H})$	See Sec. 2.10.5
$W^{k,p}(a,b; \mathcal{V}, \mathcal{V}^*)$	See Sec. 2.10.5
$\tilde{W}^{k,p}(a,b; \mathcal{V}, \mathcal{V}^*)$	See Sec. 2.10.5
$g_{|A}$	Restriction of $\varphi : X \to Y$ to A
A^*	Adjoint of A (linear)
$D(A)$	Domain of A
$Ker(A)$	Kernel of A
$R(A)$	Range of A
$\sigma(A)$	Spectrum of A (linear)
id_X	Identity mapping from X onto X
Δ	Laplace operator
$\Delta\Delta$	Biharmonic operator
$\sigma = \{\sigma_{ij}\}$	Cauchy stress tensor

$n = \{n_i\}$	Outward unit normal to Γ
$S = \{S_i\}$	Stress vector on Γ
h_N, h_T	Normal and tangential component of h with respect to Γ
$C = \{C_{ijkl}\}$	Hooke's tensor
$\varepsilon = \{\varepsilon_{ij}\}$	Strain tensor
$[\zeta]^{(\alpha)}$	Relative deflection $= v_\alpha - v_{\alpha+1}$ ($v \in \mathbb{R}^N$) see (2.10.21)
$[u]$	Relative displacement, see Sec. 2.11.3
δ_{ij}	Kronecker's delta

MIX
Papier aus verantwortungsvollen Quellen
Paper from responsible sources
FSC® C105338

If you have any concerns about our products,
you can contact us on
ProductSafety@springernature.com

In case Publisher is established outside the EU,
the EU authorized representative is:
Springer Nature Customer Service Center GmbH
Europaplatz 3, 69115 Heidelberg, Germany

Printed by Libri Plureos GmbH
in Hamburg, Germany